Strategic Studies Institute
and
U.S. Army War College Press

MOVING BEYOND PRETENSE:
NUCLEAR POWER AND NONPROLIFERATION

Henry Sokolski
Editor

June 2014

CONTENTS

FOREWORD

The President of the United States and nearly all his critics agree that the spread of nuclear weapons and the possibility of their seizure and potential use is the greatest danger facing the United States and the world. Looking at the way government and industry officials downplay the risks of civilian nuclear technology and materials being diverted to make bombs, though, a person would get almost the opposite impression. In fact, most governments have made the promotion of nuclear power's growth and global development a top priority. Throughout, they have insisted that the dangers of nuclear weapons proliferation are manageable either by making future nuclear plants more "proliferation-resistant" or by strengthening International Atomic Energy Agency (IAEA) safeguards and acquiring more timely intelligence on proliferators.

How sound is this view, though? How useful might civilian nuclear programs be for states that want to get nuclear weapons quickly? Are current IAEA nuclear safeguards sufficient to block military nuclear diversions from civilian programs? Are there easy fixes to upgrade these controls? How much can we count on more timely intelligence on proliferators to stem the further spread of nuclear weapons?

This volume taps the insights and analyses of 13 top security and nuclear experts to get the answers. What emerges is a comprehensive counternarrative to the prevailing wisdom and a series of innovative reforms to tighten existing nuclear nonproliferation controls. For any official, analyst, or party concerned about the spread of nuclear technology, this book is essential reading.

DOUGLAS C. LOVELACE, JR.
Director
Strategic Studies Institute and
    U.S. Army War College Press

CHAPTER 1

INTRODUCTION:
NUCLEAR ENERGY'S SECURITY STORY

Henry Sokolski

Governments have funded most of the nuclear industry's research and development, financed or guaranteed loans for its construction and export of nuclear plants, capped its liability for off-site damages in the case of nuclear accidents, and promoted its development internationally. Throughout, officials have insisted that the dangers of nuclear weapons proliferation attendant to the further spread of nuclear energy programs are manageable.

This view is fortified with a narrative. A sharp line is drawn between boiling water and making nuclear fuel. The nuclear weapons risks attendant to nuclear power are marginalized, whereas those related to making nuclear fuel are recognized. Nuclear supplier states, it is argued, though, should be able to persuade their nonweapons state customers not to make nuclear fuel because this activity is expensive and complex. Future nuclear plants also can be made more proliferation-resistant. The narrative then emphasizes the utility of International Atomic Energy Agency (IAEA) safeguards. These can be strengthened to deter and detect most of what matters, and what they cannot deter or detect, it is argued, can be countered. How? This can be accomplished with more timely intelligence to support covert national operations, which the United States and other like-minded states can be counted to take against Nuclear Nonproliferation Treaty (NPT) violators. Finally, although several more countries are certain to acquire nuclear weapons, this will not matter, it is argued, since nuclear weapons are not militarily useful except to deter use, a mission that these weapons can easily accomplish.[1]

This is the complete proliferation narrative. Although not every nuclear proponent makes it, all of these arguments are now, in one fashion or another, made by industry, academics, and public officials. It is powerful and bolsters nuclear power's further expansion. Indeed, so far, no real counternarratives, only counterpoints (i.e., qualifications), have yet been offered to challenge this upbeat view. Some have noted that truly proliferation-resistant reactors and fuel cycles are not yet at hand, that what has been proposed is unlikely to work, and that nuclear power is currently too expensive to be practical.[2] Yet, the rejoinder to these counterpoints—that in time, affordable proliferation-resistant systems will be developed—has been easy to make. A true counternarrative would be more difficult to deflect. It would show how the truth is actually the opposite of each of the nuclear security points made herein. Again, such a counternarrative has not yet been offered.

This volume is designed to do so. It features research my center commissioned over a 2-year period to reassess the assumptions currently driving U.S. and international nonproliferation policies. It spotlights the analysis and insights of some of the world's top security scholars.

The first section, "Nuclear Proliferation Matters," features the work of François Heisbourg, chairman of the International Institute for Strategic Studies (IISS), Matthew Kroenig of Georgetown University and the Council on Foreign Relations, and Matthew Fuhrmann of Texas A&M University. They contend that nuclear weapons proliferation is more likely to occur with the spread of civilian nuclear technology and that such nuclear proliferation constitutes a threat to international security—certainly if there is nuclear weapons use, but even if there is not.

The volume's second section, "Nuclear Power, Nuclear Weapons—Clarifying the Links," makes the case that civilian nuclear power programs actually afford a major leg up for any nation seeking development of a nuclear weapons option. It showcases four studies. The first, by former U.S. Nuclear Regulatory Commissioner and RAND Science Division director Victor Gilinsky, explains just how useful and quickly transformable power-reactor plutonium is to making a proliferator's first nuclear weapons. His analysis includes plutonium generated in the most proliferation resistant of power plants, the light water reactor. Susan Voss, formerly with Los Alamos National Laboratory, adds to this argument by detailing just how much intangible nuclear weapons material production-related technology and training is imparted along with any "peaceful" nuclear power program. Taken together, these studies more than suggest that nuclear power programs present nuclear weapons proliferation risks for any state but those we are certain to have forsworn making nuclear weapons and their key ingredients—highly enriched uranium and separated plutonium.

Optimism that we can easily persuade states to forswear making these nuclear fuels because of the cost and complexity doing so, moreover, is misplaced. Here, Richard Cleary of the American Enterprise Institute's account of previous American failures to get Iran, Brazil, South Korea, and Pakistan to stop making enriched uranium or separated plutonium is a cautionary tale. Compounding this sad historical record are the technical facts that nuclear fuel-making is not as complex or daunting as generally portrayed. Victor Gilinsky makes this point in regard to quick and dirty plutonium reprocessing schemes, at least one that is available in the unclassified literature. As for uranium enrichment, Scott Kemp of MIT's nuclear engineering department documents how basic centrifuge uranium enrichment technology is good enough to make bombs and is actually a relatively easy and affordable hurdle for states to climb over. Worse, it is an activity that can be hidden relatively easily from IAEA inspections until a state chooses to break out quickly to acquire nuclear weapons.

This, then, brings us to the book's third section, "How Well Can We Safeguard the Peaceful Atom?" and the question of how well the IAEA and the United Nations (UN) are likely to do their job enforcing the NPT in the future. The short answer is mixed. The key concern here is how well these institutions will be able to cope with the likely spread of nuclear energy programs to new states. In his analysis, Patrick S. Roberts of Virginia Tech raises a number of worrisome questions. What are the risks associated with simply scaling up the IAEA's current inspection system even assuming it had the funds to do so? Would it tolerate the inevitable increase in false alarms that must come with more inspections or would it tune the system to filter out such alarms even further than it already has? What, moreover, should be the metrics for IAEA success or failure in conducting its inspections? Would we know when and if the IAEA was failing at its mission and be able to take timely corrective action?

We get worrisome, partial answers from the analyses of two of the IAEA's best known deputy directors general for safeguards, Olli Heinonen and Pierre Goldschmidt. Dr. Heinonen notes that the IAEA could have the authority to conduct more special short-notice inspections and that it could do so without necessarily securing the unanimous consent of the agency's governing board. He argues that it would be most useful in the future for the IAEA's safeguards department to exercise such authority. It remains to be seen if it ever will.

Will there be clear consequences for those that violate the nuclear rules? Pierre Goldschmidt homes in on this question and recommends that the enforcement of safeguard agreements be made mandatory for any IAEA member and that they remain in force whether or not the country in question remains a member of the NPT. He also recommends that a set of country-neutral sanctions be agreed to by the UN Security Council in advance so that any country the IAEA finds in breach of its IAEA or NPT obligations will be certain to be sanctioned. The prospects for these recommendations' adoption are doubtful to unclear.

Until and unless they are adopted, whatever enforcement there might be will be taken by the major states relying on their own intelligence. But how much should we rely on such actions? This issue is examined in the book's fourth section, "Ignoring Nuclear Weapons Proliferation Intelligence."

Conventional wisdom presumes governments want to collect all the intelligence they can on proliferating states and that they are eager to act on this intelligence, especially if the proliferators are violating the rules. All that is lacking, according to this view, is sufficient, timely intelligence. With more situational awareness, it is argued, the United States and like-minded states can do much more to combat nuclear proliferation.

This supply side view of countering proliferation, however, ignores significant demand problems Washington and other states have for such timely proliferation information. Certainly, the United States sat on intelligence regarding A. Q. Khan in Pakistan and acted only very belatedly regarding intelligence concerning North Korea's uranium enrichment program. Such reticence, moreover, is hardly new.

Consider the case of Israel's acquisition of U.S. nuclear weapons material in the 1960s. Israel promised Presidents John Kennedy, Lyndon Johnson, and Richard Nixon that it would not acquire nuclear weapons. When intelligence emerged that Israel had illicitly acquired U.S. weapons-grade uranium and developed nuclear arms, Nixon and, to a lesser extent, Johnson glossed over or excused it. This sad history is detailed in Victor Gilinsky's history and backgrounder to the now famous meeting between Nixon and Golda Meir in 1969.

In yet another chapter, Leonard Weiss, formerly chief of staff of the Senate Governmental Affairs Committee, which had oversight of nuclear proliferation matters, details how the U.S. Government did all it could to deny the possibility that the Israelis, who ratified the Limited Test Ban Treaty, conducted a nuclear test off the coast of South Africa in 1979, even though the evidence clearly suggests they did.

Israel, however, was not the only country to receive such treatment. In his historical analysis, Robert Zarate of the Foreign Policy Initiative details how American policymakers either ignored or distorted proliferation intelligence on Iran's and North Korea's nuclear programs in order to avoid taking timely action against either state.

Finally, there is the problem with how we interpret the intelligence we get. Today, most officials would like to believe that there is still time to prevent Iran from developing nuclear weapons. This has encouraged the view that Iran is still far from getting its first bomb. Gregory Jones, Nonproliferation Policy Education Center's (NPEC) senior researcher, though, details how, in fact, Iran's nuclear weapons capability is so advanced that it no longer is a problem to be solved so much as a fact with which to be reckoned. That many intelligence officials cannot bring themselves to agree to this in public suggests how uncertain relying on their intelligence findings would be to assure timely counterproliferation actions to manage nuclear weapons proliferation.

How, then, are we to prevent more Irans? Mr. Jones suggests that we tighten the nuclear rules. This, then, brings us to the most important part of this volume's offerings: the nonproliferation principles and steps recommended in "Serious Rules for Nuclear Power without Proliferation." Victor Gilinsky and I developed this chapter initially as a thought exercise. What would a proper set of nonproliferation rules look like if one did not put nuclear power sales and promotion first, but instead emphasized security?

As we see it, this question has only been seriously tackled twice before: in 1946 with the Acheson-Lilienthal Report on the international control of nuclear power, and in 1976 with the Ford-Carter executive branch decisions to defer the use and production of commercial plutonium-based nuclear fuels. The Acheson-Lilienthal proposals were rejected by the Soviets. Shortly thereafter, the Dwight Eisenhower administration decided to share U.S. civilian nuclear energy internationally in the hopes that the control issues raised in the Acheson-Lilienthal Report could be solved later. This gave rise to the Atoms for Peace program, the creation of a loose set of nuclear controls administered by the IAEA, and the wholesale export of nuclear technology internationally. Atoms for Peace remained U.S. policy until 1974, when this approach was literally blown away by India's "peaceful" nuclear explosion of a bomb made of plutonium that was produced using "peaceful" U.S. and Canadian civilian nuclear assistance. Shortly thereafter, the London Suppliers Group secretly agreed to restrict the export of nuclear fuel-making technologies to nonweapons states, and Presidents Gerald Ford and Jimmy Carter announced U.S. efforts to defer the commercial use of plutonium-based fuels both domestically and abroad.

That was over 40 years ago. Now, after Iraq, Iran, North Korea, Libya, A. Q. Khan, and Syria, there is cause to review the bidding once again. Certainly, the experience of the last 3 and a half decades has challenged the assumptions that drove the nuclear policies of Presidents Ford and Carter. These policies presumed that we could detect nuclear fuel-making. Uranium enrichment centrifuges, which are relatively easy to hide, were not yet readily available then, nor had much thought been given to just how small one could make a dedicated, covert reprocessing plant. It also was presumed that if illicit nuclear activities were detected, swift, effective international enforcement would follow. Our experience with Iran and North Korea, though, has jilted many of these notions. In fact, the United States and others now find it challenging just to maintain existing nuclear nonproliferation controls, much less to tighten them.

Much of this nonproliferation defensiveness is reflected in how the United States and others view the

NPT. This view is encapsulated in a diplomatic formulation known as "the three-pillars of the NPT." According to this view, the NPT and the nuclear nonproliferation regime rest on three objectives that must be balanced against one another. The first is nonproliferation (as manifested by Articles I, II, and III of the NPT). This roughly translates into IAEA safeguards and UN Security Council enforcement measures against NPT violators. The second is nuclear disarmament (as manifested by Article VI of the NPT). It focuses on reducing the NPT nuclear weapons states' atomic arsenals (almost exclusively the United States and Russia). The third is sharing "peaceful" nuclear technology (as manifested by Article IV of the NPT). This can range, depending on who is defining "peaceful," from the sharing of benign medical isotopes to transferring proliferation-prone nuclear fuel-making technologies.

Putting aside how little of the NPT's diplomatic history actually supports this popular diplomatic interpretation,[3] the key problem with this three-pillar formulation is how intellectually self-defeating it is. First, if the nuclear-armed states are judged not to have sufficiently disarmed their nuclear stockpiles, why or how should this be used as the pretext for promoting **less** nonproliferation? Would not backing off necessary nonproliferation controls only increase the prospects for more proliferation and, therefore, increase demands for **more** nuclear armament?

Similarly, how is supplying nonweapons states with ever more "peaceful" nuclear technology a prerequisite for securing more or tighter nonproliferation controls? If the technology in question is truly peaceful and benign, by definition, it ought to be safe to share without any apprehensions that it might be diverted easily in order to make bombs. Also, if it could not be used to make bombs, nuclear supplier states would hardly need a nonproliferation incentive to share it. If, on the other hand, a specific civilian nuclear technology was particularly proliferation-prone and therefore not clearly safe to share, why would any state wanting to promote nonproliferation believe it was under an NPT obligation to transfer it?

Again, does not the promotion of nonproliferation presume the sharing of only truly "peaceful" nuclear goods and technology and the general encouragement of nuclear restraint? Why would any state want to bargain away achieving the goal of nonproliferation with its presumed benefits? How much sense does any of this make? The short answer is: not much. At the very least, sounder thought ought to drive our nonproliferation policies.

To pursue any sound undertaking to promote nuclear power without proliferation, Victor and I suggest five guiding principles:

1. **Locking down the NPT**. It is not consistent with the NPT's purpose for members to exercise the withdrawal provision after gaining technology of relevance to weapons — whether by importing it or developing it domestically — as this was done under the assumption by other members that it was for peaceful uses. Treaty members cannot exercise the withdrawal clause without squaring accounts. As a practical matter, this would mean membership in the Treaty was essentially permanent. Under this interpretation, North Korea's 2003 announcement of "withdrawal" while in noncompliance of IAEA inspection requirements left that country in a state of Treaty violation.

2. **Assuring a technological margin of safety**. The Treaty cannot be a vehicle for a state to legally come

overly close to a weapons capability. There has to be a technological safety margin between genuinely peaceful and potentially military applications. As a consequence, the "inalienable right" language in the Treaty has to be interpreted in terms of the Treaty's overriding objective, and thus there have to be restrictions on the kinds of technology that are acceptable for nonmilitary use. Nuclear power needs to develop in a way that does not provide easy access to nuclear explosives. Where to draw the line is now coming to a head in the context of Iran's nuclear program.

3. **Adjusting nuclear sovereignty for greater security**. Countries involved with nuclear energy must accept that the inherent international security dangers such involvement implies require them to relinquish a considerable degree of sovereignty to international security organizations, in particular the IAEA inspectorate. In view of the concerns about clandestine facilities, both with respect to enrichment and reprocessing, countries have to agree to essentially unlimited inspection rights for international inspectors if the circumstances warrant. The Additional Protocol is a good start toward expanding inspectors' rights, but this unfortunately goes along with a reduction in the frequency of normal inspections.

4. **Getting serious about enforcement**. The NPT needs an established enforcement mechanism to deal with Treaty violations in a predictable way. The foregoing rules for operating nuclear power plants in a manner that is consistent with international security are not self-enforcing. There has to be agreement among the Treaty parties concerning reasonably predictable responses to particular violations, and most particularly any effort by a state to withdraw from the Treaty, so as to remove the notion that violators can escape with impunity.

5. **Applying nuclear limitations and reductions to all nuclear weapons states**. All nuclear weapons states have to participate in weapons reductions. This is essential for gaining the cooperation of the other NPT members in restrictive measures. In the first instance, this includes Britain, France, and China, which up to now have not participated in the reduction process that has involved the United States and Russia. But it also has to include India, Israel, Pakistan, and North Korea. With 190 nations adhering to the NPT, its obligations should be regarded as universal, thus applying to all countries whether or not they formally joined the Treaty. From this point of view, North Korea and the three countries that never joined would be regarded as members who are out of compliance. But by participating in a suitably monitored weapons reduction process, they could be viewed as members in the process of coming into compliance.

Of course, pushing these principles in policy is sure to create considerable friction. Some have argued that it simply is impractical to push such policies. In the end, this sadly may be the case. But if so, it suggests the urgency of curbing our own enthusiasm and that of other nuclear supplier states for the international spread of nuclear energy programs where they currently do not exist. At the very least, until governments have tougher nonproliferation controls in place, they ought not be spending more to promote the export of this technology.

ENDNOTES - CHAPTER 1

1. See, e.g., Steven Kidd, "Nuclear Proliferation Risk—Is It Vastly Overrated?" *Nuclear Engineering International*, July 23, 2010, available from *www.neimagazine.com/story.asp?storyCode=2056931*.

2. See, e.g., Richard Lester, "New Nukes," *Issues in Science and Technology*, Summer 2006, available from *www.issues.org/22.4/lester.html*; Jungmin Kang, Frank von Hippel, "Limited Proliferation-Resistance Benefits from Recycling Unseparated Transuranics and Lanthanides from Light-Water Reactor Spent Fuel," *Science and Global Security*, No. 3, 2005, pp. 169-181; James M. Aton, "The Myth of Proliferation-Resistant Technology," *The Bulletin of Atomic Scientists*, November/December 2009, available from *cybercemetery.unt.edu/archive/brc/20120621022618/http://brc.gov/sites/default/files/meetings/presentations/james_m_acton-the_myth_of_proliferation-resistant_technology.pdf*; Jerry Taylor, "Nuclear Energy: Risky Business," *Reason*, October 22, 2008, available from *www.cato.org/publications/commentary/nuclear-energy-risky-business*; "Nuclear Power: Fracked Off," *The Economist*, June 1, 2013, available from *www.economist.com/news/united-states/21578690-thanks-cheap-natural-gas-americas-nuclear-renaissance-hold-fracked*; and Mycle Schneider and Anthony Froggett, *The World Nuclear Industry Status Report 2013*, Paris, France/London, UK: July 11, 2013, available from *www.worldnuclearreport.org/IMG/pdf/20130716msc-worldnuclearreport2013-lr-v4.pdf*.

3. On reading the NPT, see Albert Wohlstetter, "Spreading the Bomb without Quite Breaking the Rules," *Foreign Policy*, No. 25, Winter 1976-77, pp. 88-96, 145-179; Arthur Steiner, "Article IV and the 'Straightforward Bargain,'" PAN Heuristics Paper 78-832-08, in Albert Wohlstetter *et al.*, *Towards a New Consensus on Nuclear Technology*, Vol. II, Supporting Papers, ACDA Report No. PH-78-04-832-33, Marina del Rey, CA: PAN Heuristics, 1978, pp. 1-8; Eldon V.C. Greenberg, *The NPT and Plutonium: Application of NPT Prohibitions to "Civilian" Nuclear Equipment, Technology and Materials Associated with Reprocessing and Plutonium Use*, Washington, DC: The Nuclear Control Institute, 1993, available from *www.npolicy.org/article.php?aid=292*; Henry Sokolski, "The Nuclear Nonproliferation Treaty and Peaceful Nuclear Energy," Testimony before "Assessing 'Rights' under the Nuclear Nonproliferation Treaty," a hearing of the U.S. House of Representatives, Committee on International Relations, Subcommittee on International Terrorism and Nonproliferation, March 2, 2006, available from *www.npolicy.org/article.php?aid=392&rtid=8*; Robert Zarate, "The Three Qualifications of Article IV's 'Inalienable Right'," and Christopher Ford, "Nuclear Technology Rights and Wrongs: The

NPT, Article IV, and Nonproliferation," in Henry Sokolski, ed., *Reviewing the NPT*, Carlisle, PA: Strategic Studies Institute, U.S. Army War College, 2010, pp. 219-384, available from *www.npolicy. org/thebook.php?bid=2.*

PART I:

NUCLEAR PROLIFERATION MATTERS

CHAPTER 2

NUCLEAR PROLIFERATION — LOOKING BACK, THINKING AHEAD: HOW BAD WOULD THE FURTHER SPREAD OF NUCLEAR WEAPONS BE?

François Heisbourg

A re-visit of past proliferation helps us understand the dangers of the further spread of nuclear weapons. Notwithstanding the establishment of an international nonproliferation regime and occasional, selective, and sometimes vigorous country-specific nonproliferation policies, the fight against the spread of nuclear weapons has not been recognized in the past as an overriding policy objective by the international community, jointly or severally. It will be argued that it is largely due to an overly sanguine assessment of the consequences of past proliferation, which has been less benign than is suggested by the reassuring persistence of the taboo on the use of nuclear weapons. Future proliferation's consequences appear all the more dire as a result of a misunderstanding of the past, which meshes in with new and worrying technical, operational, and strategic developments. "Proliferation futures" will be examined in this combined light of a flawed narrative and new developments, which may lead eventually to the deliberate or inadvertent use of nuclear weapons. In order to avoid such an outcome, policy recommendations will be flagged.

A LESS THAN OVERRIDING OBJECTIVE

At first blush, the prevention of the spread of nuclear weapons appears as a rare and important feature of global consensus spanning close to half a century.

17

This is clearly the case in multilateral declaratory undertakings such as the 1978 *Final Document of the United Nations (UN) Special Session on Disarmament*,[1] notable for its universal nature committing all member states of the UN at the time, which states *inter alia* that "Non proliferation of nuclear weapons is a matter of universal concern" (§36) and "It is imperative . . . to prevent the proliferation of nuclear weapons" (§65). Previously, and more operationally, the Nuclear Nonproliferation Treaty (NPT), opened to signature on May 1, 1968, laid out the elements of an international regime that, over the years, has acquired quasi-universal status, with only India, Israel, and Pakistan holding out, and only one state (North Korea) opting out. The NPT in turn built upon an initially modest set of safeguards, established by the International Atomic Energy Agency (IAEA) after its creation in 1957, that have developed into an extensive, more intrusive system of inspections, materialized in particular by the so-called Additional Protocol, formalized in 1997, which has been acceded to by 115 states and which another 25 have signed.[2] Out of the 44 countries[3] possessing at least one operational nuclear reactor, 35 have ratified the Protocol, and three others (India, Iran, and Israel) have signed it. Even the three countries that never joined the NPT have not signaled their intent to act against the nonproliferation aims of the NPT. Only North Korea breaks what is in effect a universal declaratory pattern to which countries pay collective and individual obeisance in words, if not always in deeds.

However, this doesn't amount to an overriding policy imperative at either the multilateral or national levels. On occasion, the UN as a whole has given an overriding importance to nonproliferation as opposed to other aspects of international relations, but on a highly selective basis. Such was the case of the imposition of mandatory UN Security Council sanctions against South Africa when that country's work on a nuclear test site was uncovered in 1977,[4] and again in the wake of the Gulf War of 1991 when the Security Council mandated the nuclear, biological, and chemical disarmament of Iraq.[5] But these policies were country specific, not general in nature. Similarly, nonproliferation only rarely, and usually selectively, takes precedence over other elements of bilateral relations between given states. Israel takes firm exception to nuclear "wannabes" insofar that they deny their right to exist, but is little interested beyond that. American militancy against Pakistan's nuclear ambitions withered when Islamabad's help was required after the Soviet invasion of Afghanistan. In 2005, Washington spectacularly conceded to India privileges that are normally reserved to bona fide NPT signatories when it signed a bilateral nuclear agreement with that country, a precedent that China is now tempted to grant Pakistan. Russia, France, Britain, or other industrialized states take a "pick and choose" approach. Despite the misgivings and reservations of some, the 45 member states of the Nuclear Suppliers Group did not prevent the lifting of restrictions vis-à-vis India flowing from the U.S.-India nuclear agreement.[6] The weakening of the NPT entailed by that agreement took second place to other considerations, such as India's economic or strategic importance.

There is thus a substantial contrast between non-proliferation as an objective and its actual level of priority. In itself, this is neither unusual in international relations (how many other lofty goals are simultaneously proclaimed and neglected?) nor readily avoidable, as the examples cited previously demonstrate. However, nuclear weapons by general acknowledgement (which rests on an all-too reliable set of unimpeachable physics and an even less debatable set of practical data from nuclear use and testing) have a unique ability to instantaneously destroy entire populations. That consideration would normally have given nonproliferation a higher rank and a broader remit in the order of international priorities, even if one takes fully into account the *Realpolitik* requirements of the Cold and post-Cold War eras. There are strong and mutually reinforcing empirical and logical reasons that explained this disconnect in the past and that continue to inform the manner in which prospective further proliferation is being approached.

In empirical terms, two facts stand out: Runaway nuclear proliferation has not occurred, and nuclear weapons have not been used, in anger or by accident, since the bombing of Hiroshima and Nagasaki, Japan. As long as proliferation had remained confined to countries that were in alliance with the United States, such as the United Kingdom (UK) and France (which tested their first devices in 1952 and 1960, respectively), there was little additional fear of breaking the taboo on nuclear use in either Washington or Moscow — although the United States was even less happy than the Union of Soviet Socialist Republics (USSR) about French nuclear ambitions. However, a go-it-alone nuclear Red China rang loud alarm bells when it was set to test in 1964, leading both to rumblings about a decapitating Soviet or Soviet-American strike[7] and, more practically, to the drafting of the NPT, which sought to limit the nuclear club to those countries that tested before January 1, 1967. This was an era in which runaway proliferation had been hitherto considered as a mainstream scenario:[8] In a world with nuclear free agents (the expression "rogue state" had not yet been coined) such as an unpredictable Red China, nuclear use would occur. Neither development has happened. Proliferation has remained restricted to a limited set of countries (the five "official," the three "de facto," the North Korean "sort-of," and the Iranian "putative" nuclear powers), and roll-back has occurred willy-nilly: Nuclear South Africa was disarmed; quasi-nuclear Sweden and the once-aspiring or potential nuclear states of Argentina, Brazil, Canada, Germany, Iraq, Italy, Libya, South Korea, Switzerland, Syria, and Taiwan eventually renounced the nuclear road; and the nuclear legacies in Belarus, Kazakhstan, and Ukraine were liquidated. The "system," however defined (from the role of the NPT to preemptive military strikes against Iraq and Syria by way of defense guarantees within the North Atlantic Treaty Organization [NATO] or to Sweden and Ukraine), has more or less worked during the last decades of the 20th century. Nor has the formal advent of India and Pakistan to nuclear military status in 1998 led to nuclear weapons use, while the prospect of Mao's China running amok has been superseded by a quiescent nuclear doctrine in the Middle Kingdom.

The power of this empirical evidence appears in the choice of our leaders' words. Dire forecasts and corresponding practical calls for concrete action are made rightly by mostly Western leaders about the possible consequences of Iran going nuclear; pie-in-the-sky speeches are made about the need to eliminate all nuclear weapons. But what is largely missing is the bridging language between these two levels of concern of the sort U.S. President John F. Kennedy used to address the perceived challenge of short-term, runaway

21

nuclear proliferation and its implied consequences. In his March 1963 press conference (see endnote 8), Kennedy linked nonproliferation to a prospective test ban treaty.[9]

Largely missing, but not entirely so, is nonstate proliferation resulting in nuclear terrorism. This was correctly seized upon after September 11, 2001 (9/11) by Presidents George W. Bush and Barack Obama, leading to the first global Nuclear Security Summit in Washington, DC, in April 2010. However necessary it may be to address that fear, identified earlier by able novelists,[10] it has not, mercifully, yet materialized in empirical terms. The empirical evidence that informs nonproliferation's policy status sustains and is sustained, in turn, by reasoning on the supposed inherent stability of deterrence in all of its declensions: unilateral, bilateral, or even multilateral.

Given their disproportionate power, nuclear weapons cannot serve to achieve limited policy goals, thus excluding their use as Clausewitzian weapons. Further, the possession of nuclear weapons may even inhibit actions that an aggressive nonnuclear power might otherwise contemplate, but a nuclear power might not. Stalin, at the head of a still clearly nonnuclear USSR, blockaded Berlin, an action that none of his nuclear-armed successors sought to emulate. As a nonnuclear power, Red China bombed Taiwan repeatedly. The worst of it ceased after Beijing acquired nuclear weapons. Possession of nuclear weapons, possibly after a learning curve, appears to self-deter escalatory aggressive behavior.

Bilateral deterrence between two nuclear powers has long been deemed to moderate direct confrontation and to deflect aggressive behavior towards proxies.[11] Although no such theoretical consensus exists vis-à-vis the possible stability of multicornered possession of nuclear weapons, the case has been made by powerful authors such as Ken Waltz and Pierre Gallois.[12] In practice, a global multipolar nuclear order was established to some extent since the 1960s, with the USSR, the United States, and China forming a strategic triangle that was perceived as such by the authors of the Nixon-to-Beijing visit. A regional multipolar dispensation arguably also exists between China, India, and Pakistan. These relationships have apparently not led to instabilities greater than (or even as great as) those that have characterized the U.S.-Soviet nuclear standoff.

In short, proliferation has been a manageable, slow-motion process. Nuclear weapons have not been used nor has the probability of their use appeared to have increased. The overall status of nuclear proliferation today is satisfactory provided some adjustments are made in terms of securing material from nonstate actors. Still, the policy mix sustaining the current situation is messy and occasionally fraught, as so many things are in international life. Difficult case-specific situations, such as Iran today, will continue to be handled as such, as Iraq was yesterday.

THE PAST IS NOT WHAT IT USED TO BE

The problem with this reassuring reading of the past is that it is not entirely true. Yes, the NPT had a major material effect by gradually making the new normal non-nuclear. Yes, again, U.S. defense guarantees weaned Germany, Italy,[13] South Korea, Taiwan, and even neutral Sweden away from the nuclear road, as did the U.S.-French-British assurances to post-Soviet Ukraine. Yes, too, various levels of coercion worked against Iraq, Libya, and Syria. But no, the practice of even the most "classical" bilateral deterrence was not nearly as reassuring as the mainstream narrative inherited from the Cold War would have one believe. Nor can we consider that our elements for empirical judgment are as methodologically satisfactory in terms of their breadth and depth as they need to be. These two negatives will be examined in turn.

Nuclear archives, as other sensitive governmental archives, open up usually after an interval of decades, and even then with varying levels of culling and redaction. Even oral histories tend to follow this pattern, as aging witnesses feel freer to speak up. Hence a paradox: When the Soviet-American nuclear confrontation was central to our lives and policies during the Cold War, we did not know how bad things really were. Now that we are beginning to know, there is little public interest, given the disappearance of the East-West contest. Yet there are lessons of general interest that can be summarized as follows.

The first lesson is that the Cuban Missile Crisis brought us much closer to the brink than we were even aware of at the time and for reasons that are germane to the current situation. These reasons include massive failures of intelligence on Soviet nuclear preparations and dispositions in Cuba (notably on tactical nukes and on the operational readiness of a number of intermediate range ballistic missiles [IRBMs] and their warheads); dysfunctional or imperfect command and control arrangements (notably vis-à-vis Soviet submarines); and unintentionally mixed signals (on each antagonist's actions). These reasons are effectively laid out in Michael Dobb's book, *One Minute to Midnight*.[14]

The second lesson relates to the safety and security of nuclear forces, which are subject to potentially

24

calamitous procedural, technical, or operational mishaps and miscalculations, somewhat along the lines of what applies to related endeavors (nuclear power and aerospace). Scott Sagan, in his *Limits of Safety*,[15] provides compelling research on the American Cold War experience. It would be interesting to have a similar treatment on the Soviet experience.

Although it can be argued that today's nuclear arsenals are much smaller and easier to manage and that the technology for their control has been vastly improved, several facts remain. First, the United States has continued to witness serious procedural lapses in the military nuclear arena.[16] Second and related, the de-emphasis of the importance of nuclear weapons in the U.S. force structure is not conducive to treating them with the respect that is due to their destructive power. Third, other nuclear powers do not necessarily benefit from the same technology and learning curves as the older nuclear states, and notably the United States. Instead, cheek-to-jowl nuclear postures, which prevailed in the Cuban Missile Crisis and which help explain why World War III nearly occurred, characterize India and Pakistan today. Indeed, despite the dearth of detail on Indian and Pakistani nuclear crisis management, we know that the stability of nuclear deterrence between India and Pakistan is by no means a given, with serious risks occurring on several occasions since the mid-1980s.[17]

At another level of analysis, we have to recognize the limits of the database on which we ground our policies on nonproliferation. The nuclear age, in terms of operationally usable devices, began in 1945, less than 70 years ago, less than the age of an old man. The fact that there has been no accidental or deliberate nuclear use during that length of time is nearly twice as reassuring as the fact that it took more than 30 years[18] for a nuclear electricity generating plant to blow up, in the form of the Chernobyl disaster of 1986. But given the destructive potential of nuclear weapons, twice as much reassurance (in the form of no use of nuclear weapons for close to 70 years) is probably not good enough.

Furthermore, the Chernobyl disaster involved the same sort of errors of judgment, procedural insufficiencies, and crisis mismanagement visible in Sagan's book, not only, or even mainly, flawed design choices. Inadvertence was at work, in other words, of the sort that could prevail in a time-sensitive, geographically constrained Indo-Pakistani or Middle Eastern conflict. Give it another 70 years to pass judgment.

The same empirical limits apply to the number of actors at play: We have simple bipolar (U.S.-USSR/Russia or India-Pakistan) and complex bipolar (U.S.-France-UK-NATO-Soviet Union/Russia) experiences; we have had U.S.-Soviet-Chinese or Sino-Indian-Pakistani tripolarity; and we have had a number of unipolar moments (one nuclear state vis-à-vis nonnuclear antagonists). But we mercifully have not had to deal with more complex strategic geometries — yet — in the Middle East or East Asia. We only know what we know; we do not know what we do not know. A historical narrative that is not reassuring and an empirical record that is less than compelling need to inform the manner in which we approach further proliferation.

PROLIFERATION PUSH AND PULL

Ongoing proliferation differs from that of the first half-century of the nuclear era in three essential ways. On the demand side, the set of putative nuclear actors is largely focused on the most strategically stressed regions of the world. On the supply side, the actual or potential purveyors of proliferation are no longer principally the first industrialized generation of nuclear powers. Instead, the technology involved in proliferation is somewhat less demanding than it was during the first nuclear age. Taken together, these changes entail growing risks of nuclear use.

Demand is currently focusing on two regions, the Middle East and East Asia (broadly defined), and involves states and, potentially, nonstate actors. In the Middle East, Iran's nuclear program is the focus of the most intense concerns. A potential consequence in proliferation terms would be to lead regional rivals of Iran to acquire nuclear weapons in turn: This concern was vividly described in 2007 by the then-President of France, Jacques Chirac[19] who specifically mentioned Egypt and Saudi Arabia. The likelihood of such a "proliferation chain reaction" may have been increased by President Obama's recent repudiation of containment as an option.[20] Short of Iran being persuaded or forced to abandon its nuclear ambitions, the neighboring states would presumably have to contemplate security options other than a Cold War-style U.S. defense guarantee. Given prior attempts by Iraq, Syria, and Libya to become nuclear powers, the probability of a multipolar nuclear Middle East has to be rated as high in case Iran is perceived as having acquired a military nuclear capability. Beyond the Middle East, there is a possibility of civil war in nuclear-armed Pakistan, leading to state failure and the possibility of nukes falling out of the hands of an effective central government.

There are historical precedents for such a risk, most notably, but not only[21] in the wake of the collapse of the Soviet Union: Timely and lasting action by outside powers, such as the United States with the Nunn-Lugar initiative, and the successor states themselves has prevented fissile material from falling into unauthorized hands in significant quantities. Pakistan could pose similar problems in a singularly more hostile domestic environment. As things stand, nonstate actors such as post-Soviet *mafiya* bosses (interested in resale potential) or al-Qaeda[22] have sought, without apparent success, to benefit from opportunities arising from nuclear disorder in the former USSR and Central Asia. Mercifully, the price al-Qaeda was ready to pay was way below the going rate (upwards of hundreds of millions of dollars) for the sorts of services provided by the A. Q. Khan network to some of his clients.

Although North Korea's nuclear ambitions appear to be both more self-centered and more containable than is the case for Iran, the possibility of state collapse in combination with regional rivalry leaves no room for complacency. More broadly, we are facing the prospect of a multipolar nuclear Middle East linked to an uncertain nuclear Pakistan, already part of a nuclear South Asia tied via China to the Korean nexus in which nuclear America and Russia also have a stake. More broadly still, such a nuclear arc-of-crisis from the Mediterranean to the Sea of Japan would presumably imply the breakdown of the NPT regime, or at least its reversion to the sort of status it had during the 1970s when many of its currently significant members had not yet joined,[23] thus unloosening both the demand and supply sides of proliferation.

On the supply side, "old style" proliferation relied on official cooperation between first-generation nuclear or nuclearizing powers, of which the Manhattan Project was a forerunner (with American, British, and Canadian national contributions and multinational scientific teams), followed *inter alia* by post-1956 French-Israeli, post-1958 U.S.-UK, and pre-1958 USSR-China cooperation. If India relied heavily on the "unwitting cooperation" of others, notably the involvement of Canada and the United States in the Atoms for Peace CIRUS research reactor, Pakistan set up the first dedicated, broad spectrum, cross-border trading network to make up for the weakness of its limited industrial base. This import-focused organization thus went beyond traditional espionage-aided efforts, as practiced by the USSR during and after the Manhattan Project, or case-by-case purloining or diversion of useful material on the global market, as practiced by Israeli operatives. Even before the Pakistani network had fulfilled its primary task of supplying the national program, it began its transformation into an export-oriented venture.

Libya, Iran, North Korea, and a fourth country that remains officially unnamed became the main outlets of what became the world's first private-sector (albeit government originated and, presumably, supported) proliferation company, which was only wound down after strong Western pressure on Pakistan after 9/11. Although the by-now richly documented A. Q. Khan network[24] appears to have ceased to function in its previous incarnation, it has powerfully demonstrated that there is an international market for proliferation that other operators can expect to exploit. Furthermore, budding, resource-weak nuclear powers have a strong incentive to cover the cost of their investment by selling or bartering their nuclear-related assets, including delivery systems. The fruits of state-to-state cooperation between Iran, North Korea, and Pakistan are clearly apparent in the close-to-identical genealogy of their nuclear-capable ballistic missiles of the No-Dong/Ghauri/Shahab families displayed in military parades and test launches. Not all such cooperation consists of televised objects.

Even in the absence of game-changing breakthroughs, technical trends facilitate both demand and supply-side proliferation. For the time being, the plutonium route towards the bomb remains essentially as easy and as difficult as from the earliest years of the nuclear era. Provided a country runs a (difficult-to-hide) research or power reactor from which low-irradiated fuel can be downloaded at will, such as CANada Deuterium Uranium (CANDU) type natural uranium reactors, reprocessing is a comparatively straightforward and undemanding task. Forging and machining a multiple-isotope metal that is notorious for its numerous physical states and chemical toxicity is a substantial challenge, along with the companion complications of devising a reliable implosion mechanism. Nuclear testing is highly desirable to establish confidence in the end-result. Opportunities for taking the plutonium-proliferation road may increase somewhat as new techniques such as pyro-processing come on line. Developments in the enriched uranium field have been more substantial in facilitating proliferation. The development of lighter and more efficient centrifuges makes it easier for a state to extract enriched uranium speedily in smaller and less visible facilities. Dealing with the resulting military-level highly enriched uranium (HEU) is a comparatively undemanding task. The long-heralded advent of industrially effective and reliable laser enrichment technology may eventually further increase ease of access. Downstream difficulties would still remain. Although implosion mechanisms are not mandatory, they are desirable in order both to reduce the critical mass of uranium-235 for a nuclear explosion and to make for a lighter, smaller, more readily deliverable weapons package.

In sum, incremental improvements increase the risk of proliferation. However, nonstate actors are not yet, and will not be on the basis of known technical trends, in a position to master the various steps of the two existing military nuclear fuel cycles, which remain the monopoly of states. Nonstate actors would need the active complicity from (or from accomplices within) states, or benefit from the windfall of state collapse, to acquire a military nuclear capability. The threat of nuclear terrorism continues to be subordinated to developments involving state actors, a remark that is not meant to be reassuring since such developments are increasingly likely as proliferation spreads to new states and as state failure threatens in the "arc of proliferation" extending from the Mediterranean to Northeast Asia. Furthermore, nonstate actors can be satisfied with levels of nuclear reliability and performance that states could not accept. A difficult-to-deliver or fizzle-prone nuclear device would not provide a state with the level of deterrence needed to shield it from pre-emptive or retaliatory action, whereas a terrorist group would not be seeking such immunity. A road or ship-delivered imperfect device, which would be closer to a radiological bomb than to a fully fledged atomic weapon, would provide its nonstate owners with immense potential. The road to a nonstate device does not need to be as well-paved.

NUCLEAR FUTURES

"New" lessons from a revisited past and current trends in nuclear proliferation will tie into a number of characteristics of contemporary international relations with potentially destabilizing consequences leading to an increasing likelihood of nuclear use. Four such characteristics will be singled out both because of their relevance to nuclear crisis management and because of their growing role in the world system in the age of globalization:

      1. Strategic upsets;

      2. Limits of imagination;

      3. Unsustainable strains; and,

      4. Radical aims.

The 2008 French *Defense and National Security White Paper*[25] developed the concept of *ruptures stratégiques* (strategic upsets) to describe the growing tendency of the world system to generate rapid, unexpected, morphing upsets of international security as a consequence of globalization broadly defined against the backdrop of urbanizing populations generating economic growth and environmental and resource constraints. In themselves, such upsets are not novel (see *inter alia*, a pandemic such as the Black Death in 1348-49, the Great Depression, and not to mention the World Wars or the major and benign strategic upset of 1989-91), but the very nature of globalization and the relationship between human activity and the Earth's ability to sustain them mean more frequent, as well as more complex upsets. If this reading is correct—and the Great Financial Crisis, the Arab revolutions, the accession of China to superpower status can be mentioned as examples that followed the publication of the white paper—then the consequences in the nuclear arena will be two-fold. First, nuclear doctrines and dispositions that were conceived under a set of circumstances (such as the Cold War or the India-Pakistan balance of power) may rapidly find themselves overtaken by events. For instance, it is easier to demonstrate that U.S. and Russian nuclear forces still visibly bear the imprint of their 1950s template than it is to demonstrate their optimal adaptation to post-post-Cold War requirements. Second, more challenges to international security and of a largely unforeseeable nature means greater strains will be placed on the ability of nuclear powers to manage crises against the backdrop of their possession of nuclear weapons. In many, indeed most, cases, such *ruptures stratégiques* will no doubt be handled with nuclear weapons appearing as irrelevant: Hypothetical security consequences of an epidemic, such as the interhuman transmission of the H5N1 bird flu virus, or prospective conflicts resulting from climate change do not have *prima facie* nuclear aspects. But beyond the reminder that we do not know that as a fact, the probability is, under the *rupture stratégiques* hypothesis, there will be more occasions for putting all crisis management, including nuclear, to the test.

Human societies tend to lack the imagination to think through, and to act upon, what have become known as "black swan" events.[26] That which has never occurred (or which has happened very rarely and in a wholly different context) is deemed not to be in the field of reality, and to which must be added eventualities that are denied because their consequences are too awful to contemplate. The extremes of human misconduct (the incredulity in the face of evidence of the Holocaust, the failure to imagine 9/11) bear testimony to this hard-wired trait of our species. This would not normally warrant mention as a factor of growing salience, if not for the recession into time of the original and only use of nuclear weapons in August 1945. Non-use of nuclear weapons may soon be taken for granted rather than being an absolute taboo. Recent writing on the reputedly limited effects of the Hiroshima and Nagasaki bombs[27] may contribute to such a trend in the name of reducing the legitimacy of nuclear weapons. Recent, and often compelling, historical accounts of the surrender of the Japanese Empire that downplay the role of the atomic bombings in comparison to early research can produce a similar effect, even if that may not have been the intention.[28] However desirable it has been, the end of atmospheric nuclear testing[29] has removed for more than 3 decades the periodic reminders that such monstrous detonations made as to the uniquely destructive nature of nuclear weapons. There is a real and growing risk that we forget what was obvious to those who first described in 1941 the unique nature of yet-to-be produced nuclear weapons.[30] The risk is no doubt higher in those states for which the history of World War II has little relevance and that have not had the will or the opportunity to wrestle at the time or *ex post facto* with the moral and strategic implications of the nuclear bombing of Japan in 1945.

Unsustainable strains are possibly the single most compelling feature of contemporary proliferation. Examples include tight geographical constraints — with, for instance, New Delhi and Islamabad, located within 300 miles of each other; nuclear multipolarity against the backdrop of multiple, crisscrossing sources of tension in the Middle East, as opposed to the relative simplicity of the U.S.-Soviet confrontation;

the existence of doctrines, such as India's "cold start," and force postures, such as Pakistan's broadening array of battlefield nukes, that rest on the expectation of early use; and the role of nonstate actors as aggravating or triggering factors when they are perceived as operating with the connivance of an antagonist state (in the past, the assassination of the Austrian Archduke in Sarajevo in 1914, and in the future, Hezbollah operatives launching rockets with effect against Israel or Lashkar-e-Taiba commandos doing a "Bombay" redux in India). Individually or in combination, these factors test crisis management capabilities more severely than anything seen during the Cold War with the partial exception of the Cuban Missile Crisis. Even the overabundant battlefield nuclear arsenals in Cold War Central Europe, with their iffy weapons' safety and security arrangements, were less of a challenge: The U.S. and Soviet short-range nuclear weapons so deployed were not putting U.S. and Soviet territory and capitals at risk.

It may be argued that these risk factors are known to potential protagonists, and that they therefore will be led to avoid the sort of nuclear brinksmanship that characterized U.S. and Soviet behavior during the Cold War in crises such as the Korean War, Berlin, the Cuban Missile Crisis, or the Yom Kippur War. Unfortunately, the multiple nuclear crises between India and Pakistan demonstrate no such prudence, rather the contrary. Were such restraint to feed into nuclear policy and crisis planning, along the lines of apparently greater U.S. and Soviet nuclear caution from the mid-1970s onwards, the fact would remain that initial intent rarely resists the strains of a complex, multi-actor confrontation between inherently distrustful antagonists. It is also worth reflecting on the fact that during the 1980s, there was real and acute fear in Soviet ruling circles that the West was preparing an out-of-the-blue nuclear strike, a fear that in turn fed into Soviet policies and dispositions.[31]

The Cold War was a set of crises and misunderstandings that came within a whisker of a nuclear holocaust. India and Pakistan's nuclear standoff is deeply unstable, not least as a result of the interaction with nonstate actors. A multipolar nuclear Middle East would make the Cuban Missile Crisis look easy in comparison.

Great conflicts tend to occur when one or several of the antagonists views the status quo as sufficiently undesirable and/or unsustainable to prompt forceful pro-action. Notwithstanding widespread perceptions to the contrary, this was not the case of the USSR and the United States during the Cold War. The United States had chosen a policy of containment, as opposed to roll-back, of the Soviet Empire within the limits established as a result of World War II. The Soviet Union seized targets of opportunity outside of its 1945 area of control but avoided direct confrontation with U.S. forces. Messianic language from the USSR on the global victory of communism or from the United States about the end of the "Evil Empire" did not take precedence over the prime Soviet concern of preserving the Warsaw Pact and the U.S. pursuit of containment, or, no less crucially, their mutual confidence that they could achieve these aims without going to war one with the other.

No such generalization can be made about the Middle East, a region in which the very existence of a key state, Israel, is challenged while other states have gone to war with each other (e.g., Iran-Iraq war, and the Gulf War of 1990-91) or are riven by deep internal conflicts. Actors such as Hezbollah, with its organic and functional links with Islamic Iran and Alawite Syria, add to the complexities and dangers. Extreme views and actions vis-à-vis the strategic status quo are widely prevalent. Although the India-Pakistan relationship corresponds to something akin to the U.S.-Soviet "adversarial partnership," that does not apply to radical nonstate actors prevalent in Pakistan with more or less tight links to that country's military intelligence services (Inter-Services Intelligence). The potential for danger is compounded by the variety of such groups: the Pashtu-related Pakistani Taliban, Kashmiri-related groups, and Jihadi militants from the core provinces of Punjab and Sind. Their common characteristics are extreme radicalism, high levels of operational proficiency, and shared enmity of India. Their potential for triggering a conflict between the two countries is substantial, above and beyond the intentions of government officials.

In summary, some 70 years after the launch of the Manhattan Project, there is every reason to upgrade and reinforce nonproliferation policies if nuclear use is to be avoided during the coming decades. Some markers to that end will be laid in the concluding section.

WHAT IS TO BE DONE?

In light of the preceding analysis, the most obvious short run implication is the absolute need to secure a satisfactory conclusion of the Iranian file. Anything that feeds the perception of less-than-full compliance of Iran with the strictest international safeguards or, worse, that creates the impression that recessed deterrence is in place, would lead to further proliferation in the Middle East and beyond. What happens to the Iranian nuclear program will be essential to the future of proliferation and nonproliferation prospects.

In the medium term, those states that share the view that current proliferation trends would have catastrophic outcomes must display greater readiness to make those concessions that could reinforce the nonproliferation regime. Since the vast majority of countries subscribe to the proposition that reinforced nonproliferation norms imply determined moves towards nuclear disarmament by nuclear weapons states, a serious attempt has to be made to test that linkage. In practice, this means the polar opposite of the sort of linkage that led to a vacuous consensus at the 2010 NPT review conference. On that occasion, there was a link between the industrialized states, including the Western nuclear weapons states, suspending their pursuit of the universalization of the IAEA Additional Protocol in exchange for the nonaligned states dropping their insistence on a calendar for nuclear disarmament. No nonproliferation in exchange for no nuclear disarmament. At the 2015 NPT Review Conference, the opportunity will exist to turn that sort of linkage inside out. The recommendations of the International Commission on Nuclear Nonproliferation and Disarmament (ICNND) in 2009[32] offer practical goals in terms of nuclear disarmament of the sort that could be implemented in synergy with a reinforced nonproliferation regime. In particular, the ICNND's report suggests a so-called vantage point of nuclear disarmament compatible with prevailing strategic circumstances but that leads in a 15-year timeframe to a reduction of some 90 percent of the world's nuclear weapons and the capping of the arsenals of the smaller nuclear powers. Such progress, however desirable, cannot be achieved without strong political will, which is unlikely to be on call in the absence of either a successful resolution of the Iran file or an unexpected proliferation wake-up call.

In parallel, multilateral and unilateral policies limiting the spread of reprocessing and enrichment facilities should be pursued, a task that overcapacity in the global market readily justifies in economic terms. Similarly, the entry into industrial service of new technologies that could facilitate proliferation needs to be discouraged (here again, market forces provide some leverage). A strengthening of the control on, and the recycling of, weapons-grade fissile material, along the lines of what has been successfully done during the last 2 decades in the former Soviet Union; the tracking and securing of radioactive sources as promoted by the Nuclear Security Summits; and the reinforcement of the Proliferation Security Initiative's work, notably on the trafficking of proliferation-relevant material and knowledge, are all necessary, not least in reducing the risk of nonstate access to something approaching a nuclear capability. However, such necessary technical measures will only serve their purpose if the political causes of proliferation are also addressed. At heart, the decision to proliferate is political and strategic in nature, and nonproliferation policy needs to provide a broader response than a narrow, technical one. This was the particular genius of the NPT and its ability to generate a bandwagon effect over time; this explains the effectiveness of defense guarantees and related blandishments as nonproliferation tools, and this also means that in certain circumstances broad-spectrum coercion, sometimes including the use of military force, may be required. This policy mix remains entirely relevant. It is the associated doses that need to be reconsidered: tougher nonproliferation norms, a greater readiness to reward the virtuous and act against the wayward, and the acceptance by the nuclear powers and their allies that it is in their interest to accept the trade-offs that may be required for such an outcome to be achieved. The Western powers may and should lead by example here, as they have been trying to do in their handling of the Iran dossier.

ENDNOTES - CHAPTER 2

1. Available from *www.un.org/disarmament/HomePage/SSOD/A-S-10-4.pdf*.

2. Figures derived from *www.iaea.org/safeguards/documents/es2008-tables.pdf*.

3. The so-called Annex 2 countries singled out in the Comprehensive Nuclear Test Ban Treaty.

4. Notably UN Security Council Resolution 418, November 4, 1977.

5. Notably UN Security Council Resolution 707, September 19, 1991.

6. The so-called 123 agreement between India and the United States became effective on December 10, 2008, after the Nuclear Suppliers Group (NSG) granted its waiver.

7. Thomas C. Reed and Danny Stillman, *The Nuclear Express: a Political History of the Bomb and Its Proliferation*, Minneapolis, MN: Zenith, 2009, p. 104.

8. President Kennedy's press conference of March 21, 1963, on "a world in which 15 or 20 or 25 nations may have these weapons (in the 1970s)," available from *www.jfklibrary.org/Research/Research-Aids/Ready-Reference/JFK-Quotations.aspx*.

9. The Partial Test Ban Treaty was signed on October 7, 1963.

10. Such as Larry Collins and Dominique Lapierre, *The Fifth Horseman*, New York: Simon & Schuster, 1980.

11. For instance, in Thomas Schelling's work on the strategy of conflict, Thomas C. Schelling, *The Strategy of Conflict*, Cambridge, MA: Harvard University Press, 1960.

12. See Ken Waltz, "The Spread of Nuclear Weapons: More May be Better," *Adelphi Paper* No. 171, London, UK: International Institute for Strategic Studies (IISS), 1981; and Pierre Gallois, *Stratégie de l'âge nucléaire* (Strategy of the Nuclear Age), Paris, France: Calman-Lévy, 1960.

13. In the framework of the secret French-German-Italian agreement of November 1957 (revealed and declassified after the Cold War). See Georges-Henri Soutou, *L'Alliance incertaine: les rapports politico-stratégiques franco-allemands 1954-1996* (The Uncertain Alliance: Franco-German Political and Strategic Relations, 1954-1996), Paris, France: Fayard, 1996.

14. Michael Dobbs, *One Minute to Midnight: Kennedy, Khrushchev and Castro on the Brink of Nuclear War*, New York: Knopf, 2008.

15. Scott Sagan, *The Limits of Safety: Organizations, Accidents, and Nuclear Weapons*, Princeton, NJ: Princeton University Press, 1995.

16. Such as the unauthorized flight of a B-52H bomber loaded with six nuclear-tipped AGM-129 ACM cruise missiles in August 2009 (leading to the resignation of the Secretary of the Air Force and of the Chief of Staff of the Air Force); or the mistaken shipment to Taiwan of fuses for the nuclear payload of Minuteman intercontinental ballistic missiles in 2006 (leading to the disciplining of 17 senior officers). On these incidents, see *inter alia* Lolita Baldur, "Report slams Pentagon nuke oversight," Associated Press, January 7, 2009, available from *www.nbcnews.com/id/28547375/ns/us_news-military/t/report-slams-pentagon-nuke-oversight/*.

17. Beginning with the Operation BRASS TACKS, Indian military exercises in 1986-87; the Kargil conflict in Kashmir

in 1999; and the very tense standoff that followed for close to half a year the terrorist attack against the Indian Parliament in December 2001.

18. The Soviet reactor at Obninsk was the first to feed power into the electricity grid in 1954; the Chernobyl nuclear accident occurred 32 years later.

19. Elaine Sciolino and Katrin Bennhold, "Chirac Strays from Assailing a Nuclear Iran," *The New York Times*, February 1, 2007 (quote: "Why wouldn't Saudi Arabia do it? Why won't it help Egypt to do it as well?").

20. "I do not have a policy of containment; I have a policy of preventing Iran from obtaining nuclear weapons," Barack Obama, Remarks by the President at American Israel Public Affairs Committee (AIPAC) Policy Conference, speech at the Washington Convention Center, Washington, DC, March 4, 2012, available from *www.whitehouse.gov/the-press-office/2012/03/04/remarks-president-aipac-policy-conference-0*.

21. During the attempted military coup by French generals in Algeria in April 1961, there was a question about the exercise of control on the French nuclear test site in the Sahara and of a nuclear device that was being prepared for testing there. See Bruno Tertrais, "A 'Nuclear Coup'? France, the Algerian War, and the April 1961 Nuclear Test," Henry D. Sokolski and Bruno Tertrais, eds., *Nuclear Weapons Security Crises: What Does History Teach?*, Carlisle, PA: Strategic Studies Institute, U.S. Army War College, 2013, pp. 25-64, available from *www.strategicstudiesinstitute.army. mil/pubs/display.cfm?pubID=1156*.

22. "Al Qaeda has tried to acquire or make nuclear weapons for at least 10 years," quoted from National Commission on Terrorist Attacks Upon the United States, "The 9/11 Commission Report: Final Report of the National Commission on Terrorist Attacks Upon the United States," Washington, DC: U.S. Government Printing Office, 2004, p. 386.

23. China, France, Argentina, Brazil, South Africa, Algeria, among others, only acceded to the NPT after 1990.

24. See "Nuclear Black Markets: Pakistan, A.Q. Khan and the rise of proliferation networks. A Net Assessment," London, UK: IISS, 2007. For a narrower, racier read, see Catherine Collins and Douglas Frantz, *Fallout: the True Story of the CIA's Secret War on Nuclear Trafficking*, New York: Free Press, 2011.

25. For an English-language presentation, see *www.archives. livreblancdefenseetsecurite.gouv.fr/IMG/pdf/white_paper_press_ kit.pdf*.

26. An expression coined, in this meaning, by Nassim Taieb in his *The Black Swan: The Impact of the Highly Improbably*, New York: Random House, 2007.

27. Ken Berry *et al.*, "Delegitimizing Nuclear Weapons: Examining the Validity of Nuclear Deterrence," Monterey, CA: Center for Nonproliferation Studies, May 2010, available from *cns.miis. edu/opapers/pdfs/delegitimizing_nuclear_weapons_may_2010.pdf*.

28. See Tsuyoshi Hasegawa, *Racing for the Enemy: Stalin, Truman, and the Surrender of Japan*, Cambridge, MA: Harvard University Press, 2005; and Richard B. Frank, *Downfall: The End of the Imperial Japanese Empire*, New York: Random House, 1999.

29. The last atmospheric test was conducted by China on October 16, 1980.

30. Otto Frisch and Rudolph Pierls, "Frisch-Peierls Memorandum, March 1940: On the Properties of a Radioactive Superbomb," available from *www.atomicarchive.com/Docs/Begin/Frisch-Peierls.shtml*.

31. For the so-called Operation RYAN (the Russian initials for nuclear missile attack), see Benjamin Fischer, "A Cold War Conundrum. The 1983 Cold War Scare," Washington, DC: CIA Center for Intelligence Studies, 1997, available from *www.cia.gov/ library/center-for-the-study-of-intelligence/csi-publications/books-and-monographs/a-cold-war-conundrum/source.htm*.

32. International Commission on Nuclear Non-proliferation and Disarmament, "Eliminating the Nuclear Threat: A Practical Agenda," Canberra, Australia: Paragon, 2009, available from *icnnd.org/Reference/reports/ent/downloads.html*.

CHAPTER 3

THE HISTORY OF PROLIFERATION OPTIMISM: DOES IT HAVE A FUTURE?

Matthew Kroenig

Should we worry about the spread of nuclear weapons? At first glance, this might appear to be an absurd question. After all, nuclear weapons are the most powerful weapons ever created by man. A single nuclear weapon could vaporize large portions of a major metropolitan area, killing millions of people, and a full-scale nuclear war between superpowers could end life on Earth as we know it. For decades during the Cold War, the public feared nuclear war, and post-apocalyptic nuclear war scenarios became a subject of fascination and terror in popular culture. Meanwhile, scholars carefully theorized the dangers of nuclear weapons and policymakers made nuclear nonproliferation a top national priority. To this day, the spread of nuclear weapons to additional countries remains a foremost concern of U.S. leaders. Indeed, in his 2012 annual threat assessment to the U.S. Congress, Director of National Intelligence James Clapper argued that nuclear proliferation poses one of the greatest threats to U.S. national security.[1]

Recently, however, academics have become more vocal in questioning the threat posed by the spread of nuclear weapons. Students of international politics known as "proliferation optimists" argue that the spread of nuclear weapons might actually be beneficial because it deters great power war and results in greater levels of international stability.[2] Other scholars, whom I label "proliferation anti-obsessionists," maintain that nuclear proliferation is neither good nor bad, but irrelevant.[3] They claim that nuclear weapons do not have any meaningful effect on international politics and that the past 70 years of world history would have been roughly the same, had nuclear weapons never been invented. Some take this line of argument even further and argue that the only real problem is not the nuclear weapons themselves, but great power nonproliferation policy.[4] They argue that the cure that countries like the United States implement in order to prevent other states from acquiring nuclear weapons is much worse than the disease of the spread of nuclear weapons itself.

While these arguments remain provocative, they are far from new. The idea that a few nuclear weapons are sufficient to deter a larger adversary and keep the peace has its origins in the early strategic thinking of the 1940s. Moreover, a critical review of this literature demonstrates that many of these arguments are much less sound than they initially appear. Indeed, both proliferation optimism and proliferation anti-obsessionism rest on internal logical contradictions.

In this chapter, I argue that the spread of nuclear weapons poses a grave threat to international peace and to U.S. national security. Scholars can grab attention by making counterintuitive arguments about nuclear weapons being less threatening than power holders believe them to be, but their provocative claims cannot wish away the very real dangers posed by the spread of nuclear weapons. The more states that possess nuclear weapons, the more likely we are to suffer a number of devastating consequences, including nuclear war, nuclear terrorism, global and regional instability, constrained U.S. freedom of action, weakened alliances, and the further proliferation of nuclear weapons. While it is important not to exaggerate these threats, it would be an even greater sin to underestimate them and, as a result, not take the steps necessary to combat the spread of the world's most dangerous weapons.

The chapter is in three parts. First, I provide a critical review of the proliferation optimism literature, including a careful consideration of the argument's historical origins. Next, I detail the various threats posed by nuclear proliferation, supported by nuclear deterrence theory and historical evidence, and frequently illustrated with a discussion of a case currently on the minds of nonproliferation officials: Iran's advanced nuclear program. I conclude with an implication of my analysis for the scholarly study of nuclear proliferation and for U.S. nonproliferation policy.

AN INTELLECTUAL HISTORY OF PROLIFERATION OPTIMISM

Many of the key pillars of proliferation optimism arguments made today can be found in early Cold War debates about nuclear strategy. These pillars include the ideas that a small nuclear arsenal capable of targeting an enemy's cities is sufficient for deterring a powerful adversary and that nuclear wars, because they would be so devastating for everyone involved, will never be fought. These ideas stood in stark contrast to other strands of deterrence thinking that emphasized counterforce targeting, nuclear vulnerability, nuclear brinkmanship, inadvertent and accidental nuclear escalation, and limited nuclear wars.[5] It is noteworthy that some of the most influential early advocates of minimum deterrence and proliferation optimism (indeed, as we will see, these ideas are mutually reinforcing) cannot truly be understood without reference to the parochial interests and resource-constrained environments in which the strategic thinkers who developed them operated.

Early Academic Writing.

Shortly after the first use of nuclear weapons on Hiroshima and Nagasaki, U.S. strategists began to grapple with the question of what the atomic bomb meant for international peace and security. The first answer given is one that presaged the contemporary proliferation optimism literature, namely, that nuclear weapons are "absolute weapons" that are terrifyingly destructive, that are invulnerable to enemy attack, and that render great power war obsolete.[6]

Perhaps the first person to articulate this position was University of Chicago economist Jacob Viner in a speech to the American Philosophical Society in Philadelphia, PA, on November 16, 1945 — just months after the first use of nuclear weapons on Hiroshima and Nagasaki, Japan.[7] In the speech, Viner argued that counterforce nuclear targeting would be useless and splendid first strikes impossible. In doing so, he laid the basis for subsequent claims about a minimum nuclear posture being sufficient to deter a more powerful adversary. Viner argued, "The atomic bomb, unlike battleships, artillery, airplanes, and soldiers, are not an effective weapon against its own kind. A superior bomb cannot neutralize the inferior bomb of an enemy." Viner went on to argue that the awesome destructive power of nuclear weapons would induce great caution in leaders and possibly produce peace among the major powers. In his words:

the universal recognition that if war does break out, there can be no assurance that the atomic bombs will not be resorted to may make statesmen and people determined to avoid war even where in the absence of the atomic bomb, they would regard it as the only possible procedure under the circumstances for resolving a dispute or a clash of interests.[8]

The proliferation optimism position received further elaboration a few months later in Bernard Brodie's classic book, *The Absolute Weapon*.[9] In great detail, Brodie explained the basic features of the minimum deterrence and proliferation optimism position. He argued that nuclear weapons are invulnerable, ruling out the possibility of an enemy launching a disarming first strike. He also claimed that nuclear weapons have such terrifying effects that they would make war too costly to wage, potentially leading to peace. In his most oft-quoted line, Brodie declared, "Thus far the chief purpose of our military establishment has been to win wars. From now on its chief purpose must be to avert them."[10]

The optimism position was quickly countered, however, in what would become the first incarnation of the optimism-pessimism debate, predating the now-famous Kenneth Waltz-Scott Sagan debate by over 30 years.[11] Beginning with a series of basing studies done for the Department of Defense (DoD), Albert Wohlstetter, an American strategist working at the Rand Corporation in Santa Monica, CA, argued that nuclear weapons are not as invulnerable as they appeared to optimists like Brodie. Rather, he argued that the "balance of terror" that optimists had written so eloquently about, was actually quite "delicate."[12] He demonstrated that U.S. nuclear forces were potentially vulnerable to a Soviet first strike, and that this vulnerability could tempt Moscow to launch a nuclear war. His study led to a number of improvements in the survivability of U.S. nuclear forces, including the moving of U.S. air bases beyond the range of Soviet bombers and the hardening of ballistic missile silos.

Wohlstetter's study also undermined a key pillar of proliferation optimism. If nuclear forces were potentially vulnerable, encouraging an enemy nuclear attack, it was not a great leap to argue that the spread of nuclear weapons would not necessarily lead to peace. Just as a belief in minimum deterrence supports the idea of a nuclear peace, attention to nuclear vulnerability and counterforce nuclear war necessarily leads to proliferation pessimism. Indeed, it is difficult to find analysts who simultaneously believe that nuclear posture matters and that the spread of nuclear weapons necessarily leads to peace. It should come as no surprise, therefore, that Albert Wohlstetter was a proliferation pessimist. In subsequent writing, Wohlstetter catalogued the potential downsides of nuclear proliferation for U.S. interests, even if nuclear weapons spread to friendly states, such as America's North Atlantic Treaty Organization (NATO) allies.[13] First, he identified nuclear war as a potential problem. A few nuclear weapons would not be enough for deterrence, but rather "The problem of deterring a major power requires a continuing effort because the requirements for deterrence will change with the counter-measures taken by the major power."[14] But, if that investment was not made, deterrence could fail and nuclear war could result. Second, Wohlstetter worried that the spread of nuclear weapons within the NATO alliance would undermine alliance cohesion by making the allied states less interdependent. Third, Wohlstetter forecasted that the spread of nuclear weapons would lead to the further spread of nuclear weapons. He criticized U.S. decisionmakers for calculating the pros and cons of nuclear proliferation to an "Nth" state without also figuring in the potential negative consequences of what he called the "N+1 problem."[15]

The optimism-pessimism debate did not remain relegated to the ivory tower for long, however. Shortly thereafter, powerful players in government began adapting the ideas of proliferation to fit their strategic circumstances and to advance their parochial interests.

The French *Force de Frappe.*

In 1960, France entered the nuclear club with its first nuclear test.[16] French leaders, including President Charles de Gaulle, did not believe that France could rely on the United States and NATO to provide for France's security. As de Gaulle would famously ask, would Washington really be willing to trade New York for Paris in a nuclear war? France, therefore, acquired an indigenous nuclear weapons capability that would allow Paris to pursue a more independent foreign policy. Having developed the bomb, French strategic and military thinkers were soon confronted with a new problem: how would they use their nuclear weapons? In the early- and mid-1960s, therefore, France began developing a nuclear doctrine.

At the same time that U.S. and Soviet thinkers began articulating the aspects of nuclear doctrine that would come to characterize the superpower nuclear competition throughout the Cold War (counterforce nuclear targeting, limited nuclear options, the importance of assured destruction, the advantages provided by nuclear superiority over rivals, and the pursuit of active and passive defenses), France, a medium power operating with fewer resources than the superpowers, was compelled to develop a more modest nuclear strategy. In large part due to its limited means, France eschewed the warfighting strategies of the superpowers and instead developed a minimal deterrent doctrine, in which French military planners aimed to be able to threaten significant damage against Soviet cities in the event of a Soviet invasion of France.[17]

Unlike the superpowers, France did not have the luxury of working down from strategy to capabilities, but instead had to work backwards, developing a strategy around given capabilities. As French strategic thinker Colonel Pierre Marie Gallois put it, France pursued a nuclear "strategy of the means."[18] In the words of de Gaulle, "We do not have the ambition to make a force as powerful as those of the Americans or Soviets, but a force proportionate to our means, our needs, and our size."[19] Accordingly, the key pillars of French doctrine reflected France's resource constraints. "Deterrence of the strong by the weak" was the belief that a small state could deter a much larger adversary as long as the smaller state had the ability to conduct a nuclear attack against the larger state's cities.[20] "Sufficiency" was the idea that a small number of nuclear weapons was sufficient for deterrence and that anything more was overkill.[21]

France's small size and lack of strategic depth prevented it from adopting the warfighting postures of the superpowers. As Gallois put it, "France has nothing to cede that would not be herself."[22] France's vulnerability, therefore, demanded that France launch an immediate and full-scale nuclear attack on an adversary at the initiation of hostilities. Unable to build a large enough arsenal to maintain an assured destruction capability against the Soviet Union, France aimed only, according to Gallois, to "tear an arm" off the aggressor.[23] While U.S. Secretary of Defense Robert McNamara famously assessed that destroying large portions of the Soviet population and economy was necessary to deter Moscow, French thinkers thought that the Soviet Union could be deterred if France could inflict damage on the Soviet Union roughly equivalent to the destruction of the entire country of France. In the words of one French official:

> French nuclear forces have been calculated to permit reaching a population of the adversary of the same order as that of our own country. If France were destroyed, our adversary would lose the equivalent of France.[24]

A lack of adequate delivery vehicles also prevented France from following a counterforce strategy. France's plans for the development of a land-based intercontinental ballistic missile (ICBM) were canceled due to their expense, leaving Paris with a countervalue option only. As strategist Raymond Barre described:

> it was the less costly option. . .France, a medium-sized nation with limited resources, cannot pretend seeking parity with the two great nuclear powers. The only way which is opened to us is that of the current strategy.[25]

Like proliferation optimists on the other side of the Atlantic, French strategists believed that if a small nuclear arsenal in France could deter the Soviet Union, then the spread of nuclear weapons elsewhere could have a pacifying effect on international politics more broadly. As Gallois argued, a nuclear arsenal:

increases the risk, counsels discretion, and conse-
quently strengthens the strategy of dissuasion. As
atomic armament grows more widespread . . . the
notion of dissuasion will also become more common,
each nation practicing it according to its means. . . . It
will not be long before we may have to give up war
altogether.[26]

Unsurprisingly, pessimists in the United States
were skeptical of French strategy and doctrine. Albert
Wohlstetter assessed that if the United States struggled to develop a survivable nuclear arsenal capable
of deterring the Soviet Union, then the French did
not stand a chance of developing a truly independent
deterrent capability. At the end of the day, thought
Wohlstetter, "The burden of deterring a general war
as distinct from limited wars is still likely to be on the
United States and therefore, so far as our allies are
concerned, on the alliance."[27]

In sum, the notion that a few nuclear weapons
would be sufficient to deter great power war was
warmly welcomed and advocated by strategic thinkers in Paris. Once it became a nuclear weapon state,
France's resource-constrained environment did not
permit it to adopt anything other than a minimum
deterrent posture. France was not the only place,
however, where nuclear doctrines emphasizing minimum deterrence were developed in response to the
available means.

Polaris.

In the late-1950s and early-1960s, a similar minimum deterrence strand was developing among U.S.
nuclear strategists.[28] Like in France, circumstances

would compel military planners, this time in the U.S. Navy, to argue that a few nuclear weapons would be sufficient to deter a more powerful foe, paving the way for proliferation optimists that would follow in their footsteps.

In the early stages of the Cold War, the U.S. Navy was the only major U.S. military service cut out of the strategic nuclear mission. This would have major implications for service budgets and interservice rivalries, as nuclear capabilities were of paramount importance in the Cold War's superpower rivalry, and the Navy desired a foothold in the nuclear game. The Navy sought to edge its way into a role by developing "super carriers," aircraft carriers large enough for nuclear-armed fighters to take off and land, but the program was cancelled by President Harry Truman in 1949 for budgetary reasons.

Then, in the mid-1950s, under the leadership of Admiral Arleigh Burke, the Navy began developing the innovative Polaris submarine launch ballistic missile (SLBM) system. Polaris provided the Navy with a nuclear role. Indeed, Burke argued that Polaris's unique advantages, such as greater survivability, made it a candidate to replace the more vulnerable fixed ICBMs operated by the Air Force.

Critics in other services soon countered, however, that SLBMs did not meet the requirements of U.S. nuclear strategy. SLBMs, unlike bombers and land-based ICBMs, were not accurate enough to engage in counterforce targeting. Moreover, submarines could not carry sufficient firepower to guarantee an assured destruction capability against the Soviet Union.

The Navy could not credibly argue that Polaris had capabilities that it did not have, but they could, and did, challenge the prevailing logic of deterrence.

In a prize-winning essay, Paul Bracken, a naval commander working under Burke, coined the term "finite deterrence." Bracken, and eventually Burke, argued that the massive nuclear attacks and counterforce targeting envisioned by the Air Force and the Army were unnecessary. Rather, they claimed that a few survivable nuclear weapons capable of destroying enemy soft targets—the precise capabilities provided by Polaris—were sufficient for deterrence.

In the end, Burke and the Navy lost the bureaucratic battle. While SLBMs became a central element of U.S. nuclear force structure, they did not replace bombers and ICBMs. Arguments about maintaining superiority across the entire spectrum of capabilities were more persuasive in the context of a heating up cold war. Nevertheless, the ideas of "finite" and "minimum deterrence" developed by Bracken and Burke, motivated in no small part to advance the Navy's position in an interservice competition, are alive and well today in the writings of academic proliferation optimists.

Proliferation Optimism.

Proliferation optimism received what may have been its clearest articulation by Kenneth Waltz in his seminal 1981 Adelphi paper, "The Spread of Nuclear Weapons: More May Be Better."[29] In this and subsequent works, Waltz argued that the spread of nuclear weapons has beneficial effects on international politics. He maintained that states, fearing a catastrophic nuclear war, will be deterred from going to war with other nuclear-armed states. As more and more states acquire nuclear weapons, therefore, there are fewer states against which other states will be willing to wage war. The spread of nuclear weapons, according to Waltz, leads to greater levels of international stability. Looking to the empirical record, he argued that the introduction of nuclear weapons in 1945 coincided with an unprecedented period of peace among the great powers. While the United States and the Soviet Union engaged in many proxy wars in peripheral geographic regions during the Cold War, they never engaged in direct combat. Despite regional scuffles involving nuclear-armed states in the Middle East, South Asia, and East Asia, none of these conflicts resulted in a major theater war. This lid on the intensity of conflict, according to Waltz, was the direct result of the stabilizing effect of nuclear weapons.

Following in the path blazed by the strategic thinkers reviewed here, Waltz argued that the requirements for deterrence are not high. He argued that, contrary to the behavior of the Cold War superpowers, a state need not build a large arsenal with multiple survivable delivery vehicles in order to deter its adversaries. Rather, he claimed that a few nuclear weapons are sufficient for deterrence. Indeed, he went even further, asserting that any state will be deterred even if it merely suspects its opponent **might** have a few nuclear weapons because the costs of getting it wrong are simply too high.

Not even a nuclear accident is a concern, according to Waltz, because leaders in nuclear-armed states understand that if they ever lost control of nuclear weapons, resulting in an accidental nuclear exchange, the nuclear retaliation they would suffer in response would be catastrophic. Nuclear-armed states, therefore, have strong incentives to maintain control of their nuclear weapons. Not even new nuclear states, which lack experience managing nuclear arsenals, would ever allow nuclear weapons to be used or to fall into the wrong hands.

Following Waltz, many other scholars advanced arguments in the proliferation optimism school. For example, Bruce Bueno de Mesquite and William Riker explore the "merits of selective nuclear proliferation."[30] John Mearsheimer made the case for a "Ukrainian nuclear deterrent" following the collapse of the Soviet Union.[31] In the run up to the 2003 Gulf War, John Mearsheimer and Steven Walt argued that we should not worry about a nuclear-armed Iraq because a nuclear-armed Iraq can be deterred.[32] In recent years, Barry Posen and other scholars have argued that nuclear proliferation in Iran does not pose an unmanageable threat, again arguing that a nuclear-armed Iran can be deterred.[33]

What's Wrong with Proliferation Optimism?

The proliferation optimist position, while having a distinguished pedigree, has several major flaws. Many of these weaknesses have been chronicled in brilliant detail by Sagan and other contemporary proliferation pessimists.[34] Rather than repeat these substantial efforts, I will use this section to offer some other critiques of the recent incarnations of proliferation optimism.

First and foremost, proliferation optimists present an oversimplified view of nuclear deterrence theory. Apart from the optimists, leading nuclear deterrence theorists believe that nuclear proliferation contributes to a real risk of nuclear war even in a situation of Mutually Assured Destruction (MAD) among rational states.[35]

In the 1940s, Viner, Brodie, and others argued that the advent of Mutually Assured Destruction (MAD) rendered war among major powers obsolete, but nuclear deterrence theory soon advanced beyond that simple understanding.[36] After all, great power political competition does not end with nuclear weapons. Nuclear-armed states still seek to threaten nuclear-armed adversaries. States cannot credibly threaten to launch a suicidal nuclear war, but they still want to coerce their adversaries. This leads to a credibility problem: how can states credibly threaten a nuclear-armed opponent? Since the 1960s, academic nuclear deterrence theory has been devoted almost exclusively to answering this question.[37] Unfortunately for proliferation optimists, the answers do not give us reasons to be optimistic.

Thomas Schelling was the first to devise a rational means by which states can threaten nuclear-armed opponents.[38] He argued that leaders cannot credibly threaten to intentionally launch a suicidal nuclear war, but they can make a "threat that leaves something to chance."[39] They can engage in a process, the nuclear crisis, which increases the risk of nuclear war in an attempt to force a less resolved adversary to back down. As states escalate a nuclear crisis, there is an increasing probability that the conflict will spiral out of control and result in an inadvertent or accidental nuclear exchange. As long as the benefit of winning the crisis is greater than the incremental increase in the risk of nuclear war, threats to escalate nuclear crises are inherently credible. In these games of nuclear brinkmanship, the state that is willing to run the greatest risk of nuclear war before backing down will win the crisis as long as it does not end in catastrophe. It is for this reason that Thomas Schelling called great power politics in the nuclear era a "competition in risk taking."[40] This does not mean that states eagerly bid up the risk of nuclear war. Rather, they face gut-wrenching decisions at each stage of the crisis. They can quit the crisis to avoid nuclear war, but only by ceding an important geopolitical issue to an opponent. Or they can escalate the crisis in an attempt to prevail, but only at the risk of suffering a possible nuclear exchange.

Since 1945, there have been many high stakes nuclear crises (by my count, there have been 20) in which "rational" states like the United States run a frighteningly real risk of nuclear war.[41] By asking whether states can be deterred, therefore, proliferation optimists ask the wrong question. The right question to ask is: What risk of nuclear war is a specific state willing to run against a particular opponent in a given crisis? Optimists are likely correct when they assert that Iran will not intentionally commit national suicide by launching a bolt-from-the-blue nuclear attack on the United States or Israel. This does not mean that Iran will never use nuclear weapons, however. Indeed, it is almost inconceivable to think that a nuclear-armed Iran would not, at some point, find itself in a crisis with another nuclear-armed power. It is also inconceivable that in those circumstances, Iran would not be willing to run any risk of nuclear war in order to achieve its objectives. If a nuclear-armed Iran and the United States or Israel have a geopolitical conflict in the future, over, for example, the internal politics of Syria, an Israeli conflict with Iran's client, Hezbollah, the U.S. presence in the Persian Gulf, passage through the Strait of Hormuz, or some other issue, do we believe that Iran would immediately capitulate? Or is it possible that Iran would push back, possibly even brandishing nuclear weapons in an attempt to coerce its adversaries? If the latter, there is a real risk that proliferation to Iran could result in nuclear war.

An optimist might counter that nuclear weapons will never be used, even in a crisis situation, because states have such a strong incentive, namely national survival, to ensure that nuclear weapons are not used. But, this objection ignores the fact that leaders operate under competing pressures. Leaders in nuclear-armed states also have very strong incentives to convince their adversaries that nuclear weapons could very well be used. Historically, we have seen that leaders take actions in crises, such as placing nuclear weapons on high alert and delegating nuclear launch authority to low level commanders, to increase purposely the risk of accidental nuclear war in an attempt to force less resolved opponents to back down.

Moreover, not even the optimists' first principles about the irrelevance of nuclear posture stand up to scrutiny. Not all nuclear wars would be equally devastating.[42] Any nuclear exchange would have devastating consequences no doubt, but if a crisis were to spiral out of control and result in nuclear war, any sane leader would rather be facing a country with five nuclear weapons than one with 35,000. Similarly, any sane leader would be willing to run a greater risk of nuclear war against the former state than against the latter. Indeed, systematic research has demonstrated that states are willing to run greater risks and are, therefore, more likely to win nuclear crises when they enjoy nuclear superiority over their opponents.[43] Proliferation optimists miss this point, however, because they are still mired in 1940s deterrence theory. It is true that no rational leader would choose to launch a nuclear war, but, depending on the context, she would almost certainly be willing to risk one.

Nuclear deterrence theorists have proposed a second scenario under which rational leaders could instigate a nuclear exchange: limited nuclear war.[44] By launching a single nuclear weapon against a small city, for example, a nuclear-armed state could signal its willingness to escalate a crisis, while leaving its adversary with enough left to lose to deter the adversary from launching a full-scale nuclear response. In a future crisis between China and the United States, for example, China could choose to launch a nuclear strike on Honolulu to demonstrate its seriousness. In that situation, with the continental United States intact, would Washington choose to launch a full-scale nuclear war on China that could result in the destruction of many more American cities? Or would it back down? China might decide to strike, calculating that Washington would prefer a humiliating retreat over a full-scale nuclear war. If launching a limited nuclear war could be rational, it follows that the spread of nuclear weapons increases the risk of nuclear use. Again, by ignoring contemporary developments in scholarly discourse and relying exclusively on understandings of nuclear deterrence theory that became obsolete decades ago, optimists fail to make a compelling case.

The optimists also err by confusing stability with the national interest. Even if the spread of nuclear weapons contributes to greater levels of international stability (which discussions herein suggest it might not), it does not necessarily follow that the spread of nuclear weapons is in the U.S. interest. There might be other national goals that trump stability, such as reducing to zero the risk of nuclear war in important geopolitical regions. Optimists might argue that South Asia is more stable because India and Pakistan both possess nuclear weapons, but certainly the risk of nuclear war is higher than if nuclear weapons did not exist on the subcontinent. In addition, it is wrong to assume that stability is always in the U.S. national interest. Sometimes it is, but sometimes it is not. If stability is obtained because Washington is deterred from using force against a nuclear-armed adversary in a situation where using force could have advanced national goals, stability harms, rather than advances, U.S. national interests.

The final gaping weakness in the proliferation optimist argument, however, is that it rests on a logical contradiction. This might come as a surprise to some, given that optimists are sometimes portrayed as hard-headed thinkers, following their premises to their logical conclusions. But, the contradiction at the heart of the optimist argument is glaring and simple to understand: either the probability of nuclear war is zero, or it is nonzero, but it cannot be both. If the probability of nuclear war is zero, then nuclear weapons should have no deterrent effect. States will not be deterred by a nuclear war that could never occur and states should be willing to launch large-scale wars against nuclear-armed states intentionally. In this case, proliferation optimists cannot conclude that the spread of nuclear weapons is stabilizing.

If, on the other hand, the probability of nuclear war is nonzero, then there is a real danger that the spread of nuclear weapons increases the probability of a catastrophic nuclear war. If this is true, then proliferation optimists cannot be certain that nuclear weapons will never be used.

In sum, either the spread of nuclear weapons raises the risk of nuclear war and, in so doing, deters large-scale conventional conflict or there is no danger that nuclear weapons will be used, and the spread of nuclear weapons does not increase international stability. But, despite the claims of the proliferation optimists, it is nonsensical to argue that nuclear weapons will never be used and to simultaneously claim that their spread contributes to international stability.

Proliferation Anti-obsessionists.

Other scholars, whom I label "anti-obsessionists," argue that the spread of nuclear weapons has neither been good nor bad for international politics, but rather irrelevant. They argue that academics and policymakers concerned about nuclear proliferation spend too much time and energy obsessing over nuclear weapons, that, at the end of the day, are not all that important.

In *Atomic Obsession*, John Mueller argues that widespread fears about the threat of nuclear proliferation are overblown.[45] He acknowledges that policymakers and experts have often worried that the spread of nuclear weapons could lead to nuclear war, nuclear terrorism, and cascades of nuclear proliferation, but he then sets about systematically challenging each of these fears. He contends that nuclear weapons have had little effect on the conduct of international diplomacy and that world history would have been roughly the same had nuclear weapons never been invented. Finally, Mueller concludes by arguing that the real problem is not nuclear proliferation but nuclear nonproliferation policy, because states do harmful things in the name of nonproliferation, such as take military action and deny countries access to nuclear technology for peaceful purposes.

Similarly, Ward Wilson argues that, despite the belief held by optimists and pessimists alike, nuclear weapons are not useful tools of deterrence.[46] In his study of the end of World War II, for example, Wilson argues that it was not the U.S. use of nuclear weapons on Hiroshima and Nagasaki that forced Japanese surrender, but a variety of other factors, including the Soviet Union's decision to enter the war. If the actual use of nuclear weapons was not enough to convince a country to capitulate to its opponent, he argues, then there is little reason to think that the mere threat of nuclear use has been important to keeping the peace over the past half-century. Leaders of nuclear-armed states justify nuclear possession by touting their deterrent benefits, but if nuclear weapons have no deterrent value, there is no reason, Wilson claims, to keep them.

Finally, Anne Harrington de Santana argues that nuclear experts "fetishize" nuclear weapons.[47] Just as capitalists, according to Karl Marx, bestow magical qualities on money, she argues that leaders and national security experts do the same thing to nuclear weapons. Nuclear deterrence as a critical component of national security strategy, according to Harrington de Santana, is not inherent in the technology of nuclear weapons themselves, but is rather the result of how leaders in countries around the world think about them. In short, she argues, "Nuclear weapons are powerful because we treat them as powerful."[48] But, she maintains, we could just as easily "defetishize" them, treating them as unimportant and therefore rendering them obsolete. She concludes that "Perhaps some day, the deactivated nuclear weapons on display in museums across the United States will be nothing more than a reminder of how powerful nuclear weapons used to be."[49]

The anti-obsessionists make some thought-provoking points and may help reign in some of the most hyperbolic accounts of the effect of nuclear proliferation. They remind us, for example, that our worst fears have not been realized, at least not yet. Yet, by taking the next step and arguing that nuclear weapons have been, and will continue to be, irrelevant, they go too far. Their arguments call to mind the story about the man who jumps to his death from the top of a New York City skyscraper and, when asked how things are going as he passes the 15th story window, replies, "So far, so good."

The idea that world history would have been largely unchanged had nuclear weapons never been invented is a provocative one, but it is also unfalsifiable. There is good reason to believe that world history would have been different, and in many ways better, had certain countries not acquired nuclear weapons.

Let us take Pakistan as an example. Pakistan officially joined the ranks of the nuclear powers in May 1998 when it followed India in conducting a series of nuclear tests. Since that time, Pakistan has been a poster child for the possible negative consequences of nuclear proliferation. Pakistan's nuclear weapons have led to further nuclear proliferation as Pakistan, with the help of rogue scientist A. Q. Khan, transferred uranium enrichment technology to Iran, Libya, and North Korea.[50] Indeed, part of the reason that North Korea and Iran possess uranium enrichment programs today is because they got help from Pakistan. Pakistan has arguably become more aggressive since acquiring nuclear weapons, displaying an increased willingness to sponsor cross-border incursions into India with terrorists and irregular forces.[51] In a number of high-stakes nuclear crises between India and Pakistan, U.S. officials worried that the conflicts could escalate and intervened diplomatically to prevent a nuclear exchange on the subcontinent. The U.S. Government also worries about the safety and security of Pakistan's nuclear arsenal, fearing that Pakistan's nukes could fall into the hands of terrorists in the event of a state collapse or a break down in nuclear security. We still have not witnessed the full range of consequences arising from Pakistani nuclear proliferation. Islamabad has only possessed the bomb for a little over a decade, but they are likely to keep it for decades to come. It is possible that we could still one day witness a nuclear war between India and Pakistan. In short, Pakistan's nuclear capability has already had deleterious effects on U.S. national security. and these threats are only likely to grow over time.

In addition, the anti-obsessionists are incorrect to argue that the cure of U.S. nuclear nonproliferation policy is worse than the disease of proliferation. Many observers would agree with Mueller that the U.S. invasion of Iraq in 2003 was a disaster, costing much in the way of blood and treasure and offering little strategic benefit. But the Iraq War is hardly representative of U.S. nonproliferation policy. For the most part, nonproliferation policy operates in the mundane realm of legal frameworks, negotiations, inspections, sanctions, and a variety of other tools. Even occasional preventive military strikes on nuclear facilities have been far less calamitous than the Iraq War. Indeed, the Israeli strikes on nuclear reactors in Iraq and Syria in 1981 and 2007, respectively, produced no meaningful military retaliation and a muted international response. Moreover, the idea that the Iraq War was primarily about nuclear nonproliferation is a contestable one, with Saddam Hussein's history of aggression, the unsustainability of maintaining the pre-war containment regime indefinitely, Saddam's ties to terrorist groups, his past possession and use of chemical and biological weapons, and the window of opportunity created by September 11, 2001 (9/11), all serving as possible prompts for U.S. military action in the spring of 2003.

The claim that nonproliferation policy is dangerous because it denies developing countries access to nuclear energy also rests on shaky ground. If anything, the global nonproliferation regime has, on balance, increased access to nuclear technology. Does anyone really believe that countries like Algeria, Congo, and Vietnam would have nuclear reactors today were it not for Atoms for Peace, Article IV of the Nuclear Nonproliferation Treaty (NPT), and other aspects of the nonproliferation regime that have provided developing states with nuclear technology in exchange for promises to forgo nuclear weapons development? Moreover, the sensitive fuel-cycle technology denied by the Nuclear Suppliers Group (NSG) and other supply control regimes is not necessary to the development of a vibrant nuclear energy program as the many countries that have fuel-cycle services provided by foreign nuclear suppliers clearly demonstrate. Finally, the notion that nuclear energy is somehow the key to lifting developing countries from third to first world status does not find support. Given the large upfront investments, the cost of back-end fuel management and storage, and the ever-present danger of environmental catastrophe exemplified most recently by the Fukushima disaster in Japan, many argue that nuclear energy is not a cost-effective source of energy (if all the externalities are taken into account) for any country, not to mention those developing states least able to manage these myriad challenges.

Taken together, therefore, the argument that nuclear nonproliferation policy is more dangerous than the consequences of nuclear proliferation, including possible nuclear war, is untenable. Indeed, it would certainly come as a surprise to the mild man-nered diplomats and scientists who staff the International Atomic Energy Agency, the global focal point of the nuclear nonproliferation regime, located in Vienna, Austria.

The anti-obsessionsists, like the optimists, also walk themselves into logical contradictions. In this case, their policy recommendations do not necessarily follow from their analyses. Wilson argues that nuclear weapons are irrelevant, and therefore we should eliminate them.[52] But if nuclear weapons are really so irrelevant, why not just keep them lying around? They will not cause any problems if they are as meaningless as anti-obsessionists claim, and it is certainly more cost effective to do nothing than to negotiate complicated international treaties and to dismantle thousands of warheads, delivery vehicles, and their associated facilities.

Finally, the idea that nuclear weapons are only important because we think they are powerful is arresting, but false. There are properties inherent in nuclear weapons that can be used to create military effects that simply cannot, at least not yet, be replicated with conventional munitions. If a military planner wants to quickly destroy a city on the other side of the planet, his only option today is a nuclear weapon mounted on an ICBM. Therefore, if the collective "we" suddenly decided to "defetishize" nuclear weapons by treating them as unimportant, it is implausible that some leader somewhere would not independently come to the idea that nuclear weapons could advance his or her country's national security and thereby re-fetishize them.

In short, the optimists and anti-obsessionists have brought an important perspective to the nonproliferation debate. Their arguments are provocative and raise the bar for those who wish to argue that the spread of nuclear weapons is indeed a problem. Nevertheless, their counterintuitive arguments are not enough to wish away the enormous security challenges posed by the spread of the world's most dangerous weapons. These myriad threats will be considered in the next section.

WHY NUCLEAR PROLIFERATION IS A PROBLEM

The spread of nuclear weapons poses a number of severe threats to international peace and U.S. national security, including nuclear war, nuclear terrorism, global and regional instability, constrained freedom of action, weakened alliances, and further nuclear proliferation. This section explores each of these threats in turn.

Nuclear War.

The greatest threat posed by the spread of nuclear weapons is nuclear war. The more states in possession of nuclear weapons, the greater the probability that somewhere, someday, there will be a catastrophic nuclear war. A nuclear exchange between the two superpowers during the Cold War could have arguably resulted in human extinction, and a nuclear exchange between states with smaller nuclear arsenals, such as India and Pakistan, could still result in millions of deaths and casualties, billions of dollars of economic devastation, environmental degradation, and a parade of other horrors.

To date, nuclear weapons have only been used in warfare once. In 1945, the United States used nuclear weapons on Hiroshima and Nagasaki, bringing World War II to a close. Many analysts point to the 65-plus-year tradition of nuclear nonuse as evidence that nuclear weapons are unusable, but it would be naïve to think that nuclear weapons will never be used again simply because they have not been used for some time. After all, analysts in the 1990s argued that worldwide economic downturns like the great depression were a thing of the past, only to be surprised by the dot-com bubble bursting in the late-1990s and the Great Recession of late-2000s.[53] This author, for one, would be surprised if nuclear weapons are not used again sometime in my lifetime.

Before reaching a state of MAD, new nuclear states go through a transition period in which they lack a secure second-strike capability. In this context, one or both states might believe that it has an incentive to use nuclear weapons first. For example, if Iran acquires nuclear weapons, neither Iran, nor its nuclear-armed rival, Israel, will have a secure second-strike capability. Even though it is believed to have a large arsenal, given its small size and lack of strategic depth, Israel might not be confident that it could absorb a nuclear strike and respond with a devastating counterstrike. Similarly, Iran might eventually be able to build a large and survivable nuclear arsenal, but, when it first crosses the nuclear threshold, Tehran will have a small and vulnerable nuclear force.

In these pre-MAD situations, there are at least three ways that nuclear war could occur. First, the state with the nuclear advantage might believe it has a splendid first strike capability. In a crisis, Israel might, therefore, decide to launch a preventive nuclear strike to disarm Iran's nuclear capabilities and eliminate the threat of nuclear war against Israel. Indeed, this incentive might be further increased by Israel's aggressive strategic culture that emphasizes preemptive action. Second, the state with a small and vulnerable nuclear arsenal, in this case Iran, might feel "use 'em or loose 'em" pressures. That is, if Tehran believes that Israel might launch a preemptive strike, Iran might decide to strike first rather than risk having its entire nuclear arsenal destroyed. Third, as Thomas Schelling has argued, nuclear war could result due to the reciprocal fear of surprise attack.[54] If there are advantages to striking first, one state might start a nuclear war in the belief that war is inevitable and that it would be better to go first than to go second. In a future Israeli-Iranian crisis, for example, Israel and Iran might both prefer to avoid a nuclear war but decide to strike first rather than suffer a devastating first attack from an opponent.

Even in a world of MAD, there is a risk of nuclear war. Rational deterrence theory assumes nuclear-armed states are governed by rational leaders who would not intentionally launch a suicidal nuclear war. This assumption appears to have applied to past and current nuclear powers, but there is no guarantee that it will continue to hold in the future. For example, Iran's theocratic government, despite its inflammatory rhetoric, has followed a fairly pragmatic foreign policy since 1979, but it contains leaders who genuinely hold millenarian religious worldviews and who could one day ascend to power and have their finger on the nuclear trigger. We cannot rule out the possibility that, as nuclear weapons continue to spread, some leader will choose to launch a nuclear war, knowing full well that it could result in self-destruction.

One does not need to resort to irrationality, however, to imagine a nuclear war under MAD. Nuclear weapons may deter leaders from intentionally launching full-scale wars, but they do not mean the end of international politics. As discussed previously, nuclear-armed states still have conflicts of interest, and leaders still seek to coerce nuclear-armed adversaries. This leads to the credibility problem that is at the heart of modern deterrence theory: How can you credibly threaten to attack a nuclear-armed opponent? Deterrence theorists have devised at least two answers to this question. First, as stated earlier, leaders can choose to launch a limited nuclear war.[55] This strategy might be especially attractive to states in a position of conventional military inferiority that might have an incentive to escalate a crisis quickly. During the Cold War, the United States was willing to use nuclear weapons first to stop a Soviet invasion of Western Europe, given NATO's conventional inferiority. As Russia's conventional military power has deteriorated since the end of the Cold War, Moscow has come to rely more heavily on nuclear weapons in its strategic doctrine. Indeed, Russian strategy calls for the use of nuclear weapons early in a conflict (something that most Western strategists would consider to be escalatory) as a way to de-escalate a crisis. Similarly, Pakistan's military plans for nuclear use in the event of an invasion from conventionally stronger India. Finally, Chinese generals openly talk about the possibility of nuclear use against a U.S. superpower in a possible East Asia contingency.

Second, as was also discussed earlier, leaders can make a "threat that leaves something to chance."[56] They can initiate a nuclear crisis. By playing these risky games of nuclear brinkmanship, states can increase the risk of nuclear war in an attempt to force a less resolved adversary to back down. Historical crises have not resulted in nuclear war, but many of them, including the 1962 Cuban Missile Crisis, have come close. Scholars have documented historical incidents when accidents could have led to war.[57] When we think about future nuclear crisis dyads, such as Iran and Israel, there are fewer sources of stability than existed during the Cold War, meaning that there is a very real risk that a future Middle East crisis could result in a devastating nuclear exchange.

Nuclear Terrorism.

The spread of nuclear weapons also increases the risk of nuclear terrorism.[58] It used to be said that "Terrorists want a lot of people watching, not a lot of people dead," but the terrorist attacks of 9/11 changed expert perceptions of the terrorist threat.[59] These attacks demonstrated that al-Qaeda and other modern terrorist groups are interested in imposing massive casualties, and there are few better ways of killing large numbers of civilians than detonating a nuclear weapon in a major metropolitan area. While 9/11 was one of the greatest tragedies in American history, it would have been much worse had Osama Bin Laden been able to acquire nuclear weapons. Osama Bin Laden declared it a "religious duty" for al-Qaeda to acquire nuclear weapons, and radical clerics have issued *fatwas* declaring it permissible to use nuclear weapons in jihad against the West.[60] Unlike states, which can be more easily deterred, there is little doubt that if terrorists acquired nuclear weapons, they would use them. Indeed, in recent years, many U.S. politicians and security analysts have agreed that nuclear terrorism poses the greatest threat to U.S. national security.[61]

Wanting nuclear weapons and actually possessing them, however, are two different things, and many analysts have pointed out the tremendous hurdles that terrorists would have to overcome to acquire nuclear weapons.[62] Nevertheless, as nuclear weapons spread, the possibility that they will eventually fall into terrorist hands increases. States could intentionally transfer nuclear weapons, or the fissile material required to build them, to terrorist groups. There are good reasons why a state might be reluctant to transfer nuclear weapons to terrorists, but as nuclear weapons spread, the possibility that a leader might someday purposely arm a terrorist group increases. Some fear, for example, that Iran, with its close ties to Hamas and Hezbollah, might be at a heightened risk of transferring nuclear weapons to terrorists. Moreover, even if no state would ever intentionally transfer nuclear capabilities to terrorists, a new nuclear state, with underdeveloped security procedures, might be vulnerable to theft, allowing terrorist groups or corrupt or ideologically motivated insiders to transfer dangerous material to terrorists. There is evidence, for example, that representatives from Pakistan's atomic energy establishment met with al-Qaeda members to discuss a possible nuclear deal.[63] Finally, a nuclear-armed state could collapse, resulting in a breakdown of law and order and a loose nuclear weapons problem. U.S. officials are currently very concerned about what would happen to Pakistan's nuclear weapons if the government were to fall. As nuclear weapons spread, this problem is only further amplified. Iran is a country with a history of revolutions and a government with a tenuous hold on power. The regime change that

Washington has long dreamed about in Tehran could actually become a nightmare if a nuclear-armed Iran suffered a break down in authority, forcing us to worry about the fate of Iran's nuclear arsenal.

Regional Instability.

The spread of nuclear weapons also emboldens nuclear powers contributing to regional instability. States that lack nuclear weapons need to fear direct military attack from other states, but states with nuclear weapons can be confident that they can deter an intentional military attack, giving them an incentive to be more aggressive in the conduct of their foreign policy. In this way, nuclear weapons provide a shield under which states can feel free to engage in lower-level aggression. Indeed, international relations theories about the "stability-instability paradox" maintain that stability at the nuclear level contributes to conventional instability.[64]

Historically, we have seen that the spread of nuclear weapons has emboldened their possessors and contributed to regional instability. Recent scholarly analyses have demonstrated that, after controlling for other relevant factors, nuclear-weapon states are more likely to engage in conflict than non-nuclear weapon states and that this aggressiveness is more pronounced in new nuclear states that have less experience with nuclear diplomacy.[65] Similarly, research on internal decisionmaking in Pakistan reveals that Pakistani foreign policymakers may have been emboldened by the acquisition of nuclear weapons, which encouraged them to initiate militarized disputes against India.[66]

Currently, Iran restrains its foreign policy because it fears a major military retaliation from the United

States or Israel, but with nuclear weapons, it could feel free to push harder. A nuclear-armed Iran would likely step up support to terrorist and proxy groups and engage in more aggressive coercive diplomacy. With a nuclear-armed Iran increasingly throwing its weight around in the region, we could witness an even more crisis-prone Middle East. In a poly-nuclear Middle East with Israel, Iran, and, in the future, possibly other states, armed with nuclear weapons, any one of those crises could result in a catastrophic nuclear exchange.

Nuclear proliferation can also lead to regional instability due to preventive strikes against nuclear programs. States often conduct preventive military strikes to prevent adversaries from acquiring nuclear weapons. Historically, the United States attacked German nuclear facilities during World War II; Israel bombed a nuclear reactor in Iraq in 1981; Iraq bombed Iran's Bushehr reactors in the Iran-Iraq War in the 1980s, and Iran returned the favor against Iraq's Osiraq reactor; a U.S.-led international coalition destroyed Iraq's nuclear infrastructure in the first Gulf War in 1991; and Israel bombed a Syrian nuclear reactor in 2007. These strikes have not led to extensive conflagrations in the past, but we might not be so lucky in the future. At the time of this writing in 2012, the United States and Israel were polishing military plans to attack Iran's nuclear program, and some experts maintain that such a strike could result in a wider regional war.

Constrained Freedom of Action.

The spread of nuclear weapons also disadvantages American's national security by constraining U.S. freedom of action. As the most powerful country on the planet with the ability to project power to every

77

corner of the globe, the United States has the ability to threaten or protect every other state in the international system. This is a significant source of strategic leverage and maintaining freedom of action is an important objective of U.S. national security policy.[67]

As nuclear weapons spread, however, America's military freedom of action is constrained. The United States can use, or credibly threaten to use, force against nonnuclear states. The threat of military action against nuclear-armed states is much less credible, however, because nuclear-armed states can deter U.S. military action with the threat of nuclear retaliation. In January 2012, for example, Iran threatened to close the Strait of Hormuz, a narrow Persian Gulf waterway through which roughly 20 percent of the world's oil flows, and the United States issued a counterthreat, declaring that Washington would use force to reopen the Strait if necessary. If Iran had had nuclear weapons, however, Washington's threats would have been much less credible. Would a U.S. President really be willing to risk nuclear war with Iran in order to reopen the Strait? Maybe. But, maybe not. While the United States might not be deterred in every contingency against a nuclear-armed state, it is clear that, at a minimum, the spread of nuclear weapons greatly complicates U.S. decisions to use force.

Undermines Alliances.

The spread of nuclear weapons also complicates U.S. alliance relationships. Washington uses the promise of military protection as a way to cement its alliance structures. U.S. allies depend on America's protection, giving Washington influence over allied states' foreign policies. Historically, the United States has offered, and threatened to retract, the security guarantee carrot to prevent allied states from acting contrary to its interests. As nuclear weapons spread, however, alliances held together by promises of military protection are undermined in two ways. First, U.S. allies may doubt the credibility of Washington's commitments to provide a military defense against nuclear-armed states, leading them to weaken ties with their patron. As Charles de Gaulle famously asked about the U.S. commitment to defend France from the Soviet Union during the Cold War, would Washington be willing to trade New York for Paris? Similarly, if Iran acquires nuclear weapons, U.S. partners in the Middle East, such as Israel and Gulf States, will question Washington's resolve to defend them from Iran. After all, if the United States proves unwilling to use force to prevent Iran from acquiring nuclear weapons, would it really be willing to fight a war against a nuclear-armed Iran? Qatar, for example, already appears to be hedging its bets, loosening ties to Washington and warming to Tehran.

Second, nuclear proliferation could encourage client states to acquire nuclear weapons themselves, giving them greater security independence and making them less dependable allies. According to many scholars, the acquisition of the *force de frappe* was instrumental in permitting the French Fifth Republic under President Charles de Gaulle to pursue a foreign policy path independent from Washington and NATO.[68] Similarly, it is possible that Turkey, Saudi Arabia, and other regional states will acquire independent nuclear capabilities to counter Iran's nuclear arsenal, greatly destabilizing an already unstable region and threatening Washington's ability to influence regional dynamics.

Further Proliferation.

Nuclear proliferation poses an additional threat to international peace and security because it causes further proliferation. As former Secretary of State George Schultz once said, "proliferation begets proliferation."[69] When one country acquires nuclear weapons, its regional adversaries, feeling threatened by its neighbor's new nuclear capabilities, are more likely to attempt to acquire nuclear weapons in response. Indeed, the history of nuclear proliferation can be read as a chain reaction of proliferation. The United States acquired nuclear weapons in response to Nazi Germany's crash nuclear program. The Soviet Union and China acquired nuclear weapons to counter the U.S. nuclear arsenal. The United Kingdom (UK) and France went nuclear to protect themselves from the Soviet Union. India's bomb was meant to counter China, and it, in turn, spurred Pakistan to join the nuclear club. Today, we worry that if Iran acquires nuclear weapons, other Middle Eastern countries, such as Egypt, Iraq, Turkey, and Saudi Arabia, might desire nuclear capabilities, triggering an arms race in a strategically important and volatile region.

Of course, reactive proliferation does not always occur. In the early-1960s, for example, U.S. officials worried that a nuclear-armed China would cause Taiwan, Japan, India, Pakistan, and other states to acquire nuclear weapons.[70] In hindsight, we now know that they were correct in some cases but wrong in others. Using statistical analysis, Philipp Bleek has shown that reactive proliferation is not automatic, but rather that states are more likely to proliferate in response to neighbors when three conditions are met: 1) there is an intense security rivalry between the two countries, 2) the potential proliferant state does not have a security guarantee from a nuclear-armed patron, and 3) the potential proliferant state has the industrial and technical capacity to launch an indigenous nuclear program.[71] In other words, reactive proliferation is real, but it is also conditional. If Iran enters the nuclear club, therefore, it is likely that some, but not all, of the countries that we currently worry about will eventually follow suit and become nuclear powers.

We should worry about the spread of nuclear weapons in every case, therefore, because the problem will likely extend beyond that specific case. As Wohlstetter cautioned decades ago, proliferation is not an N problem but an N+1 problem. Further nuclear proliferation is not necessarily a problem, of course, if the spread of nuclear weapons is irrelevant or even good for international politics as obsessionists and optimists protest. But, as the previous discussion makes clear, nuclear proliferation, and the further nuclear proliferation it causes, increases the risk of nuclear war and nuclear terrorism, threatens global and regional stability, constrains U.S. freedom of action, and weakens America's alliance relationships, giving us all good reason to fear the spread of nuclear weapons.

CONCLUSION

This chapter analyzed the past, present, and future of proliferation optimism. It began by reviewing the academic and policy origins of the pillars of proliferation optimism thinking. Next, it examined more recent work in this tradition, including a review of both proliferation optimism and proliferation anti-obsessionism. I demonstrated that this literature brings an important perspective to bear on the question of nuclear proliferation and reins in worst-case analyses of the consequences of nuclear proliferation. At the same time, I argued that, in making the case for the irrelevance of nuclear weapons, this literature swings too far in the opposite direction. Moreover, I demonstrated that too often these theorists support their arguments with contradictory logics and weak empirical evidence. Finally, I restated the argument about why the spread of nuclear weapons continues to pose a threat to international peace and security. Despite the claims of optimists, there is no getting around the fact that nuclear proliferation increases the risks of nuclear war, nuclear terrorism, regional instability, constrained U.S. freedom of action, weakened U.S. alliances, and further proliferation.

The findings of this chapter have important implications for the scholarly study of nuclear proliferation. While proliferation optimism and proliferation anti-obsessionism have made the field of nonproliferation studies more interesting in recent years, their inherent logical weaknesses means that they should remain niche, not mainstream, approaches to the study of nuclear proliferation. This chapter, therefore, aims to bring proliferation pessimism back in. The diffusion of the most powerful weapons ever invented by man is a serious problem. The burden of proof is on those who wish to claim otherwise. So far, the optimists and anti-obsessionists have made us think, but they have not made their case. It is not yet (and my guess is that it never will be) time for the discipline to shift its null hypothesis from the point of view that the spread of nuclear weapons is bad to the position that it is either good or irrelevant.

The argument of this chapter is mostly good news for U.S. nonproliferation policy. It is difficult, if not impossible, to find U.S. national security officials who believe that the spread of nuclear weapons is beneficial or irrelevant. That is not to say that proliferation optimism has not crept into the corridors of power in more subtle ways. Its influence can be found whenever national security officials too easily dismiss the problems posed by nuclear proliferation or breezily assert that a new nuclear state can be deterred. On balance, however, optimism has had more of an effect in the classroom than in the situation room. U.S. officials are correct to treat the spread of nuclear weapons as a serious threat and to go to great lengths to prevent it. Indeed, it would be downright dangerous if Washington were to follow the advice of optimists and anti-obsessionists. Would U.S. citizens (including proliferation optimists) really stand by if Washington distributed nuclear weapons to other countries in a quixotic quest for stability? Would foreign officials be able to take us seriously if U.S. officials were to announce at the next NPT Review Conference that the aim of U.S. nonproliferation policy was to "defetishize" nuclear weapons?

Of course, there are things that U.S. officials could do better, and the fine-tuning of U.S. nonproliferation policy makes an important subject for another article. For now, however, we should rest assured, knowing that U.S. policymakers are too reasonable to be anything other than proliferation pessimists.

ENDNOTES - CHAPTER 3

1. James Clapper, Director of National Intelligence, "Unclassified Statement for the Record on the Worldwide Threat Assessment of the U.S. Intelligence Community for the Senate Select Committee on Intelligence," January 31, 2012, available from *www.intelligence.senate.gov/120131/clapper.pdf.*

2. See, for example, Kenneth Waltz, *Theory of International Politics*, New York: McGraw-Hill, 1979.

3. See, for example, John Mueller, *Atomic Obsession: Nuclear Alarmism from Hiroshima to Al-Qaeda*, New York: Oxford University Press, 2009.

4. *Ibid.*

5. On the evolution of nuclear strategy, see Lawrence Freedman, *The Evolution of Nuclear Strategy*, New York: Palgrave Macmillan, 2003.

6. See, for example, Bernard Brodie, ed., *The Absolute Weapon: Atomic Power and World Order*, New York: Harcourt Brace Jovanovich, 1946.

7. Jacob Viner, "The Implications of the Atomic Bomb for International Relations," Proceedings of the American Philosophical Society, delivered November 16, 1945.

8. *Ibid.*

9. Brodie, *The Absolute Weapon.*

10. *Ibid.*

11. Scott D. Sagan and Kenneth N. Waltz, *The Spread of Nuclear Weapons: A Debate*, New York: W. W. Norton, 1997.

12. Albert Wohlstetter, "The Delicate Balance of Power," Santa Monica, CA: Rand Corporation, 1958.

13. Albert Wohlstetter, "Nuclear Sharing: NATO and the N+1 Country," *Foreign Affairs*, Vol. 39, No. 3, April 1961, pp. 355-387.

14. *Ibid.*

15. *Ibid.*

16. On France's nuclear program, see Lawrence Scheinman. *Atomic Energy Policy in France under the Fourth Republic*, Princeton, NJ: Princeton University Press, 1965.

17. For a thorough discussion of the development of French nuclear strategy, see Bruno Tertrais, "'Destruction Assurée': The Origins and Development of French Nuclear Strategy, 1945-1982," Henry D. Sokolski ed., *Getting MAD: Nuclear Mutual Assured Destruction, Its Origins and Practice*, Carlisle, PA: Strategic Studies Institute, U.S. Army War College, 2004, pp. 51-122.

18. *Ibid.*, p. 95.

19. *Ibid.*, p. 86.

20. *Ibid.*, p. 64.

21. *Ibid.*, p. 86.

22. Pierre Marie Gallois, *Le Sablier du Siecle: Mémoires* (*The Hourglass of the Century: Memoirs*), Lausanne, France: L'Âge d'homme, 1999, p. 402.

23. Tertrais, p. 83.

24. *Ibid.*, p. 82.

25. *Ibid.*, p. 96.

26. Pierre Marie Gallois, *Stratégie de l'âge nucléaire* (*Strategy of the Nuclear Age*), Paris, France: François-Xavier de Guibert, 1960.

27. Wohlstetter, "Nuclear Sharing."

28. This section draws heavily from Harvey M. Sapolsky, "The U.S. Navy's Fleet Ballistic Missile Program and Finite Deterrence," Henry D. Sokolski, ed., *Getting MAD*, pp. 123-135.

29. Kenneth Waltz, "The Spread of Nuclear Weapons: More May Be Better," *Adelphi Papers,* No. 171, London: International Institute for Strategic Studies, 1981.

30. Bruce Bueno de Mesquita and William H. Riker, "An Assessment of the Merits of Selective Nuclear Proliferation." *Journal of Conflict Resolution*, Vol. 26, No. 2, June 1982, pp. 283-306.

31. John Mearsheimer, "The Case for a Ukrainian Deterrent," *Foreign Affairs*, Vol. 72, No. 3, Summer 1993, pp. 50-66.

32. John Mearsheimer and Stephen Walt, "An Unnecessary War," *Foreign Policy*, January/February 2003, pp. 51-59.

33. Barry R. Posen, "We Can Live with a Nuclear Iran," *The New York Times*, February 27, 2006.

34. Sagan and Waltz, *The Spread of Nuclear Weapons*.

35. See, for example, Robert Powell, "Nuclear Brinkmanship with Two-Sided Incomplete Information," *American Political Science Review,* Vol. 82, No. 1, 1988, pp. 155-178; Robert Powell, "Nuclear Deterrence and the Strategy of Limited Retaliation," *American Political Science Review*, Vol. 83, No. 2, 1989, pp. 503-519; and Matthew Kroenig, "Nuclear Superiority and the Balance of Resolve: Explaining Nuclear Crisis Outcomes," *International Organization*, Vol. 67, Issue 1, January 2013, pp. 141-171.

36. Brodie, *The Absolute Weapon*.

37. See, for example, Robert Powell. *Nuclear Deterrence Theory: The Search for Credibility*, New York: Cambridge University Press, 1990.

38. Thomas Schelling, *Arms and Influence*, New Haven: Yale University Press, 1966.

39. *Ibid.*

40. *Ibid.*

41. Kroenig, "Nuclear Superiority and the Balance of Resolve."

42. See for example, Herman Kahn, *On Thermonuclear War*, New York: Greenwood Press, 1978.

43. Kroenig, "Nuclear Superiority and the Balance of Resolve."

44. Klaus Knorr, *Limited Strategic War*, New York: Praeger, 1962.

45. Mueller, *Atomic Obsession*.

46. Ward Wilson, "The Myth of Nuclear Weapons," *The Nonproliferation Review*, Vol. 15, No. 3, November 2008, pp. 421-439.

47. Anne Harrington de Santana, "Nuclear Weapons as the Currency of Power: Deconstructing the Fetishism of Force," *The Nonproliferation Review*, Vol. 16, No. 3, 2009, pp. 325-345.

48. *Ibid.*, p. 327.

49. *Ibid.*, p. 342.

50. Matthew Kroenig, *Exporting the Bomb: Technology Transfer and the Spread of Nuclear Weapons*, Ithaca, NY: Cornell University Press, 2010.

51. Paul Kapur, *Dangerous Deterrent: Nuclear Weapons Proliferation and Conflict in South Asia*, Palo Alto, CA: Stanford University Press, 2007.

52. Wilson, "The Myth of Nuclear Weapons."

53. Steven Weber, "The End of the Business Cycle?" *Foreign Affairs*, Vol. 76, No. 4, July/August 1997, pp. 65-82.

54. Thomas Schelling, "Reciprocal Fear of Surprise Attack," Santa Monica, CA: Rand Paper, 1958.

55. Knorr, *Limited Strategic War*.

56. Schelling, *Arms and Influence*.

57. See, for example, Scott Sagan, *The Limits of Safety: Organizations, Accidents, and Nuclear Weapons*, Princeton, NJ: Princeton University Press, 1993.

58. On nuclear terrorism, see, for example, Michael Levi, *On Nuclear Terrorism*, Cambridge, MA: Harvard University Press, 2007.

59. Brian Michael Jenkins, "The New Age of Terrorism," Santa Monica, CA: Rand, 2006.

60. Fissile Materials Working Group, "After Bin Laden: Nuclear Terrorism Still a Top Threat," *Bulletin of the Atomic Scientists*, May 13, 2011.

61. *Ibid*.

62. Levi, *On Nuclear Terrorism*.

63. David Albright, *Peddling Peril: How the Secret Nuclear Trade Arms America's Enemies*, New York: Free Press, 2012.

64. Glenn H. Snyder, "The Balance of Power and the Balance of Terror," Paul Seabury, ed., *The Balance of Power*, San Francisco, CA: Chandler, 1965, pp. 184-201.

65. Robert Rauchhaus, "Evaluating the Nuclear Peace Hypothesis: A Quantitative Approach," *Journal of Conflict Resolution*, Vol. 53, No. 2, April 2009, pp. 258-277; Michael Horowitz, "The Spread of Nuclear Weapons and International Conflict: Does Experience Matter?" *Journal of Conflict Resolution* Vol. 53, No. 2, April 2009, pp. 234-257.

66. Kapur, *Dangerous Deterrent*.

67. See, for example, Donald Rumsfeld, *The National Defense Strategy of the United States of America*, Washington, DC: Department of Defense, March 2005.

68. See, for example, Scheinman, Wilfred L. Kohl, *French Nuclear Diplomacy*, Princeton, NJ: Princeton University Press, 1971.

69. Philipp Bleek, *The Nuclear Domino Myth: Why Proliferation Rarely Begets Proliferation*, Ph.D. Dissertation, Washington, DC: Department of Government, Georgetown University, 2010.

70. Francis J. Gavin, "Blasts from the Past: Proliferation Lessons from the 1960s," *International Security*, Vol. 29, No. 3, Winter 2004/2005, pp. 100-135.

71. Bleek, *The Nuclear Domino Myth*.

CHAPTER 4

PREVENTIVE WAR AND THE
SPREAD OF NUCLEAR PROGRAMS

Matthew Fuhrmann

What are the consequences of nuclear proliferation?[1] Iran's alleged pursuit of nuclear weapons and North Korea's nuclear tests in 2006 and 2009 have heightened concerns about the further spread of the bomb. Yet, debates persist about the political effects of proliferation. Some argue that the spread of nuclear weapons constitutes a major threat to international security, in part because it raises the risk of nuclear war and increases the odds that terrorists will acquire the bomb.[2] Others contend that nuclear weapons can promote international peace and stability by raising the costs of armed conflict.[3] According to this perspective, the slow and deliberate spread of nuclear weapons may actually be a good thing. Still others assert that the threat posed by nuclear proliferation is overblown and that nuclear weapons have little effect on international politics. As John Mueller pithily states, "The nuclear diffusion that has transpired has proved to have had remarkably limited, perhaps even imperceptible, consequences."[4]

This chapter contributes to ongoing debates about the consequences of nuclear proliferation by analyzing the connection between nuclear programs and preventive war. Does the pursuit of nuclear weapons increase the likelihood of preventive military force? If so, under what conditions does it do so? In this chapter, I provide answers to these questions. To begin, I raise further awareness about the targeting of nuclear programs by surveying historical cases in which countries have bombed or considered bombing nuclear facilities. Although attacks of this nature are relatively rare, countries have seriously considered using military force to delay proliferation on a number of occasions. I subsequently offer an explanation for why states strike (or consider striking), based on existing scholarly research. In the end, the evidence presented here supports the view that preventive war is a potential danger associated with the spread of nuclear programs that policymakers and scholars should take seriously.

ATTACKING NUCLEAR FACILITIES: THE HISTORICAL RECORD

Nuclear facilities or materials in non-nuclear weapons states have been targeted on more than a dozen occasions since 1941.[5] The first attempted strike against a nuclear plant occurred in 1942 when British commandos targeted the Norsk-Hydro heavy water plant in German-occupied Norway. This raid was unsuccessful, but the Allies followed up with several other strikes against the same facility, which was believed to be the main chokepoint of Germany's nuclear weapons program.[6] In November 1943, for example, Allied aircraft dropped hundreds of bombs on the heavy water plant, setting back production by a few months. Frustrated by continued attacks against the facility, Germany attempted to transport heavy water and related equipment out of Norway in 1944 on the ferry *Hydro*; saboteurs intercepted and sank the ferry in Norway's Lake Tinnsjø.

The Iran-Iraq War provided the setting for a series of strikes against nuclear facilities.[7] In 1977, prior to the onset of hostilities, Israel approached Iran to discuss joint military strikes against Iraqi nuclear infrastructure.[8] Tehran was not interested in attacking Iraq's nuclear program at that time, but it naturally warmed up to the idea after Saddam Hussein invaded Iran on September 22, 1980. Days later, Iranian F-4 *Phantoms* attacked Iraq's nuclear research reactor, Osiraq, en route home from a bombing raid, although the strike caused only minor damage to the facility. Later in the war, on March 24, 1984, Iraq raided Iran's nuclear power plant that was under construction at Bushehr. Baghdad targeted this facility on multiple other occasions during the conflict, despite an Iranian-backed International Atomic Energy Agency (IAEA) resolution prohibiting strikes against nuclear installations. The Iraqi raids, which damaged the Bushehr facility to varying degrees, occurred as part of a broader campaign to destroy economic and industrial targets.

During the 1991 Persian Gulf War, the United States bombed numerous Iraqi nuclear facilities, including the Tuwaitha Nuclear Research Center near Baghdad.[9] This campaign heavily damaged some Iraqi nuclear plants, but many of the bombs that were dropped missed their intended targets.[10] Moreover, some facilities escaped the war unscathed, partially because the United States was unaware of their existence or their location.

Iraqi nuclear infrastructure was targeted on two other instances in the 1990s. On January 17, 1993, the U.S. Navy used Tomahawk land attack missiles against the Zaafaraniyah uranium enrichment plant, which was left largely intact following the Gulf War.[11] These strikes, which were intended to punish Baghdad for its refusal to fully comply with the United Nations (UN)-mandated nuclear inspections regime, significantly curtailed Iraq's electromagnetic isotope separation program.[12] Then, in December 1998, the United States and Great Britain launched Operation DESERT FOX, a campaign that was intended to degrade Iraq's weapons of mass destruction (WMD) capabilities. Despite the stated objective of the operation, it appears that only one facility relevant to Baghdad's nuclear program was targeted: a plant housing machine tools relevant for centrifuge development.[13]

Israel has conducted two "bolt from the blue" raids against nuclear programs. In 1981, after Iran failed to destroy Iraq's Osiraq reactor, the Israeli Air Force bombed the facility in a mission known as Operation OPERA. This strike was successful in the sense that it destroyed Osiraq, which was widely regarded as the centerpiece of Baghdad's nuclear program, although other aspects of this raid's effectiveness are still debated.[14] More recently, in September 2007, Israel bombed a Syrian reactor at al Kibar that had yet to become operational. The plant, being built with assistance from North Korea, was heavily damaged as a result of the Israeli strike.[15] After the raid, Syria bulldozed what was left of the site in an apparent effort to prevent others, particularly the IAEA, from obtaining additional information about the plant.[16] Few leaders condemned the Israelis for using preventive military force, leading some to conclude that the international community secretly welcomed the destruction of the Syrian nuclear facility.[17]

On a number of other occasions, countries seriously considered attacking nuclear programs but ultimately did not strike.[18] Egypt had plans to destroy Dimona, Israel's main nuclear facility, during the 1967 crisis, possibly with assistance from the Soviet Union.[19] Indian Prime Ministers Indira Gandhi (in 1982 and

1984) and Rajiv Gandhi (in 1986-87) actively sought to destroy the Pakistani enrichment plant at Kahuta in a joint operation with Israel.[20] Indira Gandhi even approved plans for a preventive strike, but the raid was called off at the "last minute."[21] Pakistani officials likewise considered attacking nuclear installations in India during the 1984 crisis.[22] In a lesser-known case, the Soviet Union considered preventive strikes against South Africa in the 1970s when Moscow detected apparent preparations for a nuclear test. The Soviet Union approached the United States and asked for assistance in attacking the Y Plant, one of South Africa's key nuclear installations.[23] Washington did not respond positively to this overture.

The United States did, however, strongly consider using military force to delay nuclear proliferation on other occasions. In the early-1960s, some in Washington feared that China would soon become the world's fifth nuclear power.[24] President John F. Kennedy, in particular, was deeply concerned about the prospect of a Chinese bomb and seriously considered using military force to frustrate Beijing's nuclear program. The options that were put on the table included the use of tactical nuclear weapons against Chinese nuclear facilities, as well as employing Taiwanese saboteurs to infiltrate the mainland and destroy key plants.[25] U.S. officials ultimately chose not to attack, and Beijing conducted its first nuclear test in 1964.

Washington likewise considered using force during the 1994 North Korean nuclear crisis.[26] This crisis began when the IAEA detected irregularities at North Korean nuclear plants and called for the UN Security Council to authorize a special inspections regime, leading Pyongyang to announce its withdrawal from the Nuclear Nonproliferation Treaty (NPT). Some U.S.

officials believed that military action could reduce the threat posed by North Korea's nuclear program. For example, Secretary of Defense William Perry later indicated, "We believed that the nuclear program on which North Korea was embarked was . . . dangerous, and were prepared to risk a war to stop it."[27] Any American operation against North Korea would likely have involved cooperation from South Korea, which had seriously considered raiding nuclear facilities at Yongbyon as early as 1991. In the end, a diplomatic bargain known as the Agreed Framework brought a (temporary) end to the crisis.[28]

Iran's nuclear program has also raised the prospect of preventive military action. Some elites in Washington and Jerusalem have recently called for military raids against Tehran's nuclear facilities. President Barack Obama has not publicly threatened to bomb Iranian nuclear facilities, but he has said that "all options are on the table," and that the option of last resort is the "military component."[29] Israeli officials, including Prime Minister Benjamin Natanyahu, have more forcefully advocated for military strikes. Speaking about the prospect of attacking Iran, Natanyahu said, "None of us can afford to wait much longer. . . . I will never let my people live in the shadow of annihilation."[30] Officials outside of the United States and Israel — including Saudi King Abdullah — have similarly voiced support for preventive raids against Iran's nuclear infrastructure. It remains to be seen, however, whether Israel or the United States will take military action against Iran.

In the cases previously mentioned, the target state had yet to acquire nuclear weapons. Does the danger of preventive war disappear once a potential target assembles a nuclear arsenal? Kenneth Waltz, a prominent proponent of "nuclear optimism," maintains that "preventive strikes against states that have, or may have, nuclear weapons are hard to imagine."[31] It is true that attacks against nuclear states are potentially more dangerous than strikes against states that are still non-nuclear. However, countries have occasionally considered raiding nuclear infrastructure in states that possessed nuclear arsenals. Some elites in the United States called for preventive strikes against the Soviet Union during the 1950s.[32] The Soviet Union seriously contemplated striking Chinese nuclear facilities during the 1969 border crisis.[33] Libya hoped to launch a retaliatory raid against Israel's Dimona plant following the 1981 Osiraq strike. Tripoli sought cooperation to implement such a strike from Iraq and the Soviet Union, both of whom expressed little interest in attacking Israel.[34] Iraq did, however, launch Scud missiles at Dimona during the 1991 Persian Gulf War, but they did not come close to hitting the target.[35]

WHY COUNTRIES ATTACK

The preceding discussion underscores that concerns about nuclear proliferation have occasionally led to preventive strikes against nuclear facilities. Why do countries attack or consider attacking nuclear plants in other states? Prior research has shown that states are more likely to target nuclear programs when they are highly threatened by the target state's potential acquisition of nuclear weapons.[36] Two main factors shape this threat perception: violent interstate conflict and the proliferator's regime type.[37]

Nuclear proliferation can be especially threatening to states that fear that they could be targeted with the bomb. The likelihood of nuclear use is generally low, and nuclear weapons have not been used in war since the bombings of Hiroshima and Nagasaki, Japan, in 1945. However, a history of bad relations among states can increase fears of a future nuclear attack, perhaps leading to the perception that a rival's acquisition of the bomb poses an existential threat. For example, some Israeli officials viewed the Iraqi nuclear program as a threat of the highest magnitude, in part because Iraq fought against Israel in the 1948 War of Independence and the 1973 Yom Kippur War.[38] As Prime Minister Menachem Begin proclaimed shortly after the strike against Osiraq in 1981:

> If we stood by idly . . . Saddam Hussein would have produced his three, four, five bombs. . . . Then, this country and this people would have been lost. . . . Another Holocaust would have happened in the history of the Jewish people.[39]

States are substantially less threatened when their nonrivals pursue nuclear weapons. Attacks against nuclear infrastructure are therefore unlikely in the absence of hostile relations — even when states are far from friendly. Algeria, for instance, may have coveted nuclear weapons,[40] and Algiers was one of the last capitals to consider normalizing relations with Israel.[41] Yet, Israel did not raid Algeria's nuclear plants, in part because the absence of major war between the two countries lessened the threat posed by an Algerian bomb.[42] Needless to say, attacks become exceedingly unlikely when the potential attacker and target are military allies. It is unthinkable, for instance, that the United States would have attacked British nuclear facilities in the early-1950s to delay London's ability to build the bomb.

A country's regime type also affects the degree to which other states are threatened by its nuclear program. Highly authoritarian proliferators are more likely than democracies to be attacked. Indeed, all of the strikes against non-nuclear weapons states had a nondemocratic target even though many democracies thought about building (or built) the bomb (e.g., Australia, Britain, France, and India). Why is this the case?

Democratic leaders are constrained by domestic institutions such as legislatures and judiciaries, which can limit capricious foreign policy decisions and promote compliance with international norms.[43] Authoritarian countries, on the other hand, often have less respect for norms because of opaque institutions and relatively little domestic accountability. Autocrats might thus be more likely to threaten other states with nuclear weapons, use the bomb first during a crisis, or engage in other provocative actions. Concerns such as these can motivate states to use military force to delay proliferation. For example, U.S. National Security Advisor Brent Scowcroft believed that Saddam Hussein's "notoriously mercurial" behavior magnified the threat of an Iraqi bomb and helped justify targeting Baghdad's nuclear program during the Persian Gulf War.[44] President George W. Bush likewise believed that the world should not allow Iran to acquire nuclear weapons because Tehran has a "nontransparent" government, implying that its regime type heightens the risk of aggressive or unpredictable behavior.[45]

Aside from the perceived threat posed by the target's nuclear program, two other general considerations may also affect the likelihood of preventive strikes.[46] First, potential attackers are likely to consider whether raids against nuclear facilities could be successful. The likelihood of success depends partially on the military capabilities of the attacker. Weak states will often be unable to destroy their enemies' nuclear programs in the absence of cooperation from their allies. For instance, although Zambia may have been threatened by the prospect of a South African bomb in the 1970s, it would have struggled mightily to successfully destroy the relevant facilities on its own, decreasing the odds that officials in Lusaka would even consider the military option.

The number of nuclear facilities that the target possesses also influences the likelihood that raids against nuclear programs will be successful. Iraq and Syria each possessed one main chokepoint facility at the time that they were attacked, and neither state was on the verge of building nuclear weapons. Israel therefore needed only to destroy a single facility to delay proliferation in these two cases. This situation becomes more complex, however, when potential targets have well-developed nuclear programs. Iran, to cite one example, has multiple facilities that would probably need to be destroyed to curtail significantly its nuclear program: the uranium enrichment facilities at Natanz and Qom, the heavy water production facility at Arak, the uranium conversion center at Isfahan, the Bushehr nuclear power plant, and the Tehran research reactor. This does not mean that it is impossible for Israel or the United States to delay successfully Iran's nuclear program using military force, but the probability of success is substantially lower relative to a scenario in which Iran possessed a single nuclear chokepoint.[47]

Second, the costs of raiding nuclear programs could deter countries from attacking. States may be unlikely to attack if they believe that a limited preventive strike would lead to a large-scale war or produce other undesirable outcomes. For example, the United

States refrained from bombing Chinese and North Korean nuclear facilities in part because officials in Washington believed that the military costs of such operations were too high. Concerns about costs have also influenced the debate about how to respond to Iran's nuclear program. U.S. officials that are considering bombing Iran's nuclear facilities today must wrestle with the possibility that Tehran could retaliate by closing the Strait of Hormuz or engage in other actions that threaten core U.S. politico-strategic interests.[48]

States may also worry about the normative costs of targeting nuclear programs. There is an international norm against the preventive use of force, and Article 56 of Protocol I Additional to the Geneva Conventions (1977) specifically prohibits the targeting of nuclear plants. Thus, states might be deterred from using military force by the prospect of political or economic isolation. One reason that India ultimately refrained from bombing Pakistan's Kahuta enrichment plant in the 1980s was because officials in New Delhi feared that "the international community would condemn us."[49] Similarly, after Israeli Prime Minister Ehud Olmert asked President George W. Bush to bomb Syria's al Kibar reactor during a 2007 phone conversation, Bush concluded that the normative and political costs were too great. As he recounted in his memoir:

> As a military matter, the bombing mission would be straightforward. The Air Force could destroy the target, no sweat. But bombing a sovereign country with no warning or announced justification would create severe blowback.[50]

DISCUSSION AND CONCLUSION

Scholars have previously argued that nuclear weapons programs are dangerous, in part because they can lead to preventive war.[51] This chapter lends credence to this argument by identifying numerous historical cases in which countries attacked or considered attacking nuclear programs. I have also articulated the conditions under which nuclear weapons programs are likely to lead to military strikes. When the potential attacker and the target have a history of violent conflict—and when the target state is authoritarian—preventive strikes are considerably more likely.[52] Other factors may also affect the use of force, but the perceived threat posed by the target's acquisition of the bomb is among the most important in triggering interest in preventive military action. This implies that nuclear weapons programs can be destabilizing, at least under certain conditions. Those interested in conflict management would therefore do well to engage in more diplomacy aimed at limiting the onset of new nuclear weapons programs.

One might dispute this conclusion, however, on the grounds that the violence caused by nuclear programs to date has been relatively minimal. Outside of ongoing interstate wars, nuclear facilities have been bombed on just a handful of occasions. During the Osiraq raid, the highest profile attack against a nuclear facility, only 10 Iraqi soldiers and one French civilian were killed.[53] Although it is important not to exaggerate the threat posed by nuclear weapons programs, the danger of preventive force should not be dismissed due to the modest amount of violence caused by the attacks discussed in this chapter. First,

102

there were a number of close calls — particularly in South Asia — where attacks were strongly considered but ultimately not conducted. Had Indira Gandhi followed through on her initial decision to attack Kahuta, it is possible, and perhaps likely, that war would have resulted between India and Pakistan. Second, attacks against nuclear programs could occur more frequently — and become deadlier — in the future, particularly if there are doubts about whether states pursuing the bomb would act as "responsible" nuclear powers.[54]

Compounding matters further, interest in nuclear energy is growing around the world — despite the March 2011 accident at Japan's Fukushima nuclear power plant — as part of a movement that some have labeled the "nuclear renaissance."[55] Although existing research tends to downplay the strategic effects of nuclear energy,[56] there is a growing recognition among scholars that nuclear programs could raise the risk of international conflict even when they are "peaceful" in nature.[57] This is in part because the development of a civilian nuclear program in one state might provide incentives for others to launch preventive strikes.

There is precedent for using military force against civilian facilities. Iran's Bushehr nuclear power plant — bombed during the Iran-Iraq War — was being built with assistance from West Germany to produce electricity. Osiraq was also technically a civilian facility. The reactor was supplied by France exclusively for peaceful purposes and placed under IAEA safeguards, meaning that it should have been difficult for Iraq to use Osiraq for military purposes. Many policymakers and analysts therefore condemned the Israeli strike and interpreted it as an indictment of the nonproliferation regime. For example, Sigvard Eklund, Director General of the IAEA, stated, "The Israeli attack on Iraq's nuclear research center was also an attack on the Agency's safeguards."[58] Why would countries have incentives to bomb civilian nuclear plants?

Nuclear facilities are dual use in nature, meaning that they can serve civilian or military purposes. Reactors can be employed to produce medical isotopes or to help meet a country's energy needs by producing electricity. These same facilities, however, also provide a potential source of plutonium for nuclear weapons. This so-called dual-use dilemma means that countries can draw on civilian nuclear programs to augment their military capabilities. India, for example, used a civilian research reactor supplied by Canada in the 1950s to conduct its first nuclear test in 1974. France similarly built between 63 and 250 nuclear weapons using plutonium that was produced in civilian power plants.[59] Examples such as these are not uncommon. Recent research shows that, on average, states that receive foreign assistance in developing peaceful nuclear programs are statistically more likely than states that do not receive atomic aid (or receive lower levels of assistance) to pursue and acquire nuclear weapons — especially if they later experience an international crisis.[60]

Therefore, when states build nuclear facilities, it is difficult for outsiders to know for certain whether the plants are meant for electricity production, the manufacture of nuclear weapons, or both. This problem is evident in the contemporary case of Iran. Many in the West suspect that Iran intends to build nuclear weapons, yet Tehran has repeatedly asserted that its program is intended only to serve peaceful ends. The oft-discussed 2007 National Intelligence Estimate (NIE) on Iran's nuclear program underscored this tension. The NIE concluded with "high confidence" that

Tehran halted its nuclear weapons program in 2003 but that it continued the civilian uranium enrichment program, and this program could be applied to nuclear weapons production if Iran decided to proliferate.[61] Countries aspiring to develop nuclear programs can use signals to convey that their intentions are peaceful.[62] For instance, willingness to subject nuclear facilities to international inspections could alleviate concerns about whether a state's plants might be used to build bombs. On the other hand, states that refuse to accept measures such as the 1997 IAEA Additional Protocol (AP), which provides the Agency with greater authority to inspect nuclear sites, are likely to create ambiguity about their intentions.[63] One reason that some believe that Iran covets nuclear weapons is that Tehran has signed, but not ratified, the AP.

Yet, even if states accept the AP and allow the IAEA to inspect their nuclear plants, they may be unable to convince others — especially their rivals — that their intentions are peaceful. Interstate rivalries, which ensue from a history of conflict, erode trust and often cause states to adopt worst-case thinking when analyzing actions taken by others.[64] For example, during the height of the Cold War, seemingly every policy adopted by Moscow was viewed suspiciously in Washington, even those that were probably innocuous. This helps explain why placing Osiraq under safeguards did not stop Israel from believing that Saddam Hussein intended to use the research reactor to produce plutonium for nuclear bombs. That said, if Iraq's intentions were peaceful, Baghdad did not help its cause by making hostile statements towards Israel and engaging in other actions that raised questions about the true purpose of Osiraq.

The current list of nuclear energy aspirants includes states that might struggle to persuade some
in the international community that they are procur-
ing technology strictly for peaceful purposes. In the
Middle East, for instance, 12 countries are considering
building nuclear power plants.[65] Many assume that
these states want nuclear energy programs as a hedge
against a possible Iranian bomb. If countries such as
Egypt, Jordan, Turkey, and Saudi Arabia expand their
civilian nuclear programs, it may be difficult for them
to convince others that their intentions are entirely
harmless, even if they sincerely have little interest in
nuclear weapons. This does not imply that these states
will have their nuclear facilities bombed in the future,
but the probability of preventive strikes may increase
if nuclear technology diffuses around the globe to the
degree that some predict.[66]

My analysis in this chapter leaves an important
question unanswered: Are countries wise to target
nuclear programs preventively? There is evidence
that prior strikes against nuclear facilities delayed the
targets' nuclear weapons programs.[67] However, looking into the future, policymakers should be exceedingly cautious when contemplating preventive strikes.
First, the conditions that led to success in the past may
not be present in the future. Importantly, many of the
conditions that facilitated success in Iraq (1981) and
Syria (2007) are notably absent in the case of Iran, as
discussed elsewhere in this chapter.[68] Second, worst-
case assessments about the consequences of nuclear
proliferation are unwarranted. Although nuclear
weapons affect international politics in some respects,
they are generally poor instruments for coercion and
intimidation. Based on an analysis of more than 200
militarized compellent threats[69] issued from 1918 to

2001, recent research shows that nuclear-armed states are not more likely than non-nuclear states to blackmail their adversaries successfully.[70] Some countries may still believe that their rival's acquisition of the bomb constitutes an existential threat, but fears about nuclear blackmail should not be used to justify preventive strikes against nuclear facilities.

In any case, history suggests that the risk of preventive war is unlikely to disappear in the future. It is therefore important for scholars to continue to examine this issue and devote greater attention to the consequences of nuclear technology diffusion more generally. Additional research in this vein could further inform enduring policy questions, such as what officials in Washington should do in response to the development of nuclear programs in countries of concern.

ENDNOTES - CHAPTER 4

1. This chapter draws partially on two articles coauthored with Sarah E. Kreps. See Matthew Fuhrmann and Sarah E. Kreps, "Targeting Nuclear Programs in War in Peace: A Quantitative Empirical Analysis, 1941-2000," *Journal of Conflict Resolution*, Vol. 35, No. 6, 2010, pp. 831-859; and Sarah E. Kreps and Matthew Fuhrmann, "Attacking the Atom: Does Bombing Nuclear Facilities Affect Proliferation?" *Journal of Strategic Studies*, Vol. 34, No. 2, 2011, pp. 161-187.

2. See, for example, Scott Sagan, "The Perils of Proliferation: Organization Theory, Deterrence Theory, and the Spread of Nuclear Weapons," *International Security*, Vol. 18, No. 4, 1994, pp. 66-107.

3. Kenneth N. Waltz, "The Spread of Nuclear Weapons: More May Be Better," *Adelphi Papers*, No. 171, London, UK: International Institute for Strategic Studies, 1981.

4. John Mueller, *Atomic Obsession: Nuclear Alarmism from Hiroshima to Al Qaeda*, New York: Oxford University Press, 2009, p. 237.

5. See Fuhrmann and Kreps, "Targeting Nuclear Programs in War and Peace." For other surveys of the historical record, see Dan Reiter, "Preventive Attacks against Nuclear, Biological, and Chemical Weapons Programs: The Track Record," William Walton Keller and Gordon R. Mitchell, eds., *Hitting First, Preventive Force in U.S. Security Strategy*, Pittsburgh, PA: Pittsburgh University Press, 2006; and Bennett Ramberg, "Preemption Paradox," *Bulletin of the Atomic Scientists*, Vol. 62, No. 4, 2006, pp. 48-56.

6. See, for example, Per F. Dahl, *Heavy Water and the Wartime Race for Nuclear Energy*, London, UK: Taylor and Francis, 1999; Knut Haukelid, *Skis against the Atom: The Exciting, First-Hand Account of Heroism and Daring Sabotage During the Nazi Occupation of Norway*, Minot, ND: North American Heritage Press, 1989; and Thomas Gallagher, *Assault in Norway: Sabotaging the Nazi Nuclear Program*, Guilford, CT: Lyons Press, 2002.

7. Ronald Bergquist, "The Air War," *The Role of Airpower in the Iran-Iraq War*, Montgomery, AL: Air University Series, 1988, pp. 41-68; Mark Hibbs, "Bushehr Construction Now Remote after Three Iraqi Air Strikes," *Nucleonics Week*, Vol. 28, No. 48, 1987, pp. 5-6; Mark Hibbs, "Iraqi Attack on Bushehr Kills West German Nuclear Official," *Nucleonics Week*, Vol. 28, No. 47, 1987; and Dilip Hiro, *The Longest War: The Iran-Iraq War*, London, UK: Grafton, 1989, pp. 129-166, 192.

8. Amos Perlmutter, Michael I. Handel, and Uri Bar-Joseph, *Two Minutes over Baghdad*, New York: Routledge, 2003, p. xxxi.

9. *Gulf War Air Power Survey, Vol. 1: Planning and Command and Control*, Washington, DC: U.S. Government Printing Office, 1993.

10. Most of these bombs were unguided. See *Gulf War Air Power Survey*.

11. Jeremy Tamsett, "The Israeli Bombing of Osiraq Reconsidered: Successful Counterproliferation?" *The Nonproliferation*

Review, Vol. 11, No. 3, 2004, pp. 70-85; and "Zaafaraniyah," Federation of American Scientists, October 9, 2000, available from *www. fas.org/nuke/guide/iraq/facility/zaafaraniyah.htm*.

12. Tamsett, "The Israeli Bombing of Osiraq Reconsidered," p. 81.

13. Anthony Cordesman, "The Lessons of Desert Fox: A Preliminary Analysis," Washington, DC: Center for Strategic and International Studies, February 16, 1999; and Charles Duelfer, Special Advisor to the Director of Central Intelligence, *Comprehensive Report of the Special Advisor to the DCI on Iraq's WMD*, Vol. 2, Washington, DC: U.S. Government Printing Office, 2004, pp. 45-46.

14. See, for example, Tamsett, "The Israeli Bombing of Osiraq Reconsidered"; Dan Reiter, "Preventive Attacks Against Nuclear Programs and the 'Success' at Osiraq," *Nonproliferation Review*, Vol. 12, No. 2, July 2005, pp. 355-371; and Kreps and Fuhrmann, "Attacking the Atom."

15. Bill Gertz and Sara Carter, "US: Syria Hid N. Korea-aided Nukes Plant," *Washington Times*, April 24, 2008.

16. Joby Warrick and Robin Wright, "Search is Urged for Syrian Nuclear Sites," *The Washington Post*, May 29, 2008.

17. Leonard S. Spector and Avner Cohen, "Israel's Airstrike on Syria's Reactor: Implications for the Nonproliferation Regime," *Arms Control Today*, July/August 2008.

18. Fuhrmann and Kreps, "Targeting Nuclear Programs in War and Peace."

19. Avner Cohen, "Cairo, Dimona, and the June 1967 War," *The Middle East Journal*, Vol. 50, No. 2, 1996, pp. 190-210; Avner Cohen, *Israel and the Bomb*, New York: Columbia University Press, 1998, pp. 243-276; Isabella Ginor and Gideon Remez, *Foxbats Over Dimona: The Soviets' Nuclear Gamble in the Six-Day War*, New Haven, CT: Yale University Press, 2007.

20. Douglas Frantz and Catherine Collins, *The Nuclear Jihadist: The True Story of the Man Who Sold the World's Most Dangerous*

Secrets and How We Could Have Stopped Him, New York: Twelve, 2007, pp. 88-89; Sumit Ganguly and Devin Hagerty, *Fearful Symmetry: India-Pakistan Crises in the Shadow of Nuclear Weapons*, Seattle, WA: University of Washington Press, 2005; and Bharat Karnad, *India's Nuclear Policy*, Westport, CT: Praeger, 2008.

21. Karnad, *India's Nuclear Policy*, p. 57.

22. Ganguly and Hagerty, *Fearful Symmetry*, p. 58.

23. David Albright, "South Africa and the Affordable Bomb," *Bulletin of the Atomic Scientists*, Vol. 50, July/August 1994, pp. 37-47; Nuclear Threat Initiative, "South Africa Profile: Nuclear Overview," May 2007, available from *www.nti.org/e_research/profiles/SAfrica/Nuclear/index.html*; and Ramberg, "Preemption Paradox," p. 56.

24. William Burr and Jeffrey Richelson, "Whether to Strangle the Baby in the Cradle: The United States and the Chinese Nuclear Program, 1960-64," *International Security*, Vol. 25, No. 3, 2000/01, pp. 54-99; Gordon Chang, "JFK, China, and the Bomb," *Journal of American History*, Vol. 74, No. 4, 1988, pp. 1289-1310; Office of National Estimates, *Chinese Communist Capabilities for Developing an Effective Atomic Weapons Program and Weapons Delivery Program*, Washington, DC: Central Intelligence Agency, June 24, 1955; Policy Planning Council (PPC) Director George McGhee to Secretary of State Dean Rusk, "Anticipatory Action Pending Chinese Demonstration of a Nuclear Capability," September 13, 1961, Washington, DC, Digital National Security Archive.

25. Burr and Richelson, "Whether to Strangle the Baby in the Cradle."

26. Ashton Carter and William Perry, *Preventive Defense: A New Security Strategy for America*, Washington, DC: The Brookings Institution, 1999, pp. 123, 131; Lyle Goldstein, *Preventive Attack and Weapons of Mass Destruction: A Comparative Historical Analysis*, Stanford, CA: Stanford University Press, 2006, pp. 133-135; David Sloss, "Forcible Arms Control: Preemptive Attacks on Nuclear Facilities," *Chicago Journal of International Law*, Vol. 4, 2003, pp. 39-58; and Joel Wit, Daniel Poneman, and Robert Gallucci, *Going Critical: The First North Korean Nuclear Crisis*, Washington, DC: The Brookings Institution, 2004, pp. 210-11, 219-220, 244.

27. Ashton Carter and William Perry, "Back to the Brink," *Washington Post,* October 20, 2002.

28. Under the terms of the deal, the United States and its allies agreed to supply North Korea with light water reactors for electricity production, and Pyongyang agreed to remain part of the NPT.

29. Quoted in Jeffrey Goldberg, "Obama to Iran and Israel: 'As President of the United States, I Don't Bluff'," *The Atlantic,* March 2, 2012.

30. Quoted in Jeffrey Heller and Matt Spetalnick, "Netanyahu Tells Obama: No Israeli Decision on Iran Attack," *Reuters,* March 6, 2012. Note, however, that this view is not universally shared among Israeli officials. For example, Mossad head Tamir Pardo has argued that a nuclear-armed Iran would not pose an existential threat. See Barak Ravid, "Mossad Chief: Nuclear Iran Not Necessarily Existential Threat to Israel," *Haaretz,* December 29, 2011.

31. Waltz, "The Spread of Nuclear Weapons," p. 15. For a critique of this argument, see Sagan, "The Perils of Proliferation."

32. "Acheson Rules out 'Preventive War'," *The New York Times,* June 14, 1950; Goldstein, *Preventive Attack and Weapons of Mass Destruction,* pp. 37-42.

33. William Burr, "Sino-American Relations, 1969: The Sino-Soviet Border War and Steps Towards Rapprochement," *Cold War History,* Vol. 1, No. 3, April 2001, pp. 73-112; and Elizabeth Wishnick, *Mending Fences: The Evolution of Moscow's China Policy from Brezhnev to Yeltsin,* Seattle, WA: University of Washington Press, 2001, pp. 34-36.

34. Ginor and Remez, *Foxbats Over Dimona,* p. 121; Ludmilla B. Herbst, "Preventive Strikes on Nuclear Facilities: An Analytic Framework," M.A. Thesis, University of British Columbia, 1995, p. 8; and George Russell, "Attack—and Fallout," *Time,* June 22, 1981.

35. Bob Hepburn, "Is Nuclear Plant Iraq's New Target in Israeli Desert?" *The Toronto Star*, February 19, 1991; "Iraq Reports 'Destructive' Attack on Israeli Reactor 'Dedicated to War Purposes'," *BBC*, February 18, 1991; Richard Owen, "Missiles Aimed at Dimona Nuclear Reactor," *The Times*, February 18, 1991; Stewart Stogel, "Iraq Fired Scuds at Israeli Reactor; '91 Attack Sought to Crack Dome," *Washington Times*, January 1, 1998.

36. Fuhrmann and Kreps, "Targeting Nuclear Programs in War and Peace."

37. Another factor that influences this threat perception is the similarity of foreign policy interests between the potential attacker and the proliferator. See Fuhrmann and Kreps, "Targeting Nuclear Programs in War and Peace," p. 840.

38. Iraq was not classified as a formal participant in the 1967 Six-Day War because it did not commit at least 1,000 troops or suffer 100 battle-related deaths. Meredith Reid Sarkees and Frank Wayman, *Resort to War: 1816-2007*, Washington, DC: CQ Press, 2010.

39. Quoted in Spector and Cohen, "Israel's Airstrike on Syria's Reactor."

40. David Albright, and Corey Hinderstein, "Algeria: Big Deal in the Desert?" *Bulletin of the Atomic Scientists*, Vol. 56, 2001, pp. 45-52.

41. Jacob Abadi, "Algeria's Policy toward Israel: Pragmatism and Rhetoric," *The Middle East Journal*, Vol. 56, No. 4, 2002, pp. 616-641.

42. Note, however, that Algeria did provide some support to Arab forces in 1967 and 1973.

43. See, for example, Matthew Fuhrmann and Jeffrey D. Berejikian, "Disaggregating Noncompliance: Predation versus Abstention in the Nuclear Nonproliferation Treaty," *Journal of Conflict Resolution*, Vol. 56, No. 3, 2012, pp. 355-381.

44. George H. W. Bush and Brent Scowcroft, *A World Transformed*, New York: Alfred Knopf, 1998, pp. 306-307.

45. Jim Garamone, "Bush: Iran Cannot Gain Nuclear Weapons," *American Forces Press Service*, January 30, 2006.

46. Note, however, that many of the factors discussed here were insignificant in the statistical analysis conducted by Fuhrmann and Kreps, "Targeting Nuclear Programs in War and Peace."

47. Matthew Fuhrmann and Sarah E. Kreps, "Why Attacking Iran Won't Stop the Nukes," *USA Today*, January 31, 2012.

48. See, for example, Caitlin Talmadge, "Closing Time: Assessing the Iranian Threat to the Strait of Hormuz," *International Security*, Vol. 33, No. 1, 2008, pp. 82-117.

49. Ramberg, "Preemption Paradox," p. 53.

50. George W. Bush, *Decision Points*, New York: Random House, 2010, p. 421.

51. For example, see Sagan, "The Perils of Proliferation."

52. This is relative to a scenario in which the target is democratic, and there is no history of violence between the potential attacker and the target.

53. "Factfile: How Osiraq Was Bombed," *BBC*, June 5, 2006, available from *news.bbc.co.uk/2/hi/5020778.stm*.

54. For a similar argument, see Dan Reiter, "The Global Nuclear Renaissance and the Spread of Violent Conflict: A Comment," Adam N. Stulberg and Matthew Fuhrmann, eds., *The Nuclear Renaissance and International Security*, Palo Alto, CA: Stanford University Press, 2013.

55. See, for example, Steven Miller and Scott Sagan, "Nuclear Power without Nuclear Proliferation?" *Daedalus*, Vol. 138, No. 4, Fall 2009; and Stulberg and Fuhrmann, *The Nuclear Renaissance and International Security*.

56. Research on the strategic effects of nuclear energy conducted to date focuses mostly on the relationship between nuclear power and nuclear proliferation. See, for example, Albert Wohlstetter, Thomas Brown, Gregory Jones, David McGarvey, Henry Rowen, Vince Taylor, and Roberta Wohlstetter, *Swords from Plowshares: The Military Potential of Civilian Nuclear Energy*, Chicago, IL: University of Chicago Press, 1979; Matthew Fuhrmann, *Atomic Assistance: How "Atoms for Peace" Programs Cause Nuclear Insecurity*, Ithaca, NY: Cornell University Press, 2012; Matthew Fuhrmann, "Spreading Temptation: Proliferation and Peaceful Nuclear Cooperation Agreements," *International Security*, Vol. 34, No. 1, Summer 2009, pp. 7-41.

57. Notable examples include Kyle Beardsley and Victor Asal, "Nuclear Weapons Programs and the Security Dilemma," Stulberg and Fuhrmann, eds., *The Nuclear Renaissance and International Security*; Michael C. Horowitz, "Nuclear Power and Militarized Conflict: Is There a Link?" Stulberg and Fuhrmann, eds., *The Nuclear Renaissance and International Security*; and Reiter, "The Global Nuclear Renaissance and the Spread of Violent Conflict."

58. Quoted in Fuhrmann, *Atomic Assistance*, p. 230.

59. Matthew Fuhrmann, "Australia's Uranium Exports and Nuclear Arsenal Expansion: Is There A Connection?" Michael Clarke, Stephan Frühling, Andrew O'Neil, eds., *Australia's Uranium Trade: The Domestic and Foreign Policy Challenges of a Contentious Export*, London, UK: Ashgate, 2011, p. 48.

60. Fuhrmann, *Atomic Assistance*.

61. *Iran: Nuclear Intentions and Capabilities*, Washington, DC: National Intelligence Council, November 2007, available from *www.dni.gov/files/documents/Newsroom/Reports%20and%20 Pubs/20071203_release.pdf*.

62. On the role of signaling in international relations, see James Fearon, "Signaling Foreign Policy Interests: Tying Hands Versus Sinking Costs," *Journal of Conflict Resolution*, Vol. 41, No. 1, 1997, pp. 68–90.

63. For a good overview of the Additional Protocol, see Theodore Hirsch, "The IAEA Additional Protocol: What It Is and Why It Matters," *The Nonproliferation Review,* Vol. 11, No. 3, Fall/Winter 2004, pp. 140-166.

64. See, for example, Paul Diehl and Gary Goertz, *War and Peace in International Rivalry*, Ann Arbor, MI: University of Michigan Press, 2000.

65. Miller and Sagan, "Nuclear Power without Nuclear Proliferation?" p. 10.

66. Of course, this does not imply that states should oppose nuclear energy. The development of nuclear power might help states meet growing energy needs, enhance their energy security, and, possibly, help address the problem of global climate change. These potential benefits must be weighed against the costs in order to determine whether the diffusion of peaceful nuclear programs is desirable. Matthew Fuhrmann, "Splitting Atoms: Why Do Countries Build Nuclear Power Plants?" *International Interactions,* Vol. 38, No. 1, 2012, pp. 29-57. See also Henry Sokolski, ed., *Nuclear Power's Global Expansion: Weighing Its Costs and Risks*, Carlisle, PA: Strategic Studies Institute, U.S. Army War College, 2010.

67. Kreps and Fuhrmann, "Attacking the Atom." For an alternative perspective, see Reiter, "Preventive Attacks against Nuclear, Biological, and Chemical Weapons Programs."

68. For further details, see Kreps and Fuhrmann, "Attacking the Atom"; Fuhrmann and Kreps, "Why Attacking Iran Won't Stop the Nukes."

69. A compellent threat is a demand to make some change to the existing status quo.

70. Todd S. Sechser and Matthew Fuhrmann, "Crisis Bargaining and Nuclear Blackmail," *International Organization,* Vol. 67, Winter 2013, pp. 173-195.

PART II:

NUCLEAR POWER, NUCLEAR WEAPONS –
CLARIFYING THE LINKS

CHAPTER 5

NUCLEAR POWER, NUCLEAR WEAPONS –
CLARIFYING THE LINKS

Victor Gilinsky

It was obvious from the beginning of the nuclear age that nuclear energy for power and nuclear energy for bombs overlapped. The 1946 Acheson-Lilienthal Report said the two were "in much of their course interchangeable and interdependent."[1] Therefore, and this was also understood from the beginning, gaining the benefits of the new energy source without spreading the bomb entailed strict international rules backed up by military force. This coupling did not diminish enthusiasm for developing nuclear energy. The United States proposed international ownership and control of what the report called intrinsically dangerous nuclear activities.[2]

The report contained powerful insights, but the proposal for international ownership was in many ways unrealistic and therefore failed. Less than a decade later, the United States, reluctant to give up the benefits of U.S. nuclear technology — at that time mostly political — reversed course to launch Atoms for Peace. The program promoted nuclear technology worldwide on the optimistic assumption that periodic international inspections would be sufficient to make sure that "peaceful" technology would not be used for weapons. This was the very arrangement the Acheson-Lilienthal Report had said would not work: "No system of inspection, we have concluded, could afford any reasonable security against the diversion of such materials to the purposes of war."[3] Not for the last time, the immediate attractions of nuclear energy overwhelmed distant security concerns.

Aside from occasional modest adjustment, we have been on that Atoms for Peace course ever since. But concerns about the weapons consequences kept intruding and stoked a continuing argument over whether occasional inspections by the International Atomic Energy Agency (IAEA) were really enough to keep the spread of nuclear electric facilities from contributing to the spread of nuclear weapons.[4] The present arguments over additional controls, especially in relation to reprocessing technology to separate plutonium and uranium enrichment, have their roots in that early history.

MAKING WAY FOR PLUTONIUM FUEL: FROM ATOMS FOR PEACE TO THE NPT

Plutonium is, of course, one of the two important nuclear explosives. Under Atoms for Peace, the United States declassified plutonium fuel technology, and the U.S. national laboratories trained foreign scientists in reprocessing technology.[5] The justification was that a shift to reliance on plutonium fuel was then considered inevitable so that reprocessing was regarded as an integral part of nuclear power operation.[6] A further rationalization, based on the scientifically incorrect argument first made in the Acheson-Lilienthal Report was that plutonium could be "denatured" to make it unusable for weapons.[7]

The other important nuclear explosive is highly enriched uranium (HEU). The United States did not release its uranium enrichment technology, then based on gaseous diffusion, and expected to monopolize it for many years. The United States, however, did ex-

120

port dozens of research reactors that were fueled with HEU, and ultimately exported over 30 tons of HEU.[8] Initially, the U.S. Government saw the various exports as too small to pose security concerns. Then, as the research and power reactor sizes increased, American and other exporters argued that IAEA inspections, or "safeguards" as they were optimistically called, were sufficient to make sure exports were not used to make material for bombs. The idea behind the inspections was that the threat of being found out and then sanctioned by the international community was sufficient to deter any would-be bomb maker from breaking the rules. The IAEA inspections were then gentlemanly affairs, with scientist-inspectors looking in on fellow scientists, and, in truth, the system was intended more to legitimize nuclear trade than to prevent wrongdoing.

In the late-1960s and early-1970s, the number and size of nuclear power installations increased rapidly. The preferred reactor type around the world was the light water reactor (LWR). Nuclear planners concluded from the large number of LWRs projected by national programs, and the limited then-known world uranium resources, that as early as 1980, the LWRs would have to be replaced by fast breeder reactors fueled with plutonium, thousands of them.

This meant many thousands of tons of plutonium in commercial channels, which gave pause to the security minded, as a bomb only requires a few kilograms. But the enthusiasms about plutonium as the fuel of the future overrode any concerns about its weapons potential. The interest in making way for the breeder reactor was strong enough to influence the negotiations over the Nonproliferation Treaty (NPT). The inspection provisions of the NPT were limited specifically to alleviate German and Japanese concerns that intrusive inspections would put them at a competitive disadvantage in supplying plutonium fuel for fast breeder reactors.[9]

Most of the less advanced non-nuclear countries saw the NPT negotiations as an opportunity to trade their signature for access to nuclear technology. They changed the NPT, which came into force in 1970, into a deal—or at least portrayed it as one—in which non-nuclear countries pledged not to make bombs in return for essentially unlimited access to "peaceful" nuclear technology. Article III stated that inspections were "to avoid hampering the economic or techno-logical development of the Parties . . . including the international exchange of nuclear material and equip-ment."[10] Article IV gave NPT members the **inalienable right** to develop and use nuclear energy, and to ben-efit from the obligation by all parties "to facilitate" the "fullest possible exchange" of nuclear technology.[11] In principle, all these activities had to conform to the overriding and fundamental prohibition on develop-ing nuclear weapons. In practice, "peaceful" came to mean whatever a country said was peaceful and sub-ject to IAEA inspections. In many quarters, that is still how the NPT is interpreted, and the phrase "inalien-able right" is still thrown back at anyone who would place restrictions on nuclear technology transfers.

SECOND THOUGHTS ON PLUTONIUM: INDIA'S BOMB TO 1976 FORD STATEMENT

It was widely believed in the early days of Atoms for Peace that, while a country might decide to make bombs on its own, no country would violate a "peace-ful uses" pledge to another.[12] This complacency was

122

punctured by India's 1974 nuclear explosion. India had obtained the plutonium for its bomb from a small Canadian reactor that used heavy water obtained from the United States. India had agreed to restrict use of the reactor and heavy water to "peaceful uses."[13] After setting off its bomb, India insisted there was no problem, as the bomb was peaceful. This was too much for the U.S. Congress and led to its rethinking of America's permissive nuclear export policy — and ultimately to the 1978 Nuclear Nonproliferation Act, which tightened the rules for U.S. nuclear export and effectively forbade nuclear fuel exports to India because it did not accept comprehensive IAEA inspections.[14]

The experience with India made clear that a country with direct access to nuclear explosives could quickly arm nuclear bombs if it wanted to. For countries with this capacity, one could no longer rely on the IAEA inspection system to provide "timely warning," that is, warning in time to stop the bomb manufacture. To keep nuclear weapons capabilities from spreading, it was necessary — despite the liberal wording of the NPT — to restrict access to fuels that were also nuclear explosives, and therefore also to reprocessing and enrichment facilities that can produce them.

Several of the chief exporting countries met secretly in London, England, in April 1975 to form the Nuclear Suppliers Group to place restrictions on the export of what they now called "sensitive" technology (as opposed to the Acheson-Lilienthal designation of "dangerous"). It appeared that France and Germany were getting ready to sell reprocessing plants to Pakistan, South Korea, Taiwan, and Brazil. The United States set out to block these projects.[15] In September 1975, U.S. Secretary of State Henry Kissinger told the United Nations (UN) General Assembly, "The greatest single danger of unrestrained nuclear proliferation resides in the spread under national control of reprocessing facilities."[16]

In the course of the 1976 presidential campaign, U.S. President Gerald Ford launched a study on the proliferation dangers of nuclear power programs and what could be done to keep them from contributing to proliferation. The President's October 1976 statement laid out the problem and announced his decisions.[17] It was the most important statement since the Acheson-Lilienthal effort.[18]

"The root of the problem," the President said, was that "the same plutonium produced in nuclear power plants can, when chemically separated, also be used to make nuclear explosives." He believed that nuclear power could proceed economically on the basis of the so-called once-through fuel cycle—without reprocessing spent fuel to extract plutonium and recycling it.

In spite of the view at the time that recycling plutonium was economically beneficial, the President declared:

> The reprocessing and recycling of plutonium should not proceed unless there is sound reason to conclude that the world community can effectively overcome the associated risks of proliferation. I believe that avoidance of proliferation must take precedence over economic interests. . . .[19]

The October 1976 statement also made a number of ancillary proposals that became part of the nonproliferation boilerplate up to the present. It urged nuclear suppliers to provide reliable fuel services instead of providing "sensitive" facilities, and proposed "suitably-sited multinational fuel-cycle centers to serve regional needs." But President Ford added the condition—often forgotten today—that any such centers had to be **economically warranted**. He raised an economic test again in asking all nations "to turn aside from pursuing nuclear capabilities which are of **doubtful economic value** and have ominous implications for nuclear proliferation and instability in the world."[20] He refused a subsidy to the still unopened Barnwell reprocessing plant and thereby ensured it would not begin operation.

The nuclear industry reacted with considerable antagonism.

THE EMPIRE STRIKES BACK:
PLUTONIUM ISN'T A PROBLEM,
AND IF IT IS, SO WHAT?

In proposing the once-through fuel cycle, an approach his successor, Jimmy Carter, would endorse, the President was trying to find a way of developing nuclear energy that preserved a safety margin for international security.[21] It was a reasonable approach, but the nuclear devotees saw it as way of postponing indefinitely their dream of moving beyond LWRs to plutonium-fueled fast breeders, in their view the ultimate objective of nuclear energy development.

In reality, Adam Smith's "Invisible Hand" had already moved that dream beyond the horizon. Fast breeder programs had fallen behind overoptimistic schedules, and their estimated costs mounted. Meanwhile, cheap uranium became plentiful and reprocessing turned out to be expensive, so there was no economic incentive to move beyond uranium-fueled LWRs. But the plutonium enthusiasts in government and industry would not relent. To keep the reprocessing efforts on track, they shifted their objective from recycling LWR plutonium in fast breeders to recycling it in LWRs. It made no economic sense — it was rationalized as a stopgap until breeder development caught up. By the time that became an unrealistic hope, fueling LWRs with a mixture of plutonium and uranium oxides, called mixed oxide fuel (MOX), had taken on a life of its own, with supportive government bureaucracies and industrial contractors.

Defenders of this substitute recycling insisted that it posed no proliferation problem because "reactor grade" plutonium, the plutonium formed in LWRs, unlike that from weapons production reactors, was contaminated with unwanted isotopes and thus unusable for weapons.[22] This echoed the Acheson-Lilienthal Report's denaturing concept, known to be incorrect at the time by those with access to weapons information but still widely believed 30 years later by those who did not. IAEA Director General Sigvard Eklund and his IAEA safeguards staff certainly believed it in 1976, as I discovered in talking to them in Vienna, Austria. Upon returning to Washington, DC, I notified the U.S. National Security Council staff, which arranged for a briefing on reactor grade plutonium at an international meeting Eklund would attend. I sat in and saw Elklund's jaw literally drop when Bob Selden, of the Los Alamos National Laboratory, made clear that the stuff could be used for bombs.[23]

LWRs can be an even more useful source for nuclear explosives than described in Selden's briefing. LWR plutonium is not necessarily heavily laden with unwanted isotopes, as would be the case if it came from spent fuel irradiated for the three fuel cycles, or about 5 years, that LWR fuel normally spends in the reactor. If the fuel is removed after one refueling, at about 18 months or earlier, the plutonium it contains is quite good even for low-technology bombs.[24]

The nuclear industry then took another tack to defend commercial reprocessing. In 1977, with the encouragement of the Electric Power Research Institute, an expert team at the Oak Ridge National Laboratory designed a small reprocessing plant that a country with minimal industrial base could build quickly and secretly. The Oak Ridge exercise's objective was to show that even if power reactor plutonium could be used for bombs, it was not going to do any good to ban commercial reprocessing, because a country could quickly build a small clandestine reprocessing plant, using essentially off-the-shelf components, and use it to produce militarily significant numbers of warheads.[25] An essential point is that an amount of plutonium (or HEU) that is commercially insignificant can be highly significant militarily.

The idea, of course, was to undermine the Ford-Carter anti-reprocessing policy. But it also undermined the Ford-Carter assumption that LWRs with no commercial reprocessing was a safe proposition. If a country with LWRs but no commercial reprocessing could secretly build a small "quick and dirty" plant to reprocess LWR spent fuel, then—contrary to conventional wisdom—it could rapidly separate enough plutonium for nuclear weapons, likely before the IAEA inspection system could set off a timely alarm.

FAST-FORWARD TO THE PRESENT: THE CENTRIFUGE AND OTHER PROBLEMS

If we fast-forward to the present, an important addition to proliferation concerns is the commercialization and wide distribution of gas centrifuge enrichment technology and the realization that centrifuge manufacturing capabilities are widespread, too. Unlike gaseous diffusion, gas centrifuge enrichment uses small amounts of electric power and lends itself to small-scale operation, which makes it easier for many states to get into small-scale enrichment. It also means that a small plant, likely difficult to spot from outside, could produce militarily significant quantities of HEU.

A country could build such a plant quite apart from any nuclear power program, but the presence of nuclear power plants would be advantageous. It would obviously provide a useful cloak to mask some of the clandestine activities and provide a source of trained personnel, but most importantly, it could provide a source of low enriched uranium fuel. The use of such feed material would reduce (either in size or duration) the enrichment effort to produce HEU by as much as a factor of five. Any such effort would also require ancillary conversion facilities for uranium compounds, which would make secrecy more difficult, but the presence of a nuclear power program would amplify the possibilities for small-scale clandestine HEU production. This provides another reason, in addition to the concern about small clandestine reprocessing, why LWRs by themselves are not necessarily a safe proposition from the point of view of proliferation.

We know that some countries, including NPT members, have cheated on their "peaceful uses" commitments, so one cannot exclude that possibility. Nor can we be confident that clandestine facilities would be found in time by the IAEA or even by national intelligence means, as it took years to find a number of secret nuclear facilities (the latest being the secret Syrian reactor).

Identifying clandestine weapons activities would become much harder if nuclear power programs expanded significantly, especially if many new countries adopted such programs, and even more so if these new countries were in the less stable parts of the globe. The IAEA bureaucracy would be faced with a larger and more complex job. It is unclear whether it could scale up effectively.

So far, the prospects are low for a large worldwide expansion in nuclear power installations. Such an increase has been held back by the nuclear power plants' extremely high cost, which is likely to be increased further by the lessons learned from the 2011 Fukushima, Japan, accident. Still, in recent years, the major nuclear bureaucracies have dedicated themselves to a worldwide nuclear "renaissance." The U.S. Congress, with the support of President Barack Obama, has voted large subsidies for U.S. nuclear plants. The nuclear vendors press their wares throughout the world, and quite a few countries, including a number in volatile regions in Asia and Africa, have expressed interest, and some of them may be willing to foot the steep bill to enter the nuclear power ranks.[26]

There is another ominous note—the nuclear "renaissance" movement includes efforts to revive commercial reprocessing. In 2007, the George Bush administration launched the Global Nuclear Energy Partnership, a crash futuristic reprocessing and recycling program. The advertised purpose was to "solve" simultaneously the nuclear waste and proliferation problems by having the United States and other major nuclear supplier countries provide a full range of fuel services. It was a poorly thought-out scheme based on exotic reprocessing and fuel technology that did not exist in practicable form.[27] The real purpose was to rekindle the nuclear dream of a fast reactor future and to start by reversing the Ford/Carter reprocessing restrictions, which always rankled the nuclear research and development (R&D) community. The enthusiasts sold President Bush on their idea, and on the occasion of signing the 2006 U.S.-India nuclear agreement, he said, "I don't see how you can advocate nuclear power . . . without advocating technological development of reprocessing."[28] The Obama administration continued a slowed down version of the Bush program with a new name, International Framework for Nuclear Energy Cooperation, but with the same basic purpose: "mainly in relation to closing the fuel cycle by reprocessing used fuel and burning actinides in fast reactors."[29] This effectively takes us back to the pre-1976 policy.

The 2006 U.S.-India agreement, proposed by former President Bush but supported by President Obama, by carving out a generous exception for India, which fought the NPT for 40 years, has seriously diminished the Treaty, and with it respect for what used to be called the nonproliferation regime. The U.S.-India agreement explicitly allows India to operate several of its nuclear power plants as part of its weapons complex.[30]

The United States also uses civilian power reactors to support its nuclear weapons program — the Department of Energy uses Tennessee Valley Authority's Watts Bar power reactor to produce tritium for warheads. When the arrangements were first announced and drew criticism, U.S. Department of Energy Assistant Secretary for Nuclear Energy Ben Rusche said the difference between civilian and weapons applications was only "psychological."

Despite these setbacks in anti-proliferation policy, there has been no letup in discussions over anti-proliferation measures because everyone knows there is a problem. Perhaps the most talked about, but also the most ineffectual, such measure is the recurring proposal for a "fuel bank" that would assure nuclear fuel supplies to countries with nuclear power plants to dissuade them from pursuing reprocessing or enrichment technology.[31] This rationale takes at face value the excuse countries give—that they worry about "security of supply"—to mask other reasons. In reality, existing commercial contracts provide a high level of assurance.[32] The talk about fuel banks allows governments to maintain the illusion of measures to control proliferation without having to incur any political costs.

A number of other proposals fall in the same category, for example, a much stricter and more intrusive IAEA inspection regime. Such an expanded and intrusive IAEA is unlikely to be realized because it would be inconsistent with industrial operations and national sovereignty and would be costly. However, an additional problem—a considerable leap from information to international action—exists. There are conflicting interests among the major states that impede a rapid response, and sometimes countries do not even want to know about illicit nuclear activities precisely because such information would force them into actions they do not want to take.[33] In other words, in an international complex of nuclear power programs, one cannot count on IAEA inspections or even national intelligence revelations leading reliably to enforcement. To maintain a decent margin of safety, there needs to be some limitation on the nuclear facilities in place.

THE LAST REFUGE: CLAIMING NUCLEAR POWER HAS LITTLE TO DO WITH PROLIFERATION

When all is said and done, the nuclear power lobby's ultimate argument against strict anti-proliferation rules for commercial nuclear facilities is that these facilities do not contribute to the proliferation problem, and so placing restrictions on commercial nuclear power programs would do little to affect proliferation. As the Nuclear Energy Institute puts it: "All nuclear weapons programs have either preceded or risen independently of civilian nuclear energy," and any future bomb makers would likely do the same because this would still be the easiest approach.[34]

As it was, of course, the nuclear age started with weapons rather than power plants, and the first five NPT weapons states did indeed start with dedicated weapons facilities. Civilian applications then piggybacked on weapons facilities and designs. The British and French built dual-purpose reactors to produce plutonium for warheads and also generate electricity.[35] The U.S. enrichment complex, built to produce highly enriched uranium for bombs, was later used to produce low enriched fuel for reactors.[36]

But would future bomb programs follow this historical pattern? Suppose, for example, that the historical sequence were reversed, and nuclear power facilities had been in place before World War II. Would the belligerents not have used them to obtain nuclear explosives for weapons? If the most readily available source of nuclear explosives will be in the commercial sector, then that is likely where bomb makers will go.

When the next group of countries—Israel, India, Pakistan, South Africa, and North Korea—decided, over the next 30 years, to build nuclear weapons, they did not yet have domestic power plants and fuel facilities that could supply nuclear explosives. They did have small nuclear research reactors that provided a focus for training a nuclear cadre.[37] At the same time, after the early-1960s, they had to cope with the fact that the international scene had become significantly less accepting of overt weapons programs. To build the larger "research" facilities they needed, the would-be bomb makers advanced secretly, or cloaked their weapons preparations in claims that they were only engaged in research directed toward "peaceful" nuclear power programs.[38]

The current situation is now different again. All the non-nuclear weapons countries are members of the NPT. A country intending to make nuclear weapons would have a choice of withdrawing from the NPT, and thus inviting a hostile reaction, or cheating under cover of the NPT. Unless we believe that this could never happen, we need to take this possibility seriously.

If a country is going to cheat—and we know that countries that were members of the NPT have cheated—it will want to limit the period of maximum vulnerability from the time its bomb program is evident, or might be discovered, to when it has bombs in its armory. The quickest future access to nuclear explosives for a country with a nuclear power program, and especially one with associated fuel facilities, is likely to be in some way related to that ongoing program.

WHAT THIS ADDS UP TO:
CURB YOUR ENTHUSIASM

It is clear that countries enriching uranium for fuel, or separating plutonium from their spent fuel in order to recycle it as nuclear fuel, have the means to produce nuclear explosives for bombs. (There is not much argument about this, although some people still cling to the notion that plutonium from commercial facilities is effectively unusable for bombs). With the general advance of technology and spread of information, the list of candidate countries that could, if they wanted, design and manufacture bombs given the necessary nuclear explosives continually expands. It is less obvious but nevertheless true that even countries lacking commercial enrichment or plutonium separation plants still have quite a leg up on making bombs if they have nuclear power plants or related research reactors.[39]

Ted Taylor, a former Los Alamos weapons designer, put it aptly:

> The connections between nuclear technology for constructive use and for destructive use are so closely tied together that the benefits of the one are not accessible without greatly increasing the hazards of the other.[40]

So far, we have not developed the technology or the international institutions to break this connection.

To cope with the hazards of proliferation in the face of weak international restraints on national nuclear programs — basically IAEA inspections and export controls — we seem to be slipping into reliance on greatly increased national intelligence operations, both to gather information and to carry out black operations to sabotage worrisome nuclear programs, and keeping open the possibility of air attacks.[41] That, at least, is what the Iran experience appears to suggest. In a sense, this is the logical consequence of expanding nuclear power around the world, one foreseen in the 1946 Acheson-Lilienthal Report. After all, someone has to enforce the NPT rules. In this respect, nuclear energy is the only electric energy source that poses major military risks if it is in the wrong hands, and that requires constant surveillance by highly alert intelligence operations with an enforcement backstop of military force. At the same time, it is difficult to imagine the current intense intelligence focus on Iran and the open option of large-scale violence as a workable model for the broader problem of proliferation. It is not even clear it will work in Iran.

It would be an especially problematic approach if the number of countries with nuclear programs, and the number of facilities, expanded significantly. Of course, there may not be any such expansion in view of nuclear power's high cost, now likely to go higher after the Fukushima accident. But increased worldwide reliance on nuclear energy remains a goal of the United States and other industrial countries. So politically powerful is this idea that President George W. Bush made a point of saying he would not object to Iranian nuclear power plants if Iran gave up enrichment. Meanwhile, Iran's example has provoked interest in nuclear power in a number of Middle Eastern and African countries.

One of the more naïve aspects of the Acheson-Lilienthal proposal was to distribute dangerous nuclear facilities owned by an international authority among the various countries with the thought that this would best dissuade countries from seizing the facilities for national weapons use. The idea was that each country would be deterred from doing so because it would know that other countries could do the same. Whatever deterrent value this arrangement had, it also had the intrinsic potential for massive failure with an avalanche of weapons decisions. Yet, that is essentially the arrangement that we are drifting toward today.

Up to now we have allowed, over and over, the interest in gaining the benefits of nuclear power to trump bomb worries. It is time to return to the principle stated 35 years ago by President Ford that, if a choice had to be made, "nonproliferation objectives must take precedence over economic and energy benefits." This would also likely mean holding up nuclear energy expansion worldwide until—to generalize President Ford's statement on plutonium use—"The world community can effectively overcome the associated risks of proliferation."

Restraining further expansion of nuclear power would not eliminate the possibilities of additional nuclear weapons countries, but it would limit the dangers—whose outlines we barely understand—inherent in further expansion, and it would be an important first step in coping with the international security implications of nuclear energy.

ENDNOTES - CHAPTER 5

1. *A Report On The International Control Of Atomic Energy,* Washington, DC, March 16, 1946. President Harry Truman appointed Under Secretary of State Dean Acheson to head a committee to set forth U.S. policy on what was then called atomic energy. The other members were scientists James Conant and Vannevar Bush, who headed the office that controlled the Manhattan Project; John McCloy; and General Leslie R. Groves, the military officer in charge of the Manhattan Project. Acheson ap-

pointed a board of consultants chaired by David Lilienthal, chairman of the Tennessee Valley Authority. J. Robert Oppenheimer, head of the Los Alamos during WWII, was the most influential member. The committee's report became known as the Acheson-Lilienthal Report. (Hereafter Report.)

2. "We were given as our starting point a political commitment already made by the United States to seek by all reasonable means to bring about international arrangements to prevent the use of atomic energy for destructive purposes and to promote the use of it for the benefit of society." Report, Sec. I.

3. Report, Chap. V.

4. The Agency was founded in 1957.

5. Soon after announcement of the Atoms for Peace program, Soviet Foreign Minister Vyacheslav Molotov asked U.S. Secretary of State John Foster Dulles why the United States wanted to spread nuclear weapons capabilities through the program. Dulles had no idea what Molotov was talking about, and when he returned to Washington, asked his assistant, Gerard Smith, to confirm that Molotov was wrong. As he later told me, Smith had to explain to the surprised Dulles that Molotov's question was a valid one. For example, Indian scientists trained in reprocessing at Oak Ridge, TN, became the nucleus for the Indian reprocessing program and hence the production of plutonium for bombs.

6. What this really means is that uranium-235, the fissionable isotope that makes up less than 1 percent of uranium, was thought to be too rare to fuel nuclear power plants for long and therefore use would have to be made of the abundant isotope, uranium-238, that with the addition of a neutron can be converted into plutonium.

7. It appeared in the Acheson-Lilienthal Report and in fact was central to the Report's conclusion that certain plutonium activities could be conducted on a national basis: "U 235 and plutonium can be denatured; such denatured materials do not readily lend themselves to the making of atomic explosives, but they can still be used with no essential loss of effectiveness for the peaceful applications of atomic energy." See Report, Chap. V. In the

case of uranium, the Report had in mind the use of low enrichment uranium fuel, as in today's LWRs. This material cannot be used for weapons without upgrading in an enrichment facility. In the case of plutonium, it meant what we would now call "reactor grade" plutonium, material that had been irradiated sufficiently to increase the fraction of unwanted isotopes. This is a much iffier concept, as plutonium of pretty much any composition can be made to explode. Robert Oppenheimer, the intellectual force behind the Report, pushed the denaturing concept, key to the whole Acheson-Lilienthal scheme, hoping it could be made to work but probably knowing it was wrong. The Report adds the following qualification: "It is not without importance to bear in mind that, although as the art now stands denatured materials are unsuitable for bomb manufacture, developments which do not appear to be in principle impossible might alter the situation."

8. That amounts to over a thousand bombs' worth. (IAEA-TECDOC-1452) Initially, the U.S. Atomic Energy Commission (AEC) did not even keep track of what was exported and where it went. The agency left this to private firms. The AEC commissioners became aware of this in the course of investigations around 1966 after the loss of about 100 kilograms of HEU at a fuel plant in Pennsylvania that could not be accounted for and was feared to have ended up in Israel. It is hard to understand why the United States was so casual about the export of HEU.

9. Wolf Haefele, the chief technical advisor to the German NPT delegation, believed that the economic opportunities were going to lie in manufacturing fast breeder fuel rather than in building the reactors themselves. ("It's the razor blade not the razor.") He thought Germany was well positioned to compete on fuel technology but worried that it would be at a disadvantage if, as a non-nuclear state, it was subject to more intrusive international inspection and was thus more vulnerable to industrial espionage. He convinced the Japanese to join in Germany's complaints. As a result, the Treaty Preamble encourages inspecting "the flow of source and special fissionable materials by use of instruments and other techniques **at certain strategic points**." (Emphasis added.)

10. Treaty on the Non-Proliferation of Nuclear Weapons, July 1, 1968. (Hereafter NPT.)

Article III 3. The safeguards required by this article shall be implemented in a manner designed to comply with Article IV of this Treaty, and to avoid hampering the economic or technological development of the Parties or international cooperation in the field of peaceful nuclear activities, including the international exchange of nuclear material and equipment for the processing, use or production of nuclear material for peaceful purposes in accordance with the provisions of this article and the principle of safeguarding set forth in the Preamble of the Treaty.

11. NPT:

Article IV 1. Nothing in this Treaty shall be interpreted as affecting the inalienable right of all the Parties to the Treaty to develop research, production and use of nuclear energy for peaceful purposes without discrimination and in conformity with articles I and II of this Treaty. 2. All the Parties to the Treaty undertake to facilitate, and have the right to participate in, the fullest possible exchange of equipment, materials and scientific and technological information for the peaceful uses of nuclear energy. Parties to the Treaty in a position to do so shall also cooperate in contributing alone or together with other States or international organizations to the further development of the applications of nuclear energy for peaceful purposes, especially in the territories of non-nuclear-weapon States Party to the Treaty, with due consideration for the needs of the developing areas of the world.

12. The IAEA inspections were not seen as performing a police function to catch wrongdoers, but rather verifying material balances to add confidence that agreements were being observed.

13. The 1956 India-U.S. heavy water contract restricted the reactor to "peaceful uses." When finally called on this, India said there was no problem because its bomb was peaceful. The American position was not as clear as it could have been. At the time, the United States had a program on so-called peaceful nuclear explosives. That was bad enough, but its real purposes were clouded. In a 1964 briefing I attended, Director of the Livermore weapons

laboratory John Foster explained that the real purpose of the program was to get the public used to nuclear explosions so that the military could get a release for battlefield use in wartime.

14. When congressional hearings raised questions, the U.S. State Department, intent on protecting nuclear exports, tried to hide the existence of the U.S.-India heavy water contract, and then lied about whether U.S. heavy water was still in the Indian reactor. It ultimately led to the Nuclear Nonproliferation Act of 1978, which required tightening nuclear cooperation agreements, including the one with India that covered the General Electric-built reactor at Tarapur. The United States essentially gave the nuclear station to India in order to introduce IAEA inspections into India. It was a unique agreement that tied Indian acceptance of IAEA inspections to the U.S. supply of fuel. The 1978 Act forbade fuel supply to countries that did not accept full-scope IAEA inspections, which India did not. The State Department scurried to find a replacement fuel supplier.

15. But neither the United States nor the other exporters ever publicly addressed the tensions in the NPT between prohibitions on bombs and liberal promises of technology, so the NPT's ambiguities remained.

16. Kissinger's apparent concern has to be put in context. When the State Department staff learned of the Indian explosion, they assumed the United States would react firmly. Kissinger cabled back from the Middle East, rejecting any strong reaction. He was apparently in the process of putting together a nuclear deal of his own that he did not want upset. In the 1975 speech to the UN, he pointed to the dangers of reprocessing conducted under national auspices and proposed a multilateral approach. This became a standard "solution" to the problem of reprocessing. It is unlikely that Kissinger understood that the activity made no economic sense at all. He would have looked at it in purely political terms. In a talk at the RAND Corporation in Santa Monica before he assumed his role in the Richard Nixon administration—I was then a department head—he said, "Never underestimate the superficiality of important people."

17. Gerald R. Ford, "Statement on Nuclear Policy," October 28, 1976.

18. And we are still waiting for the third one.

19. It subsequently became clear that plutonium reprocessing and recycling were highly uneconomic. At the time, I also accepted that plutonium recycling was marginally economic. I calculated that the advantage was about 3 percent of the cost of power, and planned to say so in a speech. I showed the draft to a fellow commissioner—Richard Kennedy. His reaction said a lot about the extent to which the nuclear community at the time was invested in the idea of plutonium recycle. He told me I had every right to deliver the speech saying the plutonium advantage was only 3 percent. But he wanted me to know that if he had wanted to kill U.S. nuclear power, that was what he would say.

20. Gerald R. Ford, "Statement on Nuclear Policy," October 28, 1976. Emphasis added.

21. The restriction would apply to HEU as well, but hardly any power reactors used this material for fuel—it was mainly used in research reactors. The United States proposed shifting these reactors to lower enrichment fuels. After a lot of foot dragging, most of these reactors have now been converted, although some, including the research reactor at the Massachusetts Institute of Technology (MIT), still resist.

22. One claim is that a bomb made of this material would "fizzle." Using the technology in the first bombs, the yield of such a fizzle would still be on the order of a kiloton.

23. The top German officials who were exporting the technology to Brazil shared the same belief that the plutonium separated from LWRs would be unusable for weapons, or at least professed to believe that. The Selden briefing was described in a memorandum for the commissioners of the U.S. Nuclear Regulatory Commission (NRC) from James Shea, director of the NRC Office of International Programs. He wrote:

> From November 15 to 19 ERDA [Energy Research and Development Administration, DOE's predecessor] conducted a series of briefings, presented by Bob Selden (Livermore) and Carson Mark (Los Alamos), directed at convincing in-

terested international VIP's attending the ANS-AIF-ENS [American Nuclear Society-Atomic Industrial Forum-European Nuclear Society] International Meetings that reactor grade plutonium is highly useful for constructing nuclear explosives.

The attendees at these meetings included Sir John Hill (UK Atomic Energy Authority), Mr. Andre Giraud (Alternative Energies and Atomic Energy Commission-France), Dr. S. Eklund and R. Romettsch (International Atomic Energy Agency), Dr. Daennert (Fundamental Research Grant-Ministry of Research and Technology), and Dr. Imai (Japan Atomic Power Company), and others from Japan. See also "ERDA says reactor grade plutonium can make powerful, reliable bombs," *Nucleonics Week* November 16, 1976. The story included the following:

> Most recently NRC commissioner Victor Gilinsky said that reactor grade plutonium would make a bomb of 1-10 kilotons yield. A source said that ERDA provided Gilinsky with the material for his statement. The ERDA source stated that such bombs would be variable [in yield] but certainly not unreliable.

24. At the end of its first refueling, a new LWR contains about 300 kilograms of plutonium that is quite suitable for weapons by any standard. That would be enough for about 50 warheads. Of course, it first would have to be separated.

25. It was promoted by Chauncey Starr, Electric Power Research Institute's head, and by Floyd Culler, one of the developers of PUREX reprocessing at Oak Ridge. See *www.npolicy.org/article. php?aid=172&rt=&key=fresh%20&sec=article*.

26. A sign that conventional opinion is changing is evident in an October 6, 2011, opinion editorial by Jim Hoagland: "In short, the proliferation of nuclear reactors across Asia is certain to facilitate and encourage nuclear weapons proliferation as well." Jim Hoagland, "Nuclear energy after Fukushima," *The Washington Post*, October 6, 2011.

27. For a discussion on the program's technical flaws, see "A Minority Opinion: Dissenting Statement of Gilinsky and Macfarlane," *Review of DOE's Nuclear Energy Research and De-*

velopment Program, Washington, DC: National Academy Press, 2008, pp. 73-76, available from *www.nap.edu/openbook.php?record_ id=11998&page=73.*

28. "President, Prime Minister Singh Discuss Growing Strategic Partnership," New Dehli, India, March 2, 2006. The U.S.-India agreement approved by Congress in October 2008 waived U.S. export restrictions on India, which has fought the NPT regime for 40 years. It makes a mockery of NPT compliance. In effect, with Democratic congressional support, Bush drove a truck through the NPT. A related U.S.-sponsored Nuclear Supplier Group waiver gave India access to the international nuclear trade. The Indian government succeeded in steamrolling the very U.S. and international criteria that were put in place in response to its initial pursuit of the bomb—without giving up anything.

29. See *www.ifnec.org.*

30. "Taking Stock of the U.S.-India Nuclear Deal," Remarks of Geoffrey Pyatt, Principal Deputy Assistant Secretary, Bureau of South and Central Asian Affairs, Mumbai, India, September 30, 2011.

> First, India agreed to draw a clear line between its civilian and military nuclear facilities, and to voluntarily place its civilian nuclear facilities under IAEA safeguards. India's 2005 Separation Plan identified 14 thermal power reactors, as well as a number of upstream and downstream facilities, and nine research facilities for the safeguarded side of India's nuclear complex. . . .

But the separation leaves several power reactors on the military side. Pyatt also makes clear the motivation for making an NPT exception for India:

> And we are open for business. In fact, U.S. companies representing the full spectrum of commercial nuclear activities have participated in six commercial trade missions to India in the past few years, including: . . . I should note— unequivocally—that all of our companies involved in these missions have the strong support of the United States government. . . .

143

31. See, for example, one of the recommended actions arising from the 2010 NPT review conference:

> Continue to discuss further, in a non-discriminatory and transparent manner under the auspices of IAEA or regional forums, the development of multilateral approaches to the nuclear fuel cycle, including the possibilities of creating mechanisms for assurance of nuclear fuel supply, as well as possible schemes dealing with the back-end of the fuel cycle without affecting rights under the Treaty and without prejudice to national fuel cycle policies, while tackling the technical, legal and economic complexities surrounding these issues, including, in this regard, the requirement of IAEA full scope safeguards.

NPT/CONF.2010/50, 2010 Review Conference of the Parties to the Treaty on the Non-Proliferation of Nuclear Weapons, Final Document, Vol. I.

32. India is mentioned as the classic example (but I believe it is the only such example) of a country that faced a halt in uranium fuel shipments from its supplier. But this happened after it exploded its 1974 bomb in violation of a peaceful uses pledge and then refused to accept comprehensive IAEA inspections as required by the 1978 U.S. export law. Even so, the U.S. State Department found an alternative supply for India, and there was no gap in its fuel shipments.

33. As was the case in 1969 at the start of the Nixon administration when Henry Kissinger decided it would be best if the United States did not know whether Israel had built nuclear weapons.

34. "Preventing the Proliferation of Nuclear Materials," Nuclear Energy Institute, October 2011, available from *www.nei.org/ Master-Document-Folder/Backgrounders/Fact-Sheets/Preventing-the-Proliferation-of-Nuclear-Materials*:

> Uranium enrichment facilities that produce fuel for commercial reactors pose no risk of proliferation . . . Used nuclear fuel, which contains plutonium generated as a byproduct of the commercial fuel cycle, poses little risk of proliferation . . . All nuclear weapons programs have either preceded or risen independently of civilian nuclear energy.

"Safeguards to Prevent Nuclear Proliferation," World Nuclear Association (WNA), April 2012, available from *www.world-nuclear.org/info/Safety-and-Security/Non-Proliferation/Safeguards-to-Prevent-Nuclear-Proliferation*:

> Civil nuclear power has not been the cause of or route to nuclear weapons in any country that has nuclear weapons, and no uranium traded for electricity production has ever been diverted for military use. All nuclear weapons programmes have either preceded or risen independently of civil nuclear power. . . . No country is without plenty of uranium in the small quantities needed for a few weapons. . . . There is no chance that proliferation will be solved by turning away from nuclear power.

The same WNA document, however, continues in a more insightful mode:

> While nuclear power reactors themselves are not a proliferation concern, enrichment and reprocessing technologies are open to use for other purposes, and have been the cause of proliferation through illicit or unsafeguarded use. . . . The NPT does not adequately deal with the issue of SNT [sensitive nuclear technology]. It refers to the 'inalienable' right to use nuclear energy, but certainly does not guarantee the right to develop SNT. Nor, however, does the it explicitly limit the development of SNT, other than by the fundamental obligations of Non-nuclear Weapons States not to acquire (or seek to acquire) nuclear weapons, and to place all their nuclear material under IAEA safeguards.

> Current approaches to control the spread of SNT have focused on measures against the transfer of equipment, components, special materials and technology, through national export controls and multilateral coordination within the Nuclear Suppliers Group (see below). However, these approaches do not fully address the problems of illicit acquisition of enrichment technology and development of indigenous enrichment technology. A way is needed to assess the international acceptability of enrichment projects.

> Concerns about SNT programs are not addressed simply by having these activities placed under safeguards. Safeguards

are an essential part of international confidence building, but safeguards alone cannot provide assurance about a country's future intent. An enrichment or reprocessing facility under safeguards today could be used as the basis for breakout from non-proliferation commitments in the future. In the case of enrichment, a large centrifuge plant, using LEU feed, could produce sufficient HEU for a nuclear weapon in a matter of days. An essential aspect of non-proliferation is minimising the risk of breakout occurring, through limiting the countries with SNT facilities to those regarded as presenting a low proliferation risk.

35. The Soviet-designed RBMK reactors were used for both plutonium and power production, but it is unclear whether the plutonium was ever used for weapons.

36. It is also significant that the U.S. light water reactors, ultimately the basis for essentially all the world's reactors, grew out of the U.S. Navy's submarine reactors developed by Admiral Hyman Rickover.

37. Israeli Research Reactor 1 (IRR1), started in 1960, was donated by the United States under Atoms for Peace. The 5-megawatt ton (MWt) reactor used HEU fuel. India claims to have built a 1-MWt research reactor that went critical in 1956. Canada supplied the CIRUS, started in 1960, which was a 40-MWt heavy water moderated and light water cooled reactor fueled by natural uranium. The United States supplied the heavy water under a contract signed in 1956. Pakistan's PARR-I Reactor was supplied by the United States. The 5-MWt reactor used HEU fuel and went critical in 1965. South Africa's SAFARI-1 20-MWt reactor was commissioned in 1965. The U.S.-supplied reactor used HEU fuel, initially operated at 6.75-MWt, and was upgraded in 1968. North Korea's IRT-2000, an 8-MWt (2-MWt from 1965-74, 4-MWt from 1974-86) heavy-water moderated research reactor, was supplied by the Union of Soviet Socialist Republics in 1965.

38. Israel was the first of these countries to develop nuclear weapons. Israel lied to President Kennedy about the purpose of the Dimona reactor, claiming it was for peaceful purposes. Had Israel tested full-scale when it built its first bombs in the late-1960s, it might have qualified as one of the original nuclear weapon

states under the NPT. India drew the plutonium for its 1974 nuclear explosion from CIRUS, a small Canadian-supplied research reactor that used U.S. heavy water. India had given both countries "peaceful uses" assurances. When challenged, India replied that its bomb was peaceful. The U.S. State Department did not press the issue. India then continued to stockpile CIRUS plutonium for weapons. Pakistan also claimed its enrichment was for peaceful uses. No one actually believed this, but the United States looked the other way to keep Pakistani assistance in Afghanistan. South Africa claimed its Valindaba enrichment plant was built to supply fuel for its research reactor. The North Korean weapons program started in the 1980s with a nuclear reactor at Yongbyon. In 1985, under Soviet pressure, Pyongyang agreed to join the NPT, but refused to sign a safeguards agreement with the IAEA. Iran claims its enrichment program is intended to supply enriched uranium for its power and research reactors. Time will tell whether this is true.

39. Victor Gilinsky, Marvin Miller, and Harmon Hubbard, "A Fresh Examination of the Proliferation Dangers of Light Water Reactors," Washington, DC: Nonproliferation Policy Education Center, October 22, 2004, available from *www.npolicy.org/files/20041022-GilinskyEtAl-LWR.pdf*.

40. Theodore B. Taylor, *Nuclear Power and Nuclear Weapons,* Santa Barbara, CA: Nuclear Age Peace Foundation, July 12, 1996, available from *www.wagingpeace.org/nuclear-power-and-nuclear-weapons/*.

41. A fuller list of the steps envisioned in current anti-proliferation doctrine would include: Count on only a few countries being interested in nuclear weapons; apply the various mild restrictions and inspections to intimidate those who worry about getting caught; make compromises on access to fuel technology that delay weapons capabilities, even if it means shaving the security safety margins; ramp up IAEA inspections and (mainly) national intelligence; concentrate on hostile states that do not yet have bombs; in the last analysis, count on sabotage, assassinations, and bombings; let the future take care of itself; and hope for the best.

CHAPTER 6

SCOPING INTANGIBLE PROLIFERATION RELATED TO PEACEFUL NUCLEAR PROGRAMS: TRACKING NUCLEAR PROLIFERATION WITHIN A COMMERCIAL NUCLEAR POWER PROGRAM

Susan Voss

INTRODUCTION

It requires an estimated 3,000 people to construct a large commercial nuclear power plant and 800 to 900 people to operate two 1,000-megawatt-electric (MWe) light water reactors. If a country has a reactor built by a commercial company or a consortium from another country, there is a large exchange of experts working together from both countries. This includes training for language skills, experts on sites, safety experts, reactor and balance-of-plant design, and plant operations, to name just a few. Construction and manufacturing of the reactor system requires that the plant construction contractor provide highly skilled labor and technical expertise on-site to ensure the work meets nuclear quality assurance standards. Therefore, the construction process brings together experts with backgrounds in many different fields and varied levels of education and skill.

One of the key proliferation concerns is that a country may choose to develop a commercial nuclear power program as a means of concealing nuclear weapons development by exploiting direct contact with nuclear experts in another country in order to obtain nuclear weapons-related information and produce nuclear weapons-useable materials and equipment. There are historic cases where nuclear power development has been used as a front for a nuclear weapons program, and technology has also been transferred within the framework of nuclear research outside of formal international agreements. The purpose of this chapter is to discuss the systematic tracking of information needed to follow and interdict nuclear proliferation and the difficulties in doing this work. A summary of technologies that have reportedly been illicitly transferred to Iran are provided as an example.

NEW CONSTRUCTION

During the past 15 years or so, there have been a number of surprises in the field of nuclear proliferation, including the discovery of the construction of the Syrian research reactor with the Democratic People's Republic of Korea's (DPRK) support; the recent reports that China provided Pakistan nuclear material and the design for a weapon; the Khan network providing uranium enrichment equipment and expertise, along with nuclear weapons design information, to Libya, the DPRK, Iran, and a still unnamed fourth nation; and the surprising advancements within the Iranian nuclear program that were revealed by a dissident group — including heavy water production, the operation of a heavy water reactor, and uranium enrichment using centrifuge technology. These examples point to the difficulties faced by the nonproliferation community in tracking illicit nuclear agreements and transfers of technology, which have steadily grown with the advent of global trade. The potential transfer of technology is a more significant issue in nuclear weapons states where nuclear institutions and education centers work on technology that can be used for nuclear weapons materials production, nuclear weapons design, and nuclear weapons testing. Therefore, to help clarify the problem of technology transfers, the issue can be divided into three categories: nuclear materials, nuclear weapons, and nuclear weapons testing.

NUCLEAR MATERIALS

The construction of a nuclear power plant requires the exchange of scientists, engineers, material scientists, and technicians between two countries unless it is being built with indigenous technology. This affords a country that may have the production of nuclear material for a nuclear weapon as its ultimate goal an opportunity to have direct contact with nuclear experts in other countries and the possibility of establishing outside contracts or agreements.

Nuclear weapons materials production is a rather broad category that includes uranium mining and conversion, uranium enrichment, fuel pin manufacturing, the operation and construction of research reactors, reprocessing spent fuel, and plutonium production. Often, knowledge of these key technical areas coexists at the institutes, companies, and educational centers involved in commercial nuclear power since the processes for producing nuclear power and nuclear weapons materials utilize complementary technologies and capabilities. Also, some of the manufacturing organizations work on many different technologies and may have expertise in manufacturing what are called dual-use capabilities. What this implies is that, given a cooperative agreement to construct a nuclear power plant, a country seeking a nuclear weapons capability will have an increased probability of having access to experts, institutions, companies, and educational centers that have expertise in the production of nuclear materials that could be used for weapons.

Given the large quantity of people, equipment, and correspondence — over 1,000 people; tens of thousands of pieces of equipment; and tens of thousands of documents, drawings, and communications — involved in the design and construction of a nuclear power plant, it is possible that a country seeking a nuclear weapons capability could use a nuclear power plant as a cover to obtain information on nuclear materials production. That said, it is ironic to note that the Syrian research reactor, reportedly being built with the help of the North Koreans, was not noticed, in part because Syria's partnership with the DPRK was unknown, and therefore Syria was not under the same scrutiny to which a country suspected of developing a nuclear weapons capability would be subjected. Examples of nuclear material pathways include the following:

1. Pakistan chose the highly enriched uranium (HEU) path based upon the illicit use of information obtained from the commercial URENCO plant. It has since built heavy water research reactors for the production of plutonium.

2. The DPRK chose a plutonium path, using spent nuclear fuel from its 5-MWe research reactor. There are recent reports that it is pursuing the HEU option, using centrifuge technology reportedly obtained through the Khan network.

3. Syria was possibly pursuing a plutonium path using the research reactor it was developing with North Korea.

4. South Africa chose an HEU path with support from Israel.

5. Iran appears to have chosen a multipronged approach for HEU production—including the use of centrifuge technology obtained through the Khan network and, early in its program, laser enrichment—while simultaneously pursuing possible plutonium production through its heavy water research reactor. Iran obtained much of its technological base through interactions with Pakistan's Khan network, China, and Russia.

6. Libya was pursuing an HEU path with centrifuge technology purchased from the Khan network.

These examples are included to illustrate the different pathways taken by some countries pursuing nuclear material production capability. Of the countries that have pursued, or appear to have been pursuing, a nuclear weapons capability during the past 15 years, Iraq, Syria, and Libya had nuclear research programs but not a civilian nuclear power program, whereas North Korea and Iran developed fuel cycle capabilities and nuclear power programs as the cornerstone for their broader nuclear research. Both the North Koreans and Iranians received nuclear training in Russia. It is reported that more than 300 North Korea nuclear specialists of various qualifications were trained at various Soviet institutes of higher education. These facilities included the Moscow Engineering Physics Institute (MEPhI), the Bauman Higher Technical School, the Moscow Energy Institute, and other educational establishments. Some of the nuclear specialists worked at the Joint Institute for Nuclear Research in Dubna and the Institute of Physics and Power Engineering in Obninsk.[1] Specific examples of Iranians training in Russia are referenced in the following text.

NUCLEAR WEAPONS AND WEAPONS TESTING

The second category of nuclear weapons expertise represents a much smaller core group of individuals, institutions, companies, and educational centers. While these groups may still be involved in the commercial nuclear power business, they are far fewer in number and generally work within a secure or classified environment.

According to Jeffrey Richelson, "While thousands were involved in India's nuclear program, only 50 to 75 scientists were actually part of the effort to design and build an explosive device."[2] (See Figure 6-1.) The number of individuals with specialized nuclear materials production and weapons knowledge is a relatively small fraction of those with knowledge in nuclear power plant construction, manufacturing, and operation.

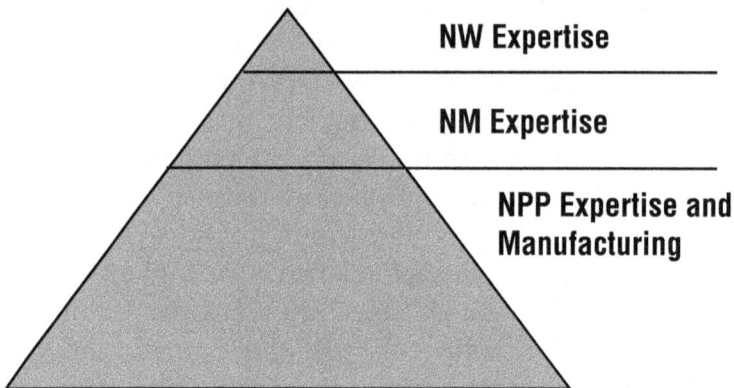

Figure 6-1. Individuals with Specialized Nuclear Materials Production and Weapons Knowledge Compared to Those with Knowledge in Nuclear Power Plant Construction, Manufacturing, and Operation.

Finally, the third category, nuclear weapons testing, also has a reduced number of people working within the civilian nuclear power business, but crossover is still possible. More difficult to trace are the affiliated organizations or front companies for the weapons laboratories that provide specialized expertise in support of nuclear weapons work and testing.

EXISTING AND NEW NUCLEAR POWER PLANT CONSTRUCTION

According to the World Nuclear Association (WNA), there are 435 nuclear power reactors operating in 30 countries, plus Taiwan, representing 2,630 billion kilowatthours (kWh) in 2010 figures, or around 14 percent of the world's electricity.[3] There are currently 60 new reactors under construction in 14 countries, including some countries with existing nuclear power capability: the United States, Canada, Finland, France, the United Kingdom (UK), Romania, Slovakia, Bulgaria, Russia, South Korea, Japan, China, Pakistan, and India. Nations that are far along in their planning for a commercial nuclear power capability include Poland, Kazakhstan, the United Arab Emirates (UAE), Jordan, Turkey, Vietnam, Indonesia, and Thailand.[4] Overall, there are "45 countries actively considering embarking upon nuclear power programs which do not currently have it."[5] According to the WNA, the following countries are actively considering nuclear power programs (NPP):

- In Europe: Italy, Albania, Serbia, Croatia, Portugal, Norway, Poland, Belarus, Estonia, Latvia, Ireland, and Turkey.

- In the Middle East and North Africa: Iran, Gulf states including the UAE, Saudi Arabia, Qatar and Kuwait, Yemen, Israel, Syria, Jordan, Egypt, Tunisia, Libya, Algeria, Morocco, and Sudan.
- In West, Central, and Southern Africa: Nigeria, Ghana, Senegal, Kenya, Uganda, and Namibia.
- In South America: Chile, Ecuador, and Venezuela.
- In Central and Southern Asia: Azerbaijan, Georgia, Kazakhstan, Mongolia, Bangladesh, and Sri Lanka.
- In Southeast Asia: Indonesia, Philippines, Vietnam, Thailand, Malaysia, Singapore, Australia, and New Zealand.
- In East Asia: North Korea.

Of the 14 countries with new reactors under construction, seven of them—the United States, France, the UK, Russia, China, Pakistan, and India—are already nuclear weapons states; and two of the seven nonweapons states, Japan and Canada, have expertise in the production of weapons-usable materials. Both countries are under International Atomic Energy Agency (IAEA) safeguards. Primary reactor design, manufacturing, and construction are from private commercial, or state-owned companies within the United States, Canada, France, Russia, South Korea, and Japan.[6]

TRACKING PROLIFERATION

Therefore, given the increase in new countries seeking a civilian nuclear power program and the potential for increased interaction with individuals, companies, organizations, institutions, and educational centers with nuclear production expertise and technology, and to a lesser extent with those with expertise in nuclear weapons and testing, we have to ask, What do we need to do to track proliferation in this environment? If we estimate that there are thousands of people working on a nuclear power plant within the country building the plant, several thousands more visiting that country, thousands of pieces of equipment being shipped, and thousands of documents and communiqués being generated, is it possible to find illicit activity between nations, institutions, companies, or even between individuals?

A "systems approach" is required to set up an infrastructure to track nuclear proliferation and identify potential gaps or discrepancies. The key to the effort is gaining an understanding of the fundamental governmental and organizational structures for each country involved in receiving or transferring nuclear technology. Without an understanding of the organizational structures, it is impossible to understand the relationship between people, companies, technologies, projects, and other important indicators of proliferation. Creating a detailed understanding of an organization and identifying the key people takes a significant amount of time and effort.

Once a model is established, it is important to keep it current as changes are made—i.e., a dynamic systems model or living document. Once organizational structures are established, additional information— such as relationships, publications, history, timelines, technologies, and projects—can be correlated. Timelines are critical for gaining a spatial understanding of the inter-relationship between technology development and a country's overall capability. Relationship charts help to map interactions between different organizations, groups, and individuals, thereby providing possible ties between military and civilian interests. Tracking past relationships can provide insights into current transactions as well. Maps, material/facility flow sheets, organization charts, and relationship charts all provide an integrated picture that incorporates thousands of pieces of information and can capture the state of knowledge at that moment.

Given the obvious need to develop not only in-depth tracking but also integrated systems thinking, why is this not accomplished more often? Interestingly, one of the primary concerns has to do with the power dynamics and the interests of the individual analyst. A nonproliferation analyst is noted for the research he or she performs on a specific topic and therefore, in a sense, the analyst owns that piece of the puzzle. When an analyst surrenders it to a larger model, it results in a loss of power and a place at the table. Therefore, one of the keys to developing extended systems thinking is finding a way to ensure that each analyst has a place in the larger picture.

Yet, even establishing an understanding of the organizational dynamics does not guarantee there will be a clear path for action. For example, it was never clear what role former President Pervez Musharraf or the military may have played in the Khan network. There are many similar examples where the line of information ends and it is indeterminable as to why; in other words, is it because there are no further connections and links, or might it simply be that we were unable to obtain additional information?

Example: Iran's Nuclear Program.

The Iranian nuclear program is used as an example of a nation whose civilian nuclear program has been identified as providing cover for illicit activity.[7] To understand their nuclear program, it is important to establish the organizational relationships of the Iranian government, the military, and the religious leadership. To gain a systems perspective of Iran's program, a model of the organizations that work with nuclear technology or nuclear policy and their inter-relationships needs to be built. The Atomic Energy Organization of Iran (AEOI) has the lead for civilian nuclear development, including uranium mining, conversion, fuel fabrication, research reactors, the Bushehr nuclear power plant, and uranium enrichment. The AEOI has its headquarters in Tehran and operates a number of different facilities and research centers throughout Iran. A good summary of the AEOI organization, capabilities, and facilities is provided by Dr. Ghannadi-Maragheh in his 2002 paper.[8]

Since February 13, 2011, the director of the AEOI has been Dr. F. Abbasi-Davani. He has a dual role as head of the AEOI and Vice President of Iran. The dual appointment shows the importance of the nuclear program within Iran. Information on the military dimension to Iran's nuclear program has emerged in the past 10 years or so. Dr. Abbasi-Davani has been linked to the military side of Iran's government. According to the Institute for Science and International Security (ISIS),[9] "Prior to his appointment as AEOI head, he chaired the physics department at Tehran's Imam Hossein University, which is linked to the Iranian Revolutionary Guard Corps (IRGC) and work on nuclear weaponization." Dr. Abbasi-Davani's unique role of being linked to the military aspect of the weapons program and his recent placement as the head of the AEOI may indicate the Iranian government's decision to ensure closer ties between civilian and military nuclear development. Both the IRGC and regular military are subordinate to the Ministry of Defense and Armed Forces Logistics (MODAFL). MODAFL and IRGC have a number of affiliated institutions, universities, and businesses, including the Imam Hossein University in Tehran, Shahid Behreshi University, Institute of Applied Physics (IAP) (formerly the Physics Research Center [PRC]), and many others.[10] The National Council of Resistance of Iran (NCRI) provides additional background on Iran's nuclear program, and in 2004, it made the claim that the P2 centrifuge technology development program was not under the AEOI but rather the military.[11]

Developing highly advanced organization charts of both the civilian and military governmental structures with ties to companies, institutions, technologies, and key individuals makes it possible to observe patterns and relationships that would otherwise be missed. Figures 6-2 and 6-3 are organization charts prepared by the IAEA on Iran's military program. Their detail reflects a significant effort to piece together information over many years.

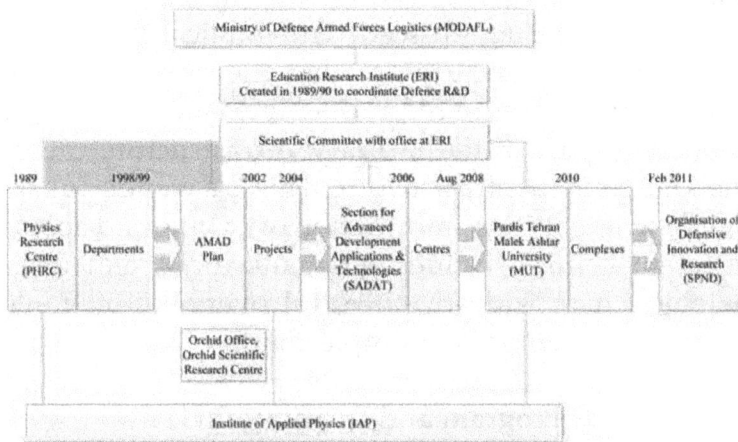

Figure 6-2. IAEA Organization and Flow Chart of the Military Side of Iran's Nuclear Program.[12]

PHRC Departments	AMAD Plan Projects	SADAT Centres
Department 01 Nuclear Physics	Project 110 Payload Design	Centre for Readiness & New Defence
Department 02 Centrifuge	Project 111 Payload Integration	Technologies
Enrichment	Project 3 Manufacture of Components	Centre for R&D (1) of Explosion & Shock
Department 03 Laser Enrichment	3.12 Explosives and EBW detonator	Technology
Department 04 Uranium Conversion	3.14 Uranium metallurgy	Centre for Industrial Research & Construction
Department 05 Geology	Project 4 Uranium Enrichment	Centre for R&T (2) of Advanced Materials –
Department 06 Health Physics	Project 5 Uranium Mining, Concentration & Conversion	Chemistry
Department 07 Workshop	5.13 Green Salt Project	Centre for R&T of Advanced Materials –
Department 08 Heavy Water	5.15 Gchine Mine Project	Metallurgy
Department 09 Analytical	Projects 8, 9 and 10	Centre for R&D of New Aerospace Technology
Laboratory	Project Health and Safety	Centre for Laser & Photonics Applications
Department 10 Computing	Project 19 Involvement of IAP	
Department 20 Analysis	Project/Group 117 Procurement and Supply	

(1) R&D = Research & Development
(2) R&T = Research & Technology

Figure 6-3. IAEA Chart of the Iranian Military Program Departments, Projects, and Centers.[13]

What is interesting about the MODAFL organization structure developed by the IAEA, as shown in Figures 6-2 and 6-3, is that it implies that nuclear research, although considered to be under the AEOI, the civilian branch of the Iran nuclear program, may fall

161

either completely or in part under the military organization, MODAFL, instead. The implications of this governmental structure are that the technologically critical infrastructure to producing weapons material is actually part of the **military infrastructure** rather than the civilian infrastructure. This is a wonderful example of integrating information over time and across different venues, including organizations and technologies. It provides a powerful overview of Iran's nuclear program in two simple charts.

Iran Nuclear Program and International Agreements.

Iran has worked primarily with Pakistan's Khan network, China, and Russia in establishing its broad based nuclear capability. To understand Iran's proliferation pathways would require establishing detailed knowledge of the Pakistani nuclear program; the Khan network with associated people and businesses; and Russia and China's civilian and military nuclear complex, institutions, companies, and technologies. This is a thoroughly complex and daunting task.

Russia's nuclear organization, Rosatom, is a highly complex organization consisting of governmental offices, technical and research institutes, design institutes, large-scale manufacturing and testing, materials development, nuclear reactor operation, nuclear weapons stockpile maintenance, materials production, universities with expertise in nuclear training, banks, and many other entities. During the 1990s, Russia was in a state of transition and economic chaos and had less control over its nuclear institutes, which opened up possibilities for technology transfer.

Its current organization chart is diverse and multifaceted. Many of the technical institutes also have affiliated private companies called limited liability companies, or LLCs. The predecessor organization to Rosatom signed the agreement with Iran to complete the unfinished Bushehr nuclear power plant back in the mid-1990s. This baseline agreement resulted in additional agreements for language training, nuclear physics and engineering education, reactor operations, nuclear safety, and others.

Training was included as part of the Bushehr Nuclear Power Plant agreement between Russia and Iran. The following list provides just a few specific examples of the sites and number of people identified for training and construction at the Bushehr Nuclear Power Plant (BNPP):

- July 22, 1989, and August 25, 1992: Russia and Iran establish training agreement with *Zarubezhatomenergostroi* (later to become part of Rosatom) to train up to 100 Iranians per year.[14]
- 1993-95: 18 Iranians reported to have completed their master's degrees at the Russian All-Union Research Institute for Power Plant, Moscow (possibly N.A. Dollezhal Scientific Research and Design Institute of Energy Technologies (NIKIET).[15]
- 1995: Within the 1995 Bushehr reactor agreement, 10-20 students trained per year in Russian schools such as the Kurchatov Institute of Atomic Energy (KIAE), MEPhI, and Novovoronezh Nuclear Power Plant.[16]
- 1998: Began training Iranians on reactor operations at the Novovoronezh Nuclear Power Plant; by 2001, a total of 342 Iranians had been trained.[17]

- 1998: Iranian student trained at the Mendelev Institute of Chemical Technology.[18]
- 2003: Up to 200 Iranians trained at Obninsk Atomic Energy University, where they learned basic skills in operating the BNPP.[19]
- Published estimates of the number of workers at the BNPP construction site:
  - 1998: 700 Russians working at the BNPP site.[20]
  - 2002: Approximately 3,900 Russians and Iranians working on the BNPP; another source reports there were 1,200 Russians and Ukrainians.[21]
- Published estimates of the amount of equipment shipped to Iran from Russia:
  - By the end of 2002: 5,000 tons of equipment had been delivered to the BNPP.[22]
  - 2003: 7,000 tons of equipment had been delivered.[23]
  - 2000-05: large pieces of equipment were shipped to Iran. This includes the reactor vessel, steam generators, turbine, piping, and other large equipment needed for the system.
- By 2008: A total of 82 megatons of fuel enriched up to 3.62 percent U-235 were delivered to the BNPP.[24]

In summary, given the number of Iranians receiving training at Russian nuclear institutes, the number of Russians traveling to and living in Iran for the construction of the BNPP, and the massive amount of equipment shipped from numerous manufacturing centers throughout Russia, it was possible for nuclear proliferation activity to go undetected and unchecked within the civilian nuclear power umbrella. Quite simply, the amount of exchange of people, equipment, and documentation between two nations engaged in the purchase and construction of a nuclear power plant is large. Tracking nuclear proliferation is highly complex, difficult to unravel, and time- and person-nel-consuming.

ISSUES WITH TECHNOLOGY TRANSFER

According to the intelligence community (IC) un-classified report to Congress in September 2001:

> These projects [referring to Russian projects with Iran] will help Iran augment its nuclear technology infrastructure, which in turn would be useful in supporting nuclear weapons research and development. The expertise and technology gained, along with the commercial channels and contacts established — **particularly through the Bushehr nuclear power plant project—could be used to advance Iran's nuclear weapons research and development program.**[25]

This document specifically flags the U.S. concern over the use of the BNPP to support Iran's nuclear weapons program.

The IC has since changed its position, stating that since 2003, it believes that Iran has ended nuclear weapons development. From the 2007 IC unclassified assessment to Congress:

> Analysis of events and activities associated with the Iranian nuclear program during the reporting period has yielded the following conclusions: We assess that Iran had been working to develop nuclear weapons through at least fall 2003, but that in fall 2003 Iran halted its nuclear weapons design and weaponization activities, and the military's covert uranium conver-

sion-and enrichment-related activities. We judge that the halt lasted at least several years, and that Tehran had not resumed these activities as of mid-2007.[26]

Since the BNPP agreement in 1995, there have been a number of projects reported in the press and by nongovernmental organizations as illicit technology transfers among Iran and Russia, China, and the Khan network for nuclear materials production and nuclear weapons design and testing. Nuclear materials production projects include: atomic vapor laser isotope separation (AVLIS), heavy water (HW) production, HW research reactor (HWRR), centrifuge enrichment technology, and uranium-hexafluoride (UF_6) gas for a centrifuge cascade. Weapon-related capabilities identified include high explosives (HE) technology and expertise, high-speed cameras, and containment vessels for weapons testing. A short summary of each is provided here.[27]

Nuclear Material Production Technologies.

- Russia-Iran AVLIS: During early-2000s, the U.S. and Russian governments engaged in discussions concerning the transfer of an AVLIS system from Russia to Iran. According to the IC unclassified report to Congress:

  > A component of the Russian Ministry of Atomic Energy (MINATOM) contracted with Iran to provide equipment clearly intended for AVLIS. The laser equipment was to have been delivered in late-1999 but continues to be held up as a result of U.S. protests. AVLIS technology could provide Iran the means to produce weapons quantities of highly enriched uranium.[28]

166

Minatom, the predecessor organization to Rosatom, had evaluated that the equipment could not be used for uranium enrichment but agreed to stop the transfer to Iran under U.S. pressure. Reportedly the AVLIS equipment was designed and manufactured at a Rosatom Institute, the Yefremov Scientific Research Institute for Electrophysical Apparatus in St. Petersburg.[29] Discussions in laser technology were part of the Russia-Iran technical agreement from 1995.

- Russia-Iran HW Production and HWRR: It was reported that prior to 1999, the Mendeleyev Russian Chemical-Technological University provided information on HW technologies to Iran[30] and in December 1998, . . . press articles citing U.S. intelligence sources reported that Russia's Research and Design Institute of Power Engineering (NIKIET) and the Mendeleyev University of Chemical Technology were negotiating to sell Iran [a] 40-MWt heavy-water research reactor and a UF_6 uranium conversion plant.[31]

  Mendeleyev University, as noted earlier, was also training an Iranian student.
  - NIKIET was identified as the U.S. focus of concern.[32] NIKIET and the Mendeleyev institute were subsequently placed under U.S. sanctions in January 1999; they have since been removed from U.S. sanctions. Since these earlier revelations in the early-1990s, ISIS released a technical note in 2009 with information on Iran's HWRR, claiming that:

Based on interviews with knowledgeable officials, NIKIET and a Russian company in Obninsk provided technology for the Arak reactor. This assistance included modifying the design of a RBMK fuel rod bundle for use in the Arak heavy water reactor. As a result of U.S. pressure, this assistance for Arak stopped in the late-1990s.[33]

This provides further confirmation of NIKIET's role in the HWRR and identifies a Russian company in Obninsk as providing technology.
— As noted earlier, in 2003, there were up to 200 Iranians trained at Obninsk Atomic Energy University, where they learned basic skills in operating the BNPP.[34] This places Iranian students in the same town as the company that ISIS identifies as providing technology for the HWRR at Arak. Both the HW production plant and HWRR project were kept hidden from the IAEA through the use of front companies until revealed by an Iranian opposition group in August 2002.[35] Tehran claims it needs the Arak facility to produce isotopes for medical purposes,[36] and that "as of October 2008, Iran had not yet installed equipment in the hot cells, including manipulators and specialized windows"[37] that could be used for isotope production.

• China-Iran Uranium-Hexafluoride: In February 2003, IAEA Director General Mohamed El-Baradei was shown the 164-centrifuge cascade uranium enrichment facility at Natanz, and it was revealed that the Iranians had received 1,000 kilograms (kg) of natural UF_6 gas from

168

China in 1991.[38] This information had previously been concealed.[39] The Chinese provided a significant amount of expertise, technology, and facility design for Iran's civilian nuclear complex within the Iranian IAEA safeguards agreement. Yet, the transfer of the UF_6 gas is an example of nuclear material transferred outside of the legal IAEA declaration.

- Pakistan Khan Network-Iran Centrifuge Uranium Enrichment: Perhaps one of the more surprising stories of proliferation was the transfer of technology and expertise from the Khan network to Iran's nuclear program, which began as early as 1985 when Dr. Khan met with the Iranians in Dubai and provided detailed construction plans and two centrifuges diverted from Kahuta.[40] This was followed by the supply of "two containers of surplus Pakistan centrifuge equipment to Iran in 1994 and 1995 for a payment of $3 million."[41] The centrifuges have been the basis for Iran's uranium enrichment program, which has continued to advance technically.

  - It is interesting to note that during nuclear negotiations with the Russians in 1995, the Iranians requested the purchase of a centrifuge uranium enrichment facility. Reportedly, the Russians had planned on providing the centrifuge technology but opted not to do so after the United States pressed them to end the agreement.
  - Later, in December 1998, NIKIET and Mendeleyev University of Chemical Technology were reportedly negotiating with Iran over the sale of a facility to convert uranium into

169

UF_6 after the Chinese canceled work on a UF_6 facility with Iran.[42] During the same time, the Iranians were purchasing an AVLIS system for uranium enrichment from Russia.

— It would appear there might have been some overlap within the Iranian nuclear program. This is where it would be interesting to trace each purchase and negotiation by organization, institution, company, and individual to see if the military and civilian sides of the nuclear program were working together. Recent studies by ISIS show that there were parts of the Iranian government purchasing equipment and materials for the centrifuge enrichment program in the late-1980s, and it is not clear how the civilian and military programs were split.[43] It is difficult to ascertain whether the technology transfers completed by the Khan network had the official involvement of the Pakistani government. The centrifuge uranium enrichment work was well hidden within Iran's larger nuclear program until it was revealed by the Iranian opposition group, NCRI, in August 2002. The IAEA director general visited the uranium enrichment plants for the first time in 2003.[44]

WEAPONS DESIGN AND NUCLEAR TESTING

- Russia-Iran Explosives Testing: According to news reports, Vyacheslav Danilenko, a former scientist from Chelyabinsk-70 (a weapons laboratory under Rosatom), contacted the Iranian embassy to inquire about possible joint ventures and later worked for the head of Iran's Physics Research Center (PRC) (see Figure 6-1). Per the IAEA organization chart, the PRC worked for the military side of Iran's nuclear program.[45] Danilenko claimed he was in Iran training on nanodiamond technology from 1995 to 2001. In the summer of 2003, Tehran engineers reportedly conducted test detonations based upon a Russian method using information obtained from Danilenko.[46]

- Russia-Iran Containment Vessel for Explosive Testing: Based upon news reports, the explosive vessel at the Iranian Parchin military site was designed with the help of Danilenko.[47] Danilenko denies he worked on the vessel.[48]

  — According to the IAEA:

  > the HE vessel, or chamber, is said to have been put in place at Parchin in 2000. A building was constructed at that time around a large cylindrical object. The IAEA was able to confirm the date of construction of the cylinder and some of its design features . . . and that it was designed to contain the detonation of up to 70 kg of HE . . .[49]

  for hydrodynamic experiments consistent with possible weapons development.

— While a former IAEA inspector has insisted that 70-kg is beyond a plausible design for a containment vessel, a report on the Comprehensive Test Ban Treaty (CTBT) states that the Russian nuclear design laboratories, which would include Arzamas-16 and Chelyabinsk-70, conducted weapon-related experiments using large containment vessels. The CTBT report continues that, based upon a paper from Arzamas-16 , the Russian Federal Nuclear Center (VNIIEF), the design laboratory in the late-1970s and early-1980s needed to develop explosive-resistant chambers:

> capable of hermetically holding inside its volume an explosive release of energy equivalent to 150 to 200 kilograms of TNT. . . . Development of these vessels–called Kolba in Russian–was completed by 1983.[50]

— It is reported that the Parchin vessel is a vacuum vessel with a diameter of 4.6 meters (m) and a length of 18.8-m. In the early-1990s, I had the opportunity to visit the Russian Research Institute of Chemical Machine Building in Zagorsk where they had large vacuum vessels for testing space equipment. One was large enough to test the Russian space shuttle with its wings clipped to fit within the chamber. This is to say that Russia has designed and built impressive facilities and large vacuum chambers consistent with the reported design of the Parchin vessel.

— Therefore, if the Iranians did receive help from the Russians in the design of containment vessels for high explosive testing of implosion systems, then 75 kg of HE is well within the design limits of Russian technology. What is not clear is whether the Russian weapons laboratories helped in the design of the Iranian vessel at Parchin.

- Russia-Iran High-speed Camera Diagnostics for HE Testing: According to ISIS,[51] the BIFO company, a subsidiary to the Russian Institute for Optical and Physical Measurement (VNIIO-FI), sold two high-speed cameras, K008 stream and "uniframe" camera and K011 "nineframe" camera to a German trading company who had purchased them for an Iranian front company. A technical description of each camera can be found in a VNIIOFI publication.[52] The cameras were purchased for $42K[53] and can be used for the diagnostics of HE implosion testing as described previously, with operating time scales in nanoseconds. It is worth noting that, according to the ISIS report, VNIIOFI has joint publications on weapons testing with Arzamas-16, one of Russia's two nuclear weapons design laboratories.

In summary, there are a number of examples where nuclear materials, nuclear materials production technology, and nuclear weapons-related technology may have been transferred to Iran while hidden within its larger civilian nuclear power program. While it is plausible that these transfers occurred as a result of contacts made while training or within the larger technical agreements, it is difficult to prove without more substantial information. During the 1990s, Russia was undergoing a significant transformation and created openings that may no longer be open. Since the mid-2000s, Rosatom management has established a clear vision for creating a professional organization aimed at meeting national and international needs in the nuclear field and implemented it through the Russian government energy program. Similarly, China has moved to the international standards for nonproliferation and export controls with the goal of becoming an international supplier of nuclear technology. The competitive economic and technical basis between nations raises the bar regarding when proliferation activities can tarnish a country's reputation. The previous examples of proliferation primarily took place during the 1990s and demonstrate what can occur within a larger commercial program if governmental oversight is not strictly enforced.

Iran's nuclear program extends over 30 years, and a timeline of its program would help provide a foundation for understanding technology developments and transfers. A timeline can be used to track interactions and changes in relationships and agreements. There are literally thousands of people and interactions within the Iranian nuclear umbrella, spanning back to the 1980s. Iran also has a large number of technology development programs where the inter-relationships between people, companies, organizations, and educational centers, if tracked, could provide insights into their activities. A systems model keeps building and building, as this is the type of complex and detailed framework needed to track potential proliferation activity.

What is clear from the previous example is that a few people can provide invaluable support to a nation seeking a nuclear weapons capability. A. Q. Khan created and directed an effort to provide centrifuge-based uranium enrichment technology to several nations, and Danilenko is reported to have provided training on implosion technology and the design of an explosive vessel. These specific examples demonstrate how difficult it can be to trace nuclear proliferation within a larger commercial nuclear power program. It is the proverbial needle in a haystack.

A FEW MORE THOUGHTS

- Technical expertise is critical to understanding proliferation pathways and how seemingly uninteresting information can be passed on without a second look by someone who lacks a deeper understanding of the materials, technologies, and types of technology that are used in materials production and nuclear weapons design and testing. Unfortunately, many of the U.S. experts who know much of the more obscure information are retiring.
- Curiosity, time, and a healthy environment for "failure" are all needed to allow analysts time to track odd information "down the rabbit hole." Failure needs to be encouraged.
- The initial question is of great importance in framing the research and in determining the final product. Questions should encourage seeking the answer from multiple views and perspective. Every question requires a new tact and approach and thereby brings in new insights.
- Physical space to do systems work is needed for sharing ideas, putting charts on the wall, and for spreading out information.

- Finally, cultural knowledge and insights, not from books, but from experience in country, is essential.

More often than not, a country chooses a civilian nuclear power program not as a cover for a nuclear weapons program, but for energy. In the cases where it does occur, a nation's pursuit of a weapons program may be accompanied by signs of noncompliance with the NPT agreement. Also, there might be a sense of a defense imbalance in the region that would become balanced if that nation obtains a nuclear weapons capability. When you consider the list of nations that are currently considering new nuclear power programs or expanding their existing program, it is clear that their trade and economic roles in the international community are of great importance to them and may serve as an incentive not to develop a nuclear weapons capability.

There are two camps on how to move forward on nuclear power when considered within the reference frame of nonproliferation: 1) Limit the transfer of any and all nuclear power technology; or 2) Provide the technology for nuclear power production, but empower the IAEA and United Nations (UN) to track and enforce any anomalies or illicit activity.

There are arguments for both perspectives, and neither has worked "perfectly." For example, in trying to limit the transfer of technology to Iran and Libya, the door was opened for the transfer of illicit technology by other nations and the establishment of a shadow organization outside of governmental purview—a much more difficult and dangerous issue to track and control. In cases where nuclear technology was provided and kept under IAEA control, it has been difficult for the IAEA to limit the growth of a nation's nuclear weapons program, including programs in North Korea, Iraq, and Iran. One of the contributing factors is the varying level of concern over proliferation activates among nations. The United States has a much more conservative idea of what limiting proliferation should be than, say, Russia or China does. Therefore, where the United States may not be willing to engage with a specific nation due to concerns of nuclear proliferation—for example, in Iran, Burma, or North Korea—Russia or China may be willing to step in.

CONCLUSIONS

Overall it takes much time and funding to develop a systems perspective on nuclear proliferation. It requires a dedicated team with a passion for digging deeper and pursuing the question longer. It also requires policy representatives who want a greater understanding of the issues of nonproliferation and who are willing to push the experts in nonproliferation to do more and perform at a higher level. It is a highly complex process, but given the complexity of the global interactions and concerns over nuclear proliferation, it is the only possible way to establish a framework to track nuclear interactions and find the inconsistencies that may signal a country is seeking more than civilian nuclear technology. This type of systems approach will be needed as the number of new countries seeking commercial nuclear technology expands.

ENDNOTES - CHAPTER 6

1. J. C. Moltz and A. Y. Monsourov, *The North Korean Nuclear Program: Security, Strategy and New Perspectives from Russia*, New York: Routledge, 1999.

2. Jeffrey Richelson, *Spying on the Bomb: American Nuclear Intelligence from Nazi Germany to Iran and North Korea*, New York: Norton, 2006.

3. "Plans for New Reactors World Wide," World Nuclear Association, May 2012, available from *www.world-nuclear.org/info/inf17.html*.

4. "Plans for New Reactors World Wide," World Nuclear Association; and "Emerging Nuclear Energy Countries," World Nuclear Association, 2012, available from *www.world-nuclear.org/info/inf102.html*.

5. "Emerging Nuclear Energy Countries," World Nuclear Association.

6. It may be possible to reduce the number of people required for planning, construction, training, and operation by moving to small or medium reactors (SMR) for nations that currently do not have a nuclear power program. Small reactors can be manufactured, built, and transported as complete units to the operating site, where they are installed in the facility. At the end of life, the complete unit is shipped back to the originating country after a few years of cool down within the reactor compartment. The other advantage to SMRs is that they allow a developing nation to build up power at the location where it is needed and expand as more power is needed within the region.

7. "Acquisition of Technology Relating to Weapons of Mass Destruction and Advanced Conventional Munitions: 1 July Through 31 December 2000," Unclassified Report to Congress from the Director of Central Intelligence, 2001, available from *www.cia.gov/library/reports/archived-reports-1/july_dec2000.htm*, hereafter DCI, Acquisition of Technology, 2000.

8. M. Ghannadi-Maragheh, "Atomic Energy Organization of Iran," in World Nuclear Association Annual Symposium, London, UK, 2002, available from *www.iranwatch.org/sites/default/files/ iran-aeoi-worldnuclearassociation-090402.pdf*.

9. David Albright, Paul Brannan, and Andrea Stricker, "Will Fereydoun Abbasi-Davani lead Iran to nuclear weapons? (Rev. 1)," Washington, DC: Institute for Science and International Security, June 24, 2011, available from *isis-online.org/isis-reports/detail/ will-fereydoun-abbasi-davani-lead-iran-to-nuclear-weapons*.

10. *Ibid.*

11. Mohammad Mohaddessin, "New Nuclear Revelations: Text of the Press Conference of Mohammad Mohaddessin, Chairman of the Foreign Affairs Committee of the National Council of Resistence of Iran," Iran Watch, Wisconsin Project on Nuclear Arms Control, September 10, 2004, available from *www.iranwatch. org/privateviews/NCRI/perspex-ncri-nuclearprogram-091004.htm*.

12. "Implementation of the NPT Safeguards Agreement and relevent provisions of Security Council resolutions in the Islamic Republic of Iran," Vienna, Austria: International Atomic Energy Association, November 8, 2011, available from *www.iaea.org/Publications/Documents/Board/2011/gov2011-65.pdf*.

13. *Ibid.*

14. Judith Perera, "Nuclear Industry of Iran," *Opensource.gov*, April 2003.

15. Mark Gorwitz, "Iranian Nuclear Science Bibliography: Open Literature References," September 2003.

16. "Specialists to Train Iranians for Bushehr Nuclear Plant," *ITAR-TASS*, NTI: Russia: Nuclear Exports to Iran: Training and Know How, 1996.

17. Perera, "Nuclear Industry of Iran"; S. Sergievskiy, "Iranskiye stazhery na Balakovskoy AE" ("Iranian trainees at Balakovo AE"), *Nezavisimaya Gazeta* online edition, 1999.

18. Perera, "Nuclear Industry of Iran."

19. Nick P. Walsh, "Russian Lessons," *The Guardian*, June 16, 2003.

20. Perera, "Nuclear Industry of Iran."

21. *Ibid.*

22. *Ibid.*

23. *Ibid.*

24. "Bushehr: Fueling the Reactor," *GlobalSecurity.org*, available from *www.globalsecurity.org/wmd/world/iran/bushehr-fuel.htm*.

25. DCI, Acquisition of Technology, 2000. Emphasis added.

26. "Acquisition of Technology Relating to Weapons of Mass Destruction and Advanced Conventional Munitions: 1 January to 31 December 2007," Unclassified Report to Congress from the Deputy Director of National Intelligence for Analysis, 2007, hereafter DDNI, Acquisition of Technology, 2007.

27. It is worth noting that the details of the stories are much richer and interesting than what I am able to provide in these short summaries.

28. DCI, Acquisition of Technology, 2000.

29. Perera, "Nuclear Industry of Iran"; Michael Knapik, "Russia tells U.S. officials it will not export lasers to Iran," *Nucleonics Week*, Vol. 42, March 8, 2001.

30. Michael R. Gordon, "Russia to Offer U.S. Deal to End Iran Nuclear Aid," *The New York Times*, March 17, 1999.

31. Perera, "Nuclear Industry of Iran."

32. Gordon, "Russia to Offer U.S. Deal."

33. David Albright, Paul Brannan, and Robert Kelley, "Mysteries Deepen Over Status of Arak Reactor Project," Washington, DC: Institute for Science and International Studies, August 11, 2009, available from *www.isisnucleariran.org/assets/pdf/Arak-FuelElement.pdf*; David Albright, Paul Brannan, and Robert Kelley, "Update on the Arak Reactor in Iran," Washington, DC: Institute for Science and International Studies, August 25, 2009, available from *www.isisnucleariran.org/assets/pdf/Arak_Update_25_August2009.pdf.*

34. Walsh, "Russian Lessons."

35. "Iran Nuclear Milestones," Iran Watch, Wisconsin Project on Nuclear Arms Control, February 2011, available from *www.iranwatch.org/wmd/wmd-nukemilestones.htm.*

36. James Phillips, "Iran's Nuclear Program: What Is Known and Unknown," Washington, DC: The Heritage Foundation, March 26, 2010, available from *www.heritage.org/research/reports/2010/03/iran-s-nuclear-program-what-is-known-and-unknown.*

37. Albright *et al.*, "Mysteries Deepen Over Status of Arak Reactor Project."

38. Mohammad Sahimi, "Iran's Nuclear Program. Part I: Its History," October 2, 2003, available from *www.payvand.com/news/03/oct/1015.html.*

39. John R. Bolton, *Surrender Is Not an Option : Defending America at the United Nations and Abroad*, New York: Threshold Editions, 2007.

40. Erich Follath and Holger Stark, "The Birth of a Bomb, A History of Iran's Nuclear Ambitions," *Der Spiegel*, June 17, 2010, available from *www.spiegel.de/international/world/0,1518,701109,00.html*; "Pakistan: Dr. Abdul Qadeer Khan Discusses Nuclear Program in TV Talk Show," Karachi Aaj News Television in Urdu, from "Islamabad Tonight" discussion program hosted by Nadeem Malik, August 31, 2009.

41. Chaim Braun and Christopher F. Chyba, "Proliferation Rings," *International Security*, Vol. 29, No. 2, Fall 2004, pp. 5-49.

42. "Russia: Nuclear Exports to Iran: Enrichment, Mining and Milling," NTI, June 22, 2011.

43. David Albright and Paul Brannan, "Iranian Smuggling for Military Nuclear Programs: From Shahid Hemmat Industrial Group to the Physics Research Center," Washington, DC: Institute for Science and International Security, May 16, 2012.

44. "Iran Nuclear Milestones," Iran Watch.

45. Joby Warrick, "Iran-Russian Scientist Vyacheslav Danilenko's Aid to Iran Offers Peek at Nuclear Program," *The Washington Post*, November 13, 2011; Mark Gorwitz, "Vyacheslav Danilenko—Background, Research, and Proliferation Concerns," Washington, DC: Institute for Science and International Security, November 29, 2011, available from *isis-online.org/isis-reports/detail/vyacheslav-danilenko-background-research-and-proliferation-concerns/*.

46. For more information on the Iranian high explosive test and a diagram of the configuration, see Follath and Stark, "The Birth of a Bomb."

47. "Parchin chamber raises nuclear test suspicions," AP, May 19, 2012, available from *gulfnews.com/news/region/iran/parchin-chamber-raises-nuclear-test-suspicions-1.1024807*.

48. *Ibid.*

49. "Implementation of the NPT Safeguards Agreement and relevent provisions of Security Council resolutions in the Islamic Republic of Iran," Vienna, Austria: International Atomic Energy Agency, November, 8, 2011.

50. Kathleen Bailey and Thomas Scheber, *The Comprehensive Test Ban Treaty: An Assessment of the Benefits, Costs, and Risks*, Fairfax, VA: National Institute for Public Policy, March 2011, available from *www.nipp.org/CTBT%203.11.11%20electronic%20version.pdf*.

51. David Albright and Christina Walrond, "The Trials of the German-Iranian Trader Mohsen Vanaki: The German Federal Intelligence Service Assesses that Iran Likely Has a Nuclear Weapons Program," Washington, DC: Institute for Science and International Security, December 15, 2009, available from *isis-online. org/isis-reports/detail/the-trials-of-the-german-iranian-trader-mohsen-vanaki-the-german-federal-in/*.

52. Vyatcheslav S. Ivanov, Yuri Zolotarevsky, Vladimir Krutikov, Vitaly B. Levedev, and Grigory G. Feldman, "High-speed image converter instrument engineering of VNIIOFI is 40 years old," presented at the 27th International Congress on High-Speed Photography and Photonics, Xi'an, China, 2007.

53. Albright and Walrond, "The Trials of the German-Iranian Trader Mohsen Vanaki."

CHAPTER 7

PERSUADING COUNTRIES TO FORGO NUCLEAR FUEL-MAKING: WHAT HISTORY SUGGESTS

Richard S. Cleary

INTRODUCTION

In recent years, there has been a resurgence of proposals designed to limit the spread of nuclear fuel-making facilities, with the understanding that ostensibly peaceful technology can allow for the production of the fissile material required for a nuclear weapon. With U.S. proposals ranging from the Global Nuclear Energy Partnership (GNEP) to a revamped, "Gold Standard" bilateral nuclear cooperation agreement, a wider array of tools has been put at the disposal of American policymakers. Prominent members of the international community have become agitated about the prospect of the proliferation of fuel-making technology as well, with numerous proposals of fuel assurances put forward by such disparate figures as Russian President Vladimir Putin and former Director of the International Atomic Energy Agency (IAEA) Mohamed El Baradei. But renewed enthusiasm for nonproliferation begs questions about how novel are the proposed instruments, and, moreover, how effective they are likely to be, particularly for the country historically at the head of nonproliferation efforts, the United States. A review of this historical record suggests that optimism about the U.S. ability to dissuade countries from this path is misplaced.

This chapter considers supply side proposals of fuel assurance and multilateral fuel-making, as well as specific interventions on both the supply and demand sides, consulting particular cases in Iran (1974-78), West Germany-Brazil (1975-77), South Korea (1974-76), and Pakistan (1972-80) to draw lessons about the effectiveness of U.S. practices under differing circumstances. The record these cases give is mixed, due to two principal causes. The first is the failure of the United States to prioritize consistently nonproliferation efforts, given Washington's global and competing interests that tend to be embraced by different factions in the federal government apparatus but whose ultimate arbiter is the President (along with his close advisors). The second is the tendency of decisions about nuclear fuel-making by the state in question to be influenced more by fundamental trends or factors than diplomatic maneuvering from Washington; diplomacy is most effective when it has the political, economic, and military backing to implicate these issues. The most important factor in U.S. efforts has tended to be the bilateral relationship between Washington and the country at hand. Decisionmakers who consider their country's relationship with the United States to be strategically vital—and believe that fuel-making would threaten this relationship—are most likely to forgo enrichment and reprocessing (ENR) technology. This calculus can be informed by a range of dynamics, some beyond U.S. control, such as security concerns, issues of prestige, and commercial and industrial interests. Domestic politics and public opinion, both in the United States and in the country considering fuel-making, can be influential.

One of the fundamental tensions of American non-proliferation efforts lies with the Nuclear Nonproliferation Treaty (NPT), the international legal framework of reference in nonproliferation matters. The prevailing interpretation of the NPT centers on what has been referred to as the "fundamental bargain": in exchange for nuclear-weapons states' movement toward disarmament and their sharing of technology and expertise for peaceful nuclear energy, non-nuclear weapons states will not pursue the bomb.[1]

One portion of the NPT, in particular, has borne on U.S. efforts to persuade countries not to pursue nuclear fuel-making technology: In Article IV, the NPT enshrines the "inalienable right . . . to develop research, production and use of nuclear energy for peaceful purposes," and pledges signatories to "undertake to facilitate . . . the fullest possible exchange of equipment, materials and scientific and technological information for the peaceful uses of nuclear energy."[2] Traditionally, the United States has elected for an ambiguous middle ground, not denying an Article IV "inalienable" right to fuel-making, but not acknowledging it either.[3] While U.S. interpretations of the NPT have not, as a practical matter, stemmed its attempts to convince countries to eschew nuclear fuel-making technology, the NPT's bargain has shaped certain stances, particularly supply side proposals such as fuel assurances.

The application of U.S. national power, on both the supply and demand sides of nuclear fuel-making, can play a role in convincing countries of the benefits of their relationship with Washington and the costs to be incurred if this relationship were fractured. The adroit use of "sticks" and "carrots" can withhold or provide incentives for cooperation, convincing countries considering ENR that the risks of doing so outweigh the benefits. The case studies examined here suggest that if the United States is to give the impression that a bilateral relationship rests in the balance, Washington may have to undertake risks of its own, perhaps compromising other policy objects for the sake of nonproliferation. When the circumstances have called for Washington to put nonproliferation goals above others, policymakers have often failed to do so.

ASSURANCES OF SUPPLY AND MULTILATERAL ARRANGEMENTS

The earliest American efforts to dissuade countries from developing nuclear fuel-making facilities were offers of international control of the atom. When the Soviet Union developed a nuclear arsenal, proposals of international control gave way to multilateral fuel-making facilities and fuel banks. There is a consistency of spirit in these proposals, grounded in the belief that a positive inducement is possible on the supply side and, more particularly, that countries will forgo nuclear fuel-making if satisfied by disarmament, or promises thereof, as well as fuel supply and other assurances.

The first systematic American attempt to grapple with the potential for nuclear weapons proliferation — and the role of fuel-making facilities in proliferation — was the so-called Acheson-Lilienthal Report. This document was the product of a committee organized by Secretary of State James Byrnes, and which included Under Secretary of State Dean Acheson, Tennessee Valley Authority chairman David Lilienthal, and a number of prominent members of the Manhattan Project, notably Robert Oppenheimer. The committee was charged with elaborating a U.S. nuclear policy to be put forward to the newly-created United Nations

Atomic Energy Commission (UNAEC).[4] The Acheson-Lilienthal Report, drafted over the course of only a few months, proposed that all fissile material—and its means of production—be concentrated in an "Atomic Development Authority." The text acknowledged explicitly the connection between fuel-making, both in enrichment and reprocessing, and the manufacture of nuclear weapons:

> Operations, like those at Hanford and Oak Ridge and their extensions and improvements, would be owned and conducted by the Authority. Reactors for producing denatured plutonium will be large installations and by the nature of the process they will yield large amounts of energy as a byproduct. . . . These production plants are intrinsically dangerous operations. Indeed they may be regarded as the most dangerous, for it is through such operations that materials can be produced which are suitable for atomic explosives.[5]

The hope of the Acheson-Lilienthal Report was that by investing an international body with authority over fuel-making facilities, uranium ore mining assets, inspections powers, and "licenses to those countries wishing to pursue peaceful nuclear research," proliferation of nuclear weapons could be averted.[6] A key portion of the text was the elimination of the "national nuclear arsenals"—that is, the only then-extant arsenal, the American one—and the report implied that U.S.-Soviet cooperation would be crucial. This grand bargain, premised on U.S. generosity in giving up its nuclear weapons and Soviet compliance in not developing them, was intended to give the world the benefits of peaceful nuclear energy, while keeping a close hold on sensitive facilities. While the hope for Soviet cooperation was misplaced—Moscow was hard in pursuit of a nuclear weapons capability — the authors did have a clear-eyed understanding of the dual nature of nuclear fuel-making technology.[7]

The next major proposal, closely based on the Acheson-Lilienthal Report, would come from Bernard Baruch, the U.S. delegate to the UNAEC. Baruch's insistence that UN Security Council members should forgo their vetoes on proliferation-related matters, as well as his modified stance on U.S. nuclear disarmament — Washington should only disarm when "guarantee[d] of safety" — were objected to by the Soviet Union, which refused to agree to the proposal. However, Baruch's plan did retain the Acheson-Lilienthal elements related to international control of fuel-making facilities, seeing their utility both in the nonproliferation of nuclear weapons and the spread of peaceful nuclear power.

Just as Acheson-Lilienthal had, Baruch envisaged an international body that "should exercise complete managerial control of the production of fissionable materials in dangerous quantities and must own and control the product of these plants." The rejection of the Baruch Plan spelled the end of U.S. proposals for international control of nuclear fuel-making. Later proposals of multilateral or international fuel-making facilities would not concentrate all fuel-making capacity in a single body. Rather, they would see multilateral facilities as enabling the spread of peaceful nuclear energy to nonweapons states, while minimizing the threat of diversion toward nuclear weapons. These facilities were understood as part of a grand bargain whereby nonweapons states would be assured of fuel for peaceful purposes in exchange for eschewing nuclear weapons.

U.S. President Dwight Eisenhower's "Atoms for Peace" speech, delivered to the UN General Assembly in 1954, outlined this basic bargain, later enshrined in the NPT. By the time of Eisenhower's speech, the Soviet Union had conducted nuclear weapons tests — including at least one hydrogen bomb test — and the United Kingdom (UK) and Canada also had access to "the secret."[8] Atoms for Peace kept the spirit of earlier proposals, which had emphasized disarmament, nonproliferation, and the spread of "peaceful" nuclear energy, but the President never touched on the danger of fuel-making facilities and their dual application. In his speech, Eisenhower called for the creation of an international body that would both conduct inspections and, by receiving contributions of fissile materials in what might be called today a fuel bank, ensure that all states would receive the benefits of peaceful nuclear energy. Eisenhower's proposal was premised in part on the idea that "the knowledge [of nuclear weapons development] now possessed by several nations will eventually be shared by others, possibly all others."[9]

Just as the Acheson-Lilienthal and Baruch plans had reflected the dynamics of a world where only one nation held nuclear weapons, Atoms for Peace was penned in a period when the peaceful benefits of nuclear power were expected to be great. To some, there was good reason to think that the trade-off between civilian and military nuclear power could be a favorable one. Eisenhower's speech resulted in a program, also called Atoms for Peace, whereby the United States transferred small-scale research reactors and fuel to developing countries around the world. The Atoms for Peace program has been criticized for spreading sensitive technology to countries that went on to develop nuclear weapons. Still, the basic bargain elaborated in Atoms for Peace would be incorporated in the 1968 NPT, which remains the legal framework of reference in matters of nonproliferation.

In recent years, the United States has again proposed a multilateral fuel facility as a means of resolving supply side concerns—and denying countries the argument that building fuel-making facilities is necessary. U.S. President Barack Obama, in a noted speech in Prague, The Czech Republic, has called for an international fuel bank.[10] Obama's proposal follows a number of appeals for multilateral, or international, facilities following Mohamed ElBaradei's 2003 proposal for putting all fuel-making facilities under "multinational control."[11] U.S. President George W. Bush also proposed, while not a multilateral facility, assurances by fuel-cycle states to those seeking fuel for peaceful purposes.[12]

The persistence of fuel-assurance proposals, particularly those related to multilateral fuel-making facilities, is striking—as is the absence of a multilateral facility in the mold proposed by Eisenhower, Obama, ElBaradei, and others. While there are a number of reasons why such a facility has not materialized, the best explanation seems to be that, other than satisfying the demand for equity as understood in the NPT grand bargain, there would be little discernible commercial advantage to one. An ample supply of nuclear fuel already exists and is priced on the international marketplace, and most countries have no difficulty in accessing this supply.[13] This supply side solution, then, has not taken hold because it has not assured the commodity in demand.

MULTILATERAL REPROCESSING
IN SHAH REZA PAHLAVI'S IRAN

The multilateral fuel-making facility has not simply been an abstract offer reserved for international fora. This concept, supported publicly by U.S. Secretary of State Henry Kissinger, was folded into negotiations between the Gerald Ford administration and Shah Reza Pahlavi's regime toward a bilateral nuclear cooperation agreement.[14] The multilateral option figured prominently in a proposed text of the agreement in May 1976, but was rejected by Iranian negotiators, and eventually discarded by U.S. negotiators.[15] Amid increasingly strict oversight from Congress, the multilateral option was absent altogether from the second proposed agreement text, presented in August 1978.[16] Indeed, negotiations between the United States and Iran never materialized in a signed, binding agreement, interrupted first by the 1976 presidential election and later by turmoil in Iran that led to the Shah's overthrow.[17] This case raises larger questions about how the United States has prioritized nonproliferation, how the policymaking process works, the means the United States has had at its disposal to combat the proliferation of sensitive fuel-making technologies, and, finally, what motivates countries to pursue nuclear fuel-making facilities.

Negotiations between Washington and Tehran, lasting from 1974 until late-summer 1978, came in a period of great nonproliferation concern internationally in the aftermath of the May 1974 Indian nuclear test. The U.S. Congress was particularly exercised about the prospect of the further spread of fuel-making technology, and multilateral ownership of ENR facilities was embraced as a potential solution. The

U.S. National Security Council's Under Secretaries Committee reached the same conclusion, advising a policy "encouraging multinational plants" (or bilateral plants involving the U.S.) in order to "restrict the spread of independent national uranium enrichment and chemical reprocessing facilities."[18] The multilateral option was attractive in part because it did not deprive countries of fuel-making technology, thereby adhering to promises made in NPT negotiations.[19]

Limiting the proliferation risk of the Shah's nuclear program would be crucial, given U.S. suspicions about Tehran's true intent. In an interview in Paris following the 1974 Indian nuclear test, the Shah was quoted as saying that Iran would have a nuclear weapon "without any doubt, and sooner than one would think."[20] This gaffe echoed the conclusions of the U.S. intelligence community that Iran was a proliferation threat, albeit not in the short-term.[21] Another indication, perhaps, of the Shah's ultimate objective came with the attempted May 1975 purchase of Lance surface-to-surface missiles, judged to be uneconomical for any nonnuclear purpose by the U.S. Department of Defense (DoD).[22]

Despite these concerns, there were several perceived advantages to reaching an agreement with the Shah, advantages that, in the eyes of some, could be compromised by over-zealous nonproliferation efforts.[23] First, the Shah's regime was a cornerstone of U.S. policy in the greater Middle East. National Security Study Memorandum (NSSM) 219, which would inform an April 1975 presidential directive on the nuclear negotiations with Iran, described the discussions as critical to the U.S.-Iran relationship:

Our [the U.S.'] ability to reach a mutually satisfactory agreement with Iran on the proposed nuclear accord is expected to have a very considerable political as well as economic importance to U.S.-Iranian relationships. . . . Conversely, failure on our part to resolve the remaining issues could have serious short, as well as long-term, adverse effects in our relations, given the Shah's sensitivity towards U.S. attitudes and Iran's strong desires to be treated in a non-discriminatory manner and as a nation that often has supported U.S. interests.

Second, the potential commercial benefits to reaching an agreement with Iran were substantial: The Shah planned a 20,000-megawatt civilian nuclear system, with 6-8 reactors presumably coming from the United States, for an estimated total of $6.8 billion.[24] Were the Shah to invest in an enrichment plant on U.S. soil, he would have brought an additional $1 billion to U.S. hands. Finally, and counterintuitively, there was a nonproliferation advantage to reaching an agreement: Should the United States demand overly stringent guidelines, it might drive the Shah away from the U.S. and toward less scrupulous nuclear suppliers, undermining U.S. goals in this area.[25] Iranian dealings with France and West Germany seemed to affirm this concern.[26]

The multilateral option figures prominently in early proposals as a way to soften Washington's "veto" on the reprocessing of U.S.-origin spent fuel. For example, U.S. negotiators indicated to their Iranian counterparts that Washington, despite its veto:

would look sympathetically on Iran's request to perform such reprocessing services. We [the U.S.] have indicated that one factor favoring U.S. approval would be a decision on the part of Iran to establish any reprocessing plant on a multinational basis with the active involvement of the country helping to establish the facility.[27]

The multinational option, however, was quickly derailed. In negotiations in Vienna, Austria, in October 1975, the U.S. veto over reprocessing, as well as the multinational concept, was rejected by Iran for both principled and practical reasons. According to Richard Helms, U.S. Ambassador to Iran:

> Iranians recognize and resent the regional reprocessing plant concept as a device to impose international control on this very sensitive stage in the nuclear fuel cycle. Iranian bruised honor aside, they believe the idea is ridiculous in the Middle East setting. The concept may have validity and a chance [of] success in a part of the world which is highly integrated economically, such as the EC [European Community], but the likelihood of Iran being able to work out close functional relationships with its neighbors for reprocessing appears remote.[28]

Meanwhile, Kissinger himself began to voice doubts in private about the effectiveness of the multinational concept, which he labeled a "fraud" — despite his public position that multilateral approaches could be effective. Furthermore, Kissinger noted the danger of placing a fuel-making facility in a proliferation-sensitive country, where the plant would be "just a cover" for the production of weapons-grade material.[29] Kissinger also put his finger on a critical weakness of the multilateral plant, that, if the foreign participant were "kicked out," it might be unwilling to protest for fear of jeopardizing other interests: Nonproliferation required fortitude.[30]

As negotiations ground to a halt and the multinational facility was found wanting, U.S. negotiators turned to another solution to Iran's quest for reprocessing: the "buy back" option, where the United States would be able to elect to purchase spent fuel of American origin while supplying fresh fuel for reactors in Iran. In May 1976, the United States proposed a draft agreement with the "buy back" option front and center, but retaining language requiring that reprocessing be "performed in facilities acceptable to the parties." Iran would be obligated to "achieve the fullest possible participation in the management and operation of such facilities of the nation or nations that serve as suppliers of technology and major equipment," but, should this fail—and should the United States not elect to buy back spent fuel—it could reprocess nationally.[31] While Kissinger insisted that the Iranian right to reprocess independently would never be realized—the United States would exercise its right to repurchase fuel in every instance—the willingness to allow, even in principle, reprocessing of U.S.-origin fuel was controversial within the Ford administration.[32]

The Shah's negotiating team, led by chairman of the Atomic Energy Organization of Iran Ahmad Etemad, rejected this proposal—despite its recognition of an Iranian right to reprocess—leading to a stalemate that persisted until 1977, when Jimmy Carter succeeded Gerald Ford in the White House following a campaign where nonproliferation policy had figured prominently.[33] With Carter promising a tougher nonproliferation stance on the campaign trail, Etemad seemed more open to the multilateral option and went so far as to eschew national reprocessing altogether.[34] With the April 27, 1977, announcement by Carter that the United States would insist on a veto of reprocessing of U.S.-origin spent fuel, the dynamics of negotiations were beginning to shift, and an agreement later initialed by negotiators in August 1978, though never signed by Carter or the Shah, would include this veto and preclude reprocessing of U.S.-origin spent fuel on Iranian soil.[35] This agreement followed the Glenn Amendment of 1977, which allowed for no aid for those countries that transferred or received reprocessing technology, multinational ownership or not; despite the sympathies of some in the executive branch toward permitting reprocessing in Iran, Congress had put its foot down. Though Iranian negotiators eventually came around to Washington's tougher line, Tehran's nuclear planning grew more suspicious. In particular, U.S. scientists from the Oak Ridge National Laboratory were wary of the proposed Esfahan Nuclear Technology Center (ENTEC) facility, noting "that the unusually large size of the planned facility makes it theoretically possible to produce weapons-grade plutonium."[36]

This episode is most notable, first, for the gradual marginalization of the multilateral option during negotiations and, second, for the flexibility that the Ford administration displayed regarding fuel-making technology. The multilateral option was shown to be intrinsically weak as a nonproliferation tool — simply to meet U.S. nonproliferation objectives, it had to be coupled with a veto and, as Kissinger suggested in private, it was hardly foolproof, with success contingent on political dynamics in the country in which the facility was located. But the balance of key interests at stake, namely the relationship with a major partner, the prospect of lucrative nuclear sales, and the wish to adhere to Article IV of the NPT for a signatory such as

Iran, was enough to drive U.S. policymakers, at least initially, to propose the multilateral reprocessing facility. Furthermore, the Ford administration's decision to sanction in principle a national reprocessing facility despite Iran's proliferation risk illustrates the way in which competing priorities can win the day over non-proliferation.

INTERVENTION WITH A SUPPLIER: WEST GERMANY'S SALE TO BRAZIL

The negotiations between the Ford and later Carter administrations and the Shah's regime took place amidst a considerable amount of activity in the nonproliferation realm. One of the most notable cases during this period was the 1975 deal between West Germany, a close U.S. ally, and Brazil, ruled by a military junta and not party to the NPT. U.S. efforts to thwart the deal, despite its close relationship with West Germany, were unsuccessful, though Washington was able to convince Brasilia, Brazil, and Bonn, Germany, to enter into a safeguards agreement with the IAEA.[37] Although the deal would not live up to expectations due to financial constraints, higher-than-expected costs, questions about the need for significant increases in energy output, and the poor performance of the West German "jet nozzle" enrichment technology, it represented the limits of Washington's ability to convince even its closest allies of the virtues of nonproliferation.[38] Because Brazil would go on to master fuel-making on its own in a secret, parallel program, this case suggests that intervention with nuclear exporters is not a definitive solution (a point the Pakistan case soon to be discussed echoes).

In what was called the "most controversial deal in the history of West German industry," Bonn agreed to sell between two and eight reactors, as well as enrichment and reprocessing technology, to Brazil.[39] This was a major sale by any measure and the first to include the full fuel cycle. It also came on the heels of spurned attempts by Brazil to secure enrichment technology from the United States and in the aftermath of the 1974 decision not to sign any new contracts for the provision of enriched uranium. According to a Central Intelligence Agency (CIA) report, Bonn had gained a decisive advantage in the sale through its willingness to share fuel-making technology: "Brazil chose West Germany as its major nuclear supplier primarily because Bonn was willing to provide a uranium enrichment plant."[40] In addition to financial compensation of between $2 and $8 billion for the reactors (depending on the number sold), West Germany stood to receive 20 percent of the uranium ore recovered in Brazil.[41]

The motives of Brazil and West Germany, and their relationships with the United States, would play a large role in the unfolding of this drama. Brazil, referred to as a "potential third-generation proliferator" by the CIA in a 1974 National Intelligence Estimate, was in a rivalry with neighbor, Argentina.[42] Brazil's military government, headed by Ernesto Geisel, held "what later proved to be very unrealistic estimates of growing energy demand" and valued the independence and prestige that mastery of the fuel cycle would bring.[43] Having fallen behind Argentina's nuclear program, the deal with West Germany would allow Brazil to "leapfrog" its competitor.[44]

Bonn, a close strategic ally of Washington, was central to American security and defense policy in Europe. West Germany was also a commercial competitor of the United States, increasingly so in the nuclear field. For Bonn, the financial advantage of the deal was considerable, ensuring a way to "pay off" German government subsidies for the nuclear industry, provide work for 300 companies in the Federal Republic, and give long-term security to Kraftwerk Union, the firm responsible for building the reactors. German banks, at the government's behest, would lend the capital for the first two reactors to Brazil's newly created *Nuclebras* organization responsible for the nuclear program.[45] The sale came at a time of economic duress in Germany, promising a much-welcomed commercial boost.[46]

The Ford administration's intervention yielded little result, except convincing Bonn to adopt safeguards on the technology to be transferred.[47] These safeguards were secured through a multilateral agreement among Brazil, West Germany and the IAEA, which did go beyond traditional Agency measures by applying safeguards to the use of know how gained from the deal in facilities across Brazil.[48] Furthermore, in the spirit of the multilateral controls that would appear in the Iranian negotiations, the Ford administration convinced Germany to play a role in the management of the enrichment project, with German nationals intimately involved.[49]

Counterintuitively, Germany's close relationship with the United States afforded Bonn latitude in this deal: The Ford administration proved unwilling to jeopardize their relationship with Bonn in order to persuade German Chancellor Helmut Schmidt to forgo the sale.[50] Although Kissinger and Ford "felt uneasy" about the deal, they refused to bring up the subject "during top-level discussions between the German and American governments."[51] In May, July, and October 1975 meetings between Kissinger, Ford, and Schmidt, the Brazil deal went untouched, with economic, oil supply, and European political and security issues taking priority.[52] Kissinger decided not to use U.S. military units in Europe, or agreements with the Soviet Union, to pressure Bonn, even reassuring German Foreign Minister Hans-Dietrich Gemscher in June 1975 that the safeguards agreement had met with Washington's approval, giving Gemscher the green light to sign the contract later that month.[53]

The Ford administration's tepid response, at least at the high level, was not shared by all in Washington; certain members of Congress and media outlets voiced their opposition. Senator Abraham Ribicoff of Connecticut noted that the demand for energy had given rise to a:

> cutthroat nuclear competition . . . leading to the spread of plutonium reprocessing and uranium enrichment facilities. The capability to produce nuclear explosives is spreading 'like the plague', in the words of the Inspector General of the International Atomic Energy Agency.[54]

The *New York Times* editorial board labeled the arrangement:

> a reckless move that could set off a nuclear arms race in Latin America, trigger the nuclear arming of half a dozen nations elsewhere and endanger the security of the United States and the world as a whole.[55]

U.S. President Jimmy Carter, who had criticized the deal in his campaign, did not improve upon the Ford administration's record.[56] A high-level diplomatic flurry between Bonn and Washington took place in

202

the early days of the Carter administration, including a visit by U.S. Vice President Walter Mondale to Bonn. [57] A stalemate had been reached by March 1977, and on April 7, 1977, West Germany confirmed that it would proceed with the deal. This announcement came the same day as Carter's own announcement on the issue of nonproliferation. Schmidt was politically invested and felt he could not back down.

The Brazil-West Germany sale exposed, similarly to Iran, unwillingness in Washington to prioritize nonproliferation if doing so might sow discord in a crucial bilateral relationship. The proliferation risks of making the technology transfer itself, however, were minimal. Jet nozzle enrichment technology had not yet been demonstrated commercially and performed poorly. The pilot reprocessing plant was delayed indefinitely as the costs of Brazil's nuclear program, and the true need for this expensive technology, came under scrutiny in Brasilia. In June 1977, West Germany announced that it would not grant any further license for the export of reprocessing facilities, following in the footsteps of a similar decision by France in 1976.[58]

The close U.S. relationship with West Germany failed to translate into an immediate nonproliferation victory. In fact, the inter-reliance of the two countries may have limited Washington's leverage, making instruments, particularly on the "stick" side, less credible. The decision not to bring up the deal in high level diplomatic exchanges, despite public and congressional pressure, does suggest that preventing fuel-making in Brazil did not fall among the administration's top priorities.

INTERVENTION ON THE BUYER:
SOUTH KOREA

Over the course of the 1970s, the Park Chung-hee regime in South Korea (Republic of Korea [ROK]) pursued reprocessing technology with the purpose of developing a nuclear weapons option in light of uncertainty over U.S. security guarantees.[59] The Ford administration, making clear that the relationship between the two countries would be jeopardized should Seoul continue down this path, convinced Park's government not to go forward with the transfer of a crucial piece of technology, a pilot reprocessing plant from France.[60] The alignment of nonproliferation with other U.S. policy objectives, namely stability in Northeast Asia, the security of South Korea, and détente with the great powers, meant that the Ford administration avoided the difficult decision it had faced in the cases of Iran and the Brazil-West Germany deal of balancing competing policy objectives.[61] Willing to apply more fully its diplomatic leverage on Seoul—and employ credible threats of deeper retrenchment—and the beneficiary of intervention in the early stages of the South Korean program, Washington tasted success.

Seoul was a close U.S. ally, bound by the 1954 Mutual Defense Treaty, occupying a critical cog in the front against communism, contributing troops to the war in Vietnam, and enjoying a high-level political relationship with Washington. It was also reliant on American security assurances, particularly after a series of arms transfers from the Soviet Union and China to North Korea gave Pyongyang significant numerical advantages in materiel.[62] As the balance of power on the Korean Peninsula tilted, however, and North Korean provocations continued, Washington embraced a policy of retrenchment, beginning with the 1969 announcement of the Nixon Doctrine, which asked that the burden for defense be increasingly shifted to U.S. allies.[63] The March 1970 disclosure to ROK President Park that the U.S. military presence on the Korean peninsula would be decreased from 63,000 to 42,000, the political fallout from the Vietnam War in the United States, and the period of détente, particularly dialogue between the United States and China, have been understood as major drivers of Korea's twin conventional forces modernization and nuclear weapons programs.[64]

The South Korean nuclear weapons program's origins can be traced to two agencies created in the aftermath of the Nixon Doctrine's announcement, the Agency for Defense Development and the Weapons Exploitation Committee, the latter recommended the pursuit of nuclear weapons to President Park.[65] With Park taking the decision in December 1974 to proceed with a nuclear weapons program, secretly of course, Seoul entered negotiations with France in 1975 to purchase a pilot reprocessing plant.[66] These talks triggered a U.S. response—unlike in the West German-Brazil deal, U.S. officials engaged their South Korean counterparts at the highest level, threatening to withdraw U.S. assistance with Korea's civilian nuclear program and, more importantly, U.S. troops from the peninsula.[67] American diplomats also pressured Seoul to sign the NPT, which it did in April 1975, leading to the adoption of IAEA safeguards in September 1975.[68] U.S. efforts, though most notable on the demand side of the deal, were also important in persuading the French to go along with the South Korean decision to cancel the reprocessing plant deal in December 1975.[69]

The stakes were high for South Korea. As Mitchell Reiss notes, had Washington withdrawn its forces from the peninsula, Seoul would have lost its ability to deter the North, at least for a period of time.[70] But the United States also, irrespective of the state of the weapons program, held significant advantages in its relationship. It was Seoul's top trading partner, holding significant amounts of South Korean debt.[71] Furthermore, in the great geopolitical game on the Korean Peninsula, the United States provided certain assurances that Seoul required, making it more likely that China and the Soviet Union would value "stability on the peninsula."[72] In an August 1975 meeting, U.S. Secretary of Defense James Schlesinger explained to Park that the United States was:

> best suited to provide nuclear deterrence on behalf of its allies. We can deal with nuclear threats against a central power in a way that smaller nuclear powers cannot. We can deter Soviet nuclear threats while the ROK could not and a ROK effort to develop its own nuclear weapons would end up providing the Soviets with justification for threatening the ROK with nuclear weapons.[73]

South Korean goals of force modernization also relied heavily on U.S. assistance; if South Korea was to achieve conventional independence, it would be best served by continued U.S. support.[74] Finally, while the Nixon Doctrine and related messaging lent credibility to U.S. threats to withdraw all troops from the peninsula, Washington continued to respond forcefully to North Korean provocations, such as the August 1976 Panmunjom incident, which led to U.S. and South Korean forces being placed on high alert, U.S. naval assets being dispatched to the area, and U.S. B-52 bombers conducting exercises.[75]

There are less auspicious aspects to the South Korean case, however. First, the pursuit by South Korea of reprocessing and a broader military nuclear program was calculated in part on half-hearted U.S. nonproliferation efforts elsewhere. Park and his close associates, looking to the example of Israel, hoped that Washington would reconcile itself to a South Korean nuclear program, even retaining significant benefits from its relationship with the United States. Second, although the Ford administration's intervention regarding the pilot reprocessing plant was successful, the Carter administration allowed the transfer of a post-irradiation examination facility, the use of which can imitate an important part of chemical reprocessing, protesting only then, asking that the size of the facility be limited.

The South Korean episode was a success, and the Ford administration pursued nonproliferation aims vigorously. But the particularity of the South Korean case raises questions about how replicable it might be. The South Korean decision to forgo the pilot reprocessing facility in 1975 was based less on U.S. diplomatic maneuvering than the fundamental economic and strategic relationship of the two countries, developed over time and bound by a bloody conflict little more than 2 decades before. American willingness to prioritize South Korea's incipient nuclear weapons program in its communication with Seoul, and to use sticks and carrots to its advantage, meant that this relationship could be exploited. But even the decision in Washington to apply pressure was an easy one: The development of a nuclear weapons program on the Korean peninsula could have been disastrous, setting off a chain of events implicating the great powers.

PAKISTAN: THE LIMITATIONS OF SUPPLIER CONTROLS

Pakistan, also a U.S. ally, embarked on a nuclear weapons program in the aftermath of its 1971 war with India. Islamabad's disastrous military defeat and partition led Pakistani Prime Minister Zulqifar Ali Bhutto, who entered office after the war, to initiate the program. These efforts would be accelerated following India's May 1974 nuclear test and would result in an established weapons program by the 1980s, with a series of tests in 1998. From the mid-1970s onward, successive U.S. administrations attempted to dissuade Pakistan from pursuing the nuclear fuel-making facilities necessary for a weapons program. While U.S. efforts on the supply side had some success, namely in convincing France to defer and later cancel the sale of a reprocessing plant to Pakistan, they failed to persuade Islamabad from developing an indigenous enrichment facility.[76] While, given Pakistan's abiding interest in developing nuclear weapons, preventing Islamabad from developing ENR technology would have been difficult, the Ford and Carter administrations' unwillingness to apply more fully and consistently diplomatic, political, economic, and military tools ensured that Pakistan's efforts would go unchecked. In this case, broader U.S. goals in South Asia won the day, precluding tougher nonproliferation measures. As Secretary of State Henry Kissinger in a July 1976 briefing said tersely, "Non-proliferation is not our only objective in South Asia."[77]

The Pakistani nuclear weapons program was understood as a way of shoring up Pakistan's defense and correcting a shifting balance of power in the subcontinent. Particularly since the 1965 war between

India and Pakistan, India had pulled ahead in conventional fighting ability, an advantage demonstrated decisively in the short 1971 war. The Pakistani military leadership, which had once seen the United States as a dependable supporter, was disenchanted with Washington after an arms embargo on both India and Pakistan during and after the 1965 conflict. Between 1965 and 1971, absent U.S. arms sales, Islamabad turned to Beijing for military support. Although the Richard Nixon administration did, contrary to Congress' wishes, facilitate arms transfers from other states to Pakistan during the 1971 war and went so far as to dispatch the U.S.S. *Enterprise* off the coast of Pakistan, it was still seen as an ally of dubious commitment. Still, Pakistan played a crucial role in enabling Nixon's outreach to China and remained in close political contact with the United States.

Thus, particularly for Bhutto, who had expressed sympathy for a Pakistani nuclear arsenal before, proclaiming in 1965, "If India builds the bomb, we will eat grass or leaves, even go hungry, but we will get one of our own," a nuclear weapons program was essential.[78] Only weeks after assuming the prime ministry, Bhutto convened Pakistan's top nuclear scientists and instructed them to build a nuclear weapon within 3 years. In 1973, Pakistan began negotiating with a French corporation regarding reprocessing technology. The Indian test of May 1974 would only serve to accelerate this drive, with the chief of Pakistan's Atomic Energy Commission commenting after the detonation that India had "opened the floodgate for nuclear weapons" and implying that Pakistan might be next to join the "nuclear club."[79] A U.S. intelligence report issued shortly after the Indian test indicates concern that Islamabad may be next.[80]

Following the Indian test, Pakistan sought a U.S. security guarantee and "arms for cash" transfers, where Pakistan would, as other allies did, compensate the United States for weapons sales.[81] The Ford administration, though unwilling to extend a security guarantee, did remove its embargo on weapons sales to Pakistan and India, after persistent requests from Islamabad in February 1975.[82] This decision, however, yielded relatively little in the way of actual arms transfers, and the balance of power on the subcontinent continued to drift in India's favor even after the embargo was lifted.[83]

Meanwhile, Islamabad continued its nuclear weapons program, turning to France for a reprocessing facility in February 1976.[84] With Pakistan having only a single reactor, a heavy water model from Canada similar to that used by the Indian program, the ambitions of Bhutto were clear: In the words of U.S. Under Secretary of State Philip Habib in an exchange with Kissinger, "What he [Bhutto] wants is to build a bomb."[85] With this in mind, the administration moved to intervene, and Ford himself sent a letter to Bhutto in March 1976 asking "that you . . . give serious consideration to foregoing present plans to acquire reprocessing and heavy water facilities. . . ."[86] When Bhutto rejected this offer, the Ford administration applied only limited pressure. For example, rather than withholding all military aid from Pakistan, Kissinger insisted on withholding only the sale of A-7 fighter aircraft, suggesting that some military aid was appropriate, given Pakistan's alliance with the United States, and that providing such assistance now would be necessary if it was to be used as leverage later.[87] Despite the promise of 100 A-7 fighter aircraft in exchange for forgoing reprocessing in August 1976 and

repeated threats after the 1976 elections that the incoming Carter administration might "make an example" of Pakistan, the Ford administration was unable to dislodge Pakistan.

In large part, the Ford administration was constrained by recent history and events beyond its immediate control, namely a lack of faith in American promises after the 1965 and 1971 wars and a deeply held belief in Pakistan of the importance of a nuclear option. The United States also was unable to address Pakistan's conventional security concerns: As Kissinger observed, because U.S. arms sales did not flow as quickly to Pakistan following the lifting of the embargo (mostly for administrative reasons) as some had hoped, the threat of not selling them implied practically the status quo. Ford's team, despite securing a commitment from France not to sell any additional reprocessing plants, was unable to convince French President Valérie Giscard d'Estaing to renege on the deal with Pakistan.

The arrival of the Carter administration opened a new chapter in U.S. negotiations with Pakistan, one where Islamabad would lose its contract with France for the reprocessing facility but, through the leadership of Abdul Qadeer Khan, open a new path to a nuclear weapon: uranium enrichment. The Carter administration, too, decided not to use its full leverage in Pakistan. When presented with a package in April 1977 that would have provided military aid, economic assistance, and a financed French reactor, Carter favored only the military sale of the A-7 and not of more advanced U.S. technology and questioned the need for U.S. financing of the "French purchase," no matter the nonproliferation gain to be had.[88] By September 1977, with Bhutto ousted in a coup by General Muhammad

211

Zia-ul-Haq, the Carter administration had eliminated development aid to Pakistan on the grounds of the reprocessing deal with France.[89]

Meanwhile, the sale of French reprocessing technology to Pakistan was foundering, thanks in part to U.S. pressure. French officials had begun to push an alternative to reprocessing, called "co-processing," which would produce a mixed oxide fuel that was understood to be proliferation-resistant.[90] After Pakistan rejected this proposal, Giscard d'Estaing's administration decided to cancel the sale altogether, informing Kissinger on Memorial Day of 1978.[91] At the time, U.S. intelligence believed that this would make the "odds favoring any sort of explosive program [in Pakistan] ... sharply diminish."[92]

Soon thereafter, despite intelligence that Pakistan was considering indigenous fuel-making options, certain State Department officials began to advocate considering a return of arms sales and economic aid.[93] But mounting evidence of a uranium enrichment program led to the invocation of the 1976 Symington Amendment of the Foreign Assistance Act, which precluded aid for countries that operated national uranium enrichment facilities outside of safeguards.[94] With the Symington Amendment enforced, the United States moved with little success to rally western allies to maintain export controls and reached out to China, a close partner of Pakistan.[95] Administration officials also reached out to India's Prime Minister, hoping to reach some kind of regional solution, such as a joint commitment not to use or develop nuclear weapons, but to no avail.[96] However, writings within the White House suggested concerns that U.S. nonproliferation efforts had come at an inopportune time with instability in Iran and Afghanistan, and that Pakistan was a more reliable partner.[97] With more disturbing news coming of the extent of Pakistan's enrichment program, U.S. officials were at a loss to find a solution. In the words of Charles van Doren at a meeting of the General Advisory Committee on Arms Control and Disarmament, "We have a great deal of talent within the U.S. government scratching its head."[98]

If the Carter administration was ambivalent about what steps to take regarding the Pakistani nuclear program, this would soon cease to be the case. The December 1979 Soviet invasion of Afghanistan fixated Washington and transformed perceived U.S. interests in the region; suddenly, Pakistani acquiescence with U.S. policy goals in Afghanistan became vital. In January 1980, Carter offered not only to restore the transfer of aid, but also to increase it dramatically to a total of $400 million per annum, split between military and economic assistance. Islamabad, well aware of its newfound leverage, spurned the offer—the American package was too modest. Within the next year, the Ronald Reagan administration had agreed upon a new assistance package extended over several years, and soon an exception to the Symington Amendment had been crafted so as to allow for U.S. aid to Pakistan.[99] While the drama surrounding the Pakistani nuclear program was far from over—it was October 1990 before the U.S. executive branch confirmed that Islamabad did not have nuclear weapons—the pattern of vacillation, mixed messaging, and half-hearted effort had been established and would continue.

CONCLUSION

The cases presented offer a common lesson: The United States, though constrained or empowered by circumstance, can exert considerable sway in nonproliferation matters but often elects not to apply the most powerful tools at its disposal for fear of jeopardizing other objectives. The persistent dilemma of how much to emphasize nonproliferation goals, and at what cost, has contributed to cases of nonproliferation failure. The inconsistent or incomplete application of U.S. power in nonproliferation cases is most harmful when it gives the impression to a nation that either sharing sensitive technology or developing it is, or will become, acceptable to Washington. U.S. reticence historically, with some exceptions, to prioritize nonproliferation—and in so doing reduce the chance of success in these cases—does not leave room for great optimism about future U.S. efforts at persuading countries to forgo nuclear fuel-making.

The most successful case, South Korea, saw the United States put in question the basis of its relationship with Seoul, its security assurance, for nonproliferation aims. The potential near-term consequences of a South Korean nuclear weapon made this bold diplomatic maneuver worth the risk. But in other cases, competing U.S. aims, often worthy, have impinged on nonproliferation goals, diluting efforts and sending mixed signals. In the case of Pakistan, for example, even well before the Soviet invasion of Afghanistan, the United States failed to use sufficiently forceful sticks or attractive carrots. U.S. efforts were bound by increasing distrust between Islamabad and Washington, a delicate geopolitical situation in the subcontinent given India's close relationship with the Soviet Union, and facing a great challenge in a Pakistani leadership that was humiliated in 1971 and keen to reestablish some power equity with India. In negotiations with Iran regarding the nuclear cooperation agreement, U.S. policymakers — hoping to reinforce the NPT after the Indian test, avoid offending the Shah, and secure civilian nuclear contracts — were initially willing to make concessions on the issue of national reprocessing. In the case of the West Germany-Brazil contract, Kissinger went so far as to tell his counterpart in Bonn that, with expanded safeguards, the deal would be acceptable to Washington despite the clear proliferation risk from Brasilia.

The previous examples show the limitations of both demand and supply side efforts. Supply side diplomatic interventions, made before the transfer of technology, have been at times effective, particularly in precluding nuclear fuel-making in the short term and buying time for more lasting solutions. However, as the Pakistan and Brazil cases illustrated, supply side interventions are no substitute for demand side solutions: Countries face political choices regarding nuclear fuel-making. A nation set upon an independent fuel-making capacity, such as Pakistan or Brazil, is unlikely to give up efforts because of supply side controls. Multilateral fuel-making arrangements, as proposed repeatedly by the United States, have not materialized and therefore seem to have had little tangible influence.

In recent years, a new nonproliferation instrument has appeared: a restructured 123 nuclear cooperation agreement, developed in the course of negotiations with the United Arab Emirates (UAE) and signed in 2009. This agreement, unlike previous bilateral nuclear cooperation agreements, offers a model for demand side nonproliferation, with the UAE vowing to forgo all enrichment and reprocessing technology on its own soil. It goes far beyond, for example, the "veto" on reprocessing of U.S.-origin spent fuel broached in the negotiations with the Shah. This "Gold Standard" agreement, much hailed at first, particularly in contrast to Iran's enrichment activities, has begun to lose its luster as, once again, competing priorities marginalize nonproliferation. In January 2012, the Obama administration announced that a "case by case" approach would be taken to the application of the Gold Standard. Countries such as Vietnam, where the United States holds out hope for a grander partnership aimed at countering China, may not be held to the UAE's standard.[100] Today, as in the 1970s with the Symington and Glenn Amendments, Congress seems most concerned about the prospect of proliferation of ENR technology.

The UAE case is a striking reminder of the lasting challenge facing American nonproliferation efforts. As a global power with ranging interests, governed by a political system where dissenting factions in Congress, the White House, and bureaucratic organs can influence policy in a number of ways, and operating in an international system with its own constraints on U.S. power, the United States has struggled to marshal its strength toward persuading countries to forgo nuclear fuel-making. While there is no guarantee that the decisive and steadfast application of "sticks and carrots" in the cases presented would have changed the outcomes—it may have brought unintended consequences of its own—a commitment to doing so would have improved the chance of persuading countries to eschew fuel-making.

ENDNOTES - CHAPTER 7

1. Norman A. Wolf, "Misinterpreting the NPT," *Arms Control Today*, September 2011, available from *www.armscontrol. org/2011_09/Misinterpreting_the_NPT*.

2. The Treaty on the Non-Proliferation of Nuclear Weapons, July 1, 1968, available from *www.fas.org/nuke/control/npt/text/ npt2.htm*.

3. There have been exceptions, most notably Ambassador John R. Bolton, who, while in office, rejected the per se right to reprocessing and enrichment technology.

4. Office of the Historian, "The Acheson Lilienthal and Baruch Plans," Washington, DC: U.S. Department of State, available from *history.state.gov/milestones/1945-1952/BaruchPlans*.

5. Department of State "A Report on the International Control of Atomic Energy," Washington, DC: U.S. Government Printing Office, March 16, 1946, pp. 35-36, available from *www.learnworld. com/ZNW/LWText.Acheson-Lilienthal.html*.

6. "The Acheson Lilienthal and Baruch Plans."

7. Bernard Baruch, "Baruch Plan," Presented to the United Nations Atomic Energy Commission, June 14, 1946, available from *www.atomicarchive.com/Docs/Deterrence/BaruchPlan.shtml*.

8. Dwight D. Eisenhower, "Address by Mr. Dwight D. Eisenhower, President of the United States of America, to the 470th Plenary Meeting of the United Nations General Assembly," New York, December 8, 1953, available from *www.iaea.org/About/ atomsforpeace_speech.html*.

9. *Ibid.*

10. Barack Obama, "Remarks by President Barack Obama," Prague, The Czech Republic, April 5, 2009, available from *www. whitehouse.gov/the_press_office/Remarks-By-President-Barack-Obama-In-Prague-As-Delivered*.

11. Mohamed ElBaradei, "Towards a Safer World," *The Economist*, October 16, 2003, available from *www.economist.com/node/2137602*.

12. Mark A. Christopher, "International Fuel Supply Guarantees," Washington, DC: Center for Strategic and International Studies, available from *csis.org/images/stories/poni/110921_Christopher.pdf*.

13. Pierre Goldschmidt, "Multilateral Nuclear Fuel Supply Guarantees and Spent Fuel Management: What Are the Priorities?" *Daedalus*, Vol. 139, No. 1, Winter 2010, available from *carnegieendowment.org/2010/01/01/multilateral-nuclear-fuel-supply-guarantees-and-spent-fuel-management-what-are-priorities/2a7s*.

14. Harold Feiveson *et al.*, "Can Future Nuclear Power Be Made Proliferation Resistant?" College Park, MD: Center for International and Strategic Studies at Maryland, July 2008, available from *www.cissm.umd.edu/papers/files/future_nuclear_power.pdf*.

15. Richard Helms, "Iranian Counterproposals for Atomic Energy Agreement," State Department Cable Tehran 7485, July 23, 1976, available from *www.gwu.edu/~nsarchiv/nukevault/ebb268/doc21.pdf*.

16. Secretary of State, "US-Iran Nuclear Agreement," State Department Cable State 125797, May 17, 1978, available from *www.gwu.edu/~nsarchiv/nukevault/ebb268/doc31a.pdf*.

17. William Burr, "US-Iran Negotiations in 1970s Featured Shah's Nationalism and U.S. Weapons Worries" The National Security Archive Electronic Briefing Book No. 268, January 13, 2009, available from *www.gwu.edu/~nsarchiv/nukevault/ebb268/index.htm*.

18. NSC Under Secretaries Committee, "US Nuclear Nonproliferation," December 4, 1974, enclosing Memorandum for the President from Robert S. Ingersoll, Chairman, December 4, 1974; and NSSM 202 Study, "Executive Summary," SECRET, available from *www.gwu.edu/~nsarchiv/nukevault/ebb268/doc03.pdf*.

19. *Ibid.*; National Security Study Memorandum 219 Working Group, "Nuclear Cooperation Agreement with Iran," April 1975, SECRET, available from *www.gwu.edu/~nsarchiv/nukevault/ebb268/doc05a.pdf*.

20. American Embassy Paris, "Interview with Shah," State Department cable 15305, June 24, 1974, available from *www.gwu.edu/~nsarchiv/nukevault/ebb268/doc01a.pdf*.

21. Director of Central Intelligence. "Prospects for Further Proliferation of Nuclear Weapons." Special National Intelligence Estimate 4-1-74, August 23, 1974, available from *www.gwu.edu/~nsarchiv/NSAEBB/NSAEBB240/snie.pdf*.

22. Live Leak, "A Brief History of Iranian Missile Technology," June 9, 2007, available from *www.liveleak.com/view?i=6a6_1181429741*.

23. American Consulate Shiraz, "The Persepolis Conference on Transfer of Nuclear Technology: A Layman's View, "State Department Cable A-16, April 18, 1977, available from *www.gwu.edu/~nsarchiv/nukevault/ebb268/doc27.pdf*.

24. National Security Study Memorandum 219 Working Group, "Nuclear Cooperation Agreement with Iran," April 1975, SECRET, available from *www.gwu.edu/~nsarchiv/nukevault/ebb268/doc05a.pdf*.

25. *Ibid.*

26. Office of Assistant Secretary of Defense for International Security Affairs, "Nuclear Energy Cooperation with Iran—Action Memorandum," enclosing Atomic Energy Commission and Department of State memoranda, CONFIDENTIAL, with handwritten note attached, June 1974, available from *www.gwu.edu/~nsarchiv/nukevault/ebb268/doc02.pdf*; and Office of the Secretary of State, "The Secretary's Meeting with FRG Ambassador Von Staden on the FRG/Iran Agreement for Nuclear Cooperation," Memorandum of Conversation, Washington, D.C. July 2, 1976, SECRET, available from *www.gwu.edu/~nsarchiv/nukevault/ebb268/doc20.pdf*.

27. National Security Study Memorandum 219 Working Group, "Nuclear Cooperation Agreement with Iran," April 1975, SECRET, available from *www.gwu.edu/~nsarchiv/nukevault/ebb268/doc05a.pdf*.

28. Richard Helms, "US/Iran Nuclear Agreement," State Department cable Tehran 11539, November 26, 1975, SECRET, available from *www.gwu.edu/~nsarchiv/nukevault/ebb268/doc09c.pdf*.

29. Department of State, "Proposed Cable to Tehran on Pakistani Nuclear Reprocessing," Memorandum of Conversation, May 12, 1976, available from *www.gwu.edu/~nsarchiv/nukevault/ebb268/doc17.pdf*.

30. *Ibid*.

31. *Ibid*.

32. Richard Helms, "Nuclear Energy Discussion," State Department cable Tehran 7886, August 3, 1976, available from *www.gwu.edu/~nsarchiv/nukevault/ebb268/doc22.pdf*.

33. "Presidential Campaign Debate between Gerald R. Ford and Jimmy Carter," San Francisco, CA, October 6, 1976, available from *www.fordlibrarymuseum.gov/library/speeches/760854.asp*; Gerald R. Ford, "Statement on Nuclear Policy," October 28, 1976, available from *www.presidency.ucsb.edu/ws/index.php?pid=6561#axzz1uo0bKAxs*.

34. American Embassy Tehran, "Nuclear Power: Comments of Head of Atomic Energy Organization of Iran (AEOI)," State Department cable Tehran 1232, February 7, 1977, available from *www.gwu.edu/~nsarchiv/nukevault/ebb268/doc25a.pdf*.

35. Jimmy Carter, "Nuclear Non-Proliferation Fact Sheet on the Proposed Nuclear Non-Proliferation Act of 1977," April 27, 1977, available from *www.presidency.ucsb.edu/ws/index.php?pid=7409#axzz1uo0bKAxs*.

36. American Embassy Tehran, "GOI/AEOI Plans for Isfahan Nuclear Technology Center, Entec," State Department cable Tehran 1437, February 14, 1977, available from *www.gwu.edu/~nsarchiv/nukevault/ebb268/doc25b.pdf*.

37. David Binder, "US Wins Safeguards in Brazil Deal," *The New York Times*, June 4, 1975.

38. Director of Central Intelligence, "Brazil's Changing Nuclear Goals: Motives and Constraints," Special National Intelligence Estimate, October 21, 1983, pp. 5-6.

39. Norman Gall, "Atoms for Brazil, Dangers for All," *Foreign Policy*, No. 23, Summer 1976, pp. 157, 160.

40. Director of Central Intelligence, "Brazil's Changing Nuclear Goals: Motives and Constraints," Special National Intelligence Estimate, October 21, 1983, p. 6.

41. Gall, p. 160.

42. Director of Central Intelligence, "Prospects for Further Proliferation of Nuclear Weapons."

43. Michael Barletta, "The Military Program in Brazil," Stanford, CA: Stanford University, Center for International Security and Arms Control, February 1985.

44. Mitchell Reiss, *Bridled Ambition: Why Countries Constrain Their Nuclear Capabilities*, Washington, DC: Woodrow Wilson Center Press, 1995, p. 50.

45. Gall, p. 158.

46. The White House, Memorandum of Conversation at the Ambassador's Residence, Brussels, Belgium, May 29, 1975, available from *www.fordlibrarymuseum.gov/library/document/0314/1553091.pdf*.

47. David Binder, "US Wins Safeguards in Brazil Deal," *The New York Times*, June 4, 1975; Karl Kaiser, "The Great Nuclear Debate: German-American Disagreements," *Foreign Policy*, No. 30, Spring 1978, p. 89.

48. Gall, p. 159.

49. Kaiser, p. 89.

50. *Ibid.*, pp. 89-90.

51. *Ibid.*

52. Memorandum of Conversation at the Ambassador's Residence, Brussels, Belgium; The White House, "Memorandum of Conversation at the Chancellery," Bonn, Germany, July 27, 1975, available from *www.fordlibrarymuseum.gov/library/document/0314/1553186.pdf*; and The White House, Memorandum of Conversation in the Oval Office, Washington, DC, October 3, 1975, available from *www.fordlibrarymuseum.gov/library/document/0314/1553253.pdf*.

53. Kaiser, pp. 89-90.

54. U.S. Congress, Senate, Congressional Record, 94th Cong., 1st sess., June 3, 1975, p. S9323, quoted in Norman Gall, "Atoms for Brazil, Dangers for All."

55. "Nuclear Madness," *The New York Times*, June 13, 1975, p. 36.

56. Kaiser, p. 97.

57. "Schmidt May Modify Rio Pact," *The New York Times*, January 27, 1977; and Kaiser, pp. 99-100.

58. Kaiser, p. 100.

59. National Foreign Assessment Center, "South Korea Nuclear Development and Strategic Decisionmaking," National Intelligence Estimate, Washington, DC: Central Intelligence Agency, June 1978, pp. 1-2; Memorandum from Richard Smyser and David Elliott of the National Security Council Staff to Secretary of State Kissinger, Washington, DC, February 28, 1975, *Foreign Relations of the United States, 1969-1976*, Vol. E-2, Documents on East and Southeast Asia, 1973-76, Document 264.

60. "South Korea Nuclear Development and Strategic Decisionmaking," pp. 1-2; Jungmin Kang and H. A. Feiveson, "South Korea's Shifting and Controversial Interest in Spent Fuel Re-

processing," *The Nonproliferation Review*, Vol. 8, No. 1, Spring 2001, p. 71.

61. Smyser and Elliott.

62. "South Korea Nuclear Development and Strategic Decisionmaking," p. 1.

63. Richard Nixon, "Informal Remarks in Guam with Newsmen," July 25, 1969, available from *www.presidency.ucsb.edu/ws/?pid=2140#ixzz1uyyKCC8j*; "South Korea Nuclear Development and Strategic Decisionmaking," p. 2; Rebecca K. C. Hersman and Robert Peters, "Nuclear U-Turns: Learning from South Korean and Taiwanese Rollback," *The Nonproliferation Review*, Vol. 13, No. 3, November 2006, available from *cns.miis.edu/npr/pdfs/133hersman.pdf*.

64. "South Korea Nuclear Development and Strategic Decisionmaking," pp. 1-2.

65. Kang and Feiveson, p. 71.

66. "South Korea Nuclear Development and Strategic Decisionmaking," pp. 5-7.

67. "Nuclear Peril," *The New York Times*, June 9, 1975, p. 30.; Hersman and Peters, "Nuclear U-Turns: Learning from South Korean and Taiwanese Rollback," p. 541.

68. Smyser and Elliott; Kang and Feiveson, p. 71.

69. "Korea's Atomic Reversal," *The New York Times*, February 15, 1976, p. E12.

70. Hersman and Peters, p. 541.

71. *Ibid.*

72. "South Korea Nuclear Development and Strategic Decisionmaking," p. 17.

73. Memorandum of Conversation, Seoul, Korea, August 27, 1975, *Foreign Relations of the United States, 1969-1976*. Vol. E-12, Documents on East and Southeast Asia, 1973-1976, Document 272.

74. "South Korea Nuclear Development and Strategic Decisionmaking," p. 17.

75. Peter Hayes and Chung-in Moon, "Park Chung Hee, the CIA & the Bomb," *Global Asia*, September 2011, p. 52.

76. Flora Lewis, "France Holds Up Proposed Sale Of a Nuclear Plant to Pakistan," *The New York Times*, June 1, 1977, p. 42.; American Embassy, Paris, France, "French Complaint Concerning US Action on Pakistan Reprocessing Plant," State Department cable Paris 24112, August 1, 1978.

77. Department of State, "Pakistani Reprocessing Issue," Memorandum of Conversation, July 9, 1976.

78. "The Spider's Stratagem: Pakistan's Nuclear Ambitions," *The Economist*, January 3, 2008.

79. Central Intelligence Agency, "Indian Test Will Spur Pakistani Effort," *National Intelligence Daily*, May 24, 1974, available from *www.foia.cia.gov/sites/default/files/document_conversions/89801/DOC_0000845822.pdf*.

80. *Ibid*.

81. "US Arms Policy Toward Subcontinent Following Indian Nuclear Test," State Department cable, June 12, 1974.

82. National Security Council, "National Security Decision Memorandum 289," March 24, 1975, available from *www.fordlibrarymuseum.gov/library/document/0310/nsdm289.pdf*.

83. "Pakistani Reprocessing Issue," pp. 6-8.

84. Central Intelligence Agency, "Pakistan Nuclear Study," Memorandum, April 26, 1978, TOP SECRET, excised copy, available from *www.gwu.edu/~nsarchiv/nukevault/ebb333/doc05.pdf*.

85. "Pakistani Reprocessing Issue," pp. 6-8.

86. Gerald Ford, Letter to Prime Minister Zulkifar Ali Bhutto of Pakistan, March 19, 1976, available from *2001-2009.state.gov/documents/organization/97435.pdf.*

87. Department of State, Memorandum of Conversation, July 9, 1976, pp. 4-5, available from *history.state.gov/historicaldocuments/frus1969-76ve08/d232.*

88. Warren Christopher, "Reprocessing Negotiations with Pakistan: A Negotiating Strategy," Memorandum for the President, April 2, 1977, SECRET, available from *www.gwu.edu/~nsarchiv/nukevault/ebb333/doc02.pdf.*

89. William Burr, "The United States and Pakistan's Quest for the Bomb," December 21, 2010, available from *www.gwu.edu/~nsarchiv/nukevault/ebb333/index.htm#12.*

90. "Pakistan Nuclear Study."

91. Secretary of State, "Reprocessing Issue," State Department cable State 136685 May 30, 1978, SECRET, available from *www.gwu.edu/~nsarchiv/nukevault/ebb333/doc06.pdf.*

92. "Pakistan Nuclear Study."

93. Memorandum to Chris [Warren Christopher] from Steve [Oxman], enclosing edits to draft cable to Islamabad and "Evening Reading" reports to President Carter on Pakistan, October 4, 1978, SECRET, available from *www.gwu.edu/~nsarchiv/nukevault/ebb333/doc18.pdf.*

94. "Pakistan," Presidential Review Committee Meeting, March 9, 1979, available from *www.gwu.edu/~nsarchiv/nukevault/ebb333/doc29.pdf.*

95. Paul H. Kreisberg to David D. Newsom, "Presidential Letter on Pakistan Nuclear Program to Western Leaders," Department of State action memorandum, March 30, 1979, SECRET, available from *www.gwu.edu/~nsarchiv/nukevault/ebb333/doc33.pdf;* Gerard Smith to the Secretary, "Consultations in Europe on Paki-

stan," November 15, 1979, available from *www.gwu.edu/~nsarchiv/ nukevault/ebb333/doc45.pdf*.

96. Smith to the Secretary, "Consultations in Europe on Pakistan."

97. *Ibid.*

98. Advisory Committee on Arms Control and Disarmament, Friday Morning Session, September 14, 1979, General SECRET, excised copy, available from *www.gwu.edu/~nsarchiv/nukevault/ ebb333/doc42.pdf*.

99. Peterson Institute for International Economics, "Case Studies in Sanctions and Terrorism: Case 79-2: US v. Pakistan (1979-Nuclear Missile Proliferation)," available from *www.iie.com/ research/topics/sanctions/pakistan.cfm*.

100. Elaine Grossman, "Nuclear Trade Reform Faces Hostile Lobbying, as Obama Team Renews Policy Review," *National Journal.com*, May 12, 2013, available from *www.nationaljournal.com/ congress/nuclear-trade-reform-bill-faces-hostile-lobbying-as-obama- team-renews-policy-review-20120504*.

CHAPTER 8

CENTRIFUGES:
A NEW ERA FOR NUCLEAR PROLIFERATION

R. Scott Kemp

The last decade has seen many new ideas for strengthening the nonproliferation regime. In 2004, the United Nations (UN) Security Council passed Resolution 1540 mandating that all UN member states adopt systems of export control to restrict transfers of sensitive technologies used in the production of weapons of mass destruction (WMD). That same year, U.S. President George W. Bush proposed to restrict the rights of states to build enrichment and reprocessing plants.[1] In 2005, the head of the International Atomic Energy Agency (IAEA), Mohamed ElBaradei, suggested that a multinational fuel cycle be established to limit national capabilities.[2] Since then, many other concepts, from fuel banks to cradle-to-grave nuclear-power leasing, have been proposed as ways to deter the proliferation of national fuel cycle capabilities. While these proposals are cast as general enhancements to the nonproliferation regime, they are all motivated by instances of centrifuge proliferation.

The uranium-enriching gas centrifuge has become one of the most coveted pieces of nuclear technology. Every aspiring nuclear-weapons state since 1975 has considered the centrifuge for its weapons program. Pakistan's first nuclear bomb was built using centrifuges, and Brazil, Iraq, Libya, Iran, South Africa, Syria, and North Korea all sought centrifuge technology for military purposes.

If centrifuges have become the proliferation technology of choice, it is not without cause. They are small, highly flexible, easy to hide, and much less resource-intensive than alternative options.[3] They produce highly enriched uranium, which is easier to handle and use in nuclear weapons than plutonium. Moreover, centrifuge programs can be deployed for ostensibly peaceful purposes and then rapidly used to make fissile material for weapons without significant modification or delay.

Policymakers have responded to recent cases of centrifuge proliferation by advocating for stronger export controls that would make it harder for states to build centrifuges. In parallel, policymakers have also considered new institutional arrangements that would make it more difficult for states to claim that their acquisition of centrifuge technology had a peaceful basis. These policies keep with a long tradition of focusing on the supply of nuclear technology rather than the demand for nuclear weapons. The newest threat to the supply-side regime has come from black-market transfers: from Germany to Iraq and South Africa; from the Netherlands to Pakistan; and from Pakistan to Libya, Iran, Syria, and North Korea.[4] By shutting down these networks and by establishing appropriate guidelines for licit transfers, many hope the centrifuge problem can be largely solved.[5] Underlying these proposals, however, is an unspoken assumption that centrifuge technology can be controlled. The proposals do not acknowledge that the centrifuge is a 50-year-old device based on straightforward principles of mechanical engineering, that essentially all of the required design information needed is in the public domain, or that basic centrifuges require no exotic tools or materials to make. If centrifuges can be indigenously produced, they cannot be effectively restrained by technology controls.

The effort needed to make basic centrifuges is, by today's standards, quite modest: Prototype centrifuges have been built by small groups of 10 to 20 engineers in 1 to 2 years, and such machines have been subsequently deployed on large scales to make nuclear weapons (particularly in the Soviet Union). Of the 20 countries that have successfully acquired centrifuges, 17 started with small, simple machines of the kind not effectively controlled by export restrictions. Fourteen of them succeeded without foreign assistance in developing these centrifuges to a level suitable for making weapons. An analysis across all 20 programs suggests that simple centrifuges are probably within the technical capability of nearly any country, including many or most developing countries.[6] Supply-side controls would not address this state of affairs; only motivations and the organizational capacity of states would restrain centrifuge proliferation. If this is indeed the case, then the nonproliferation system needs rethinking.

This chapter begins with a brief history of the effort to control the spread of fuel-cycle technologies. It explains how the centrifuge emerged from the field of other technologies as the most pernicious and most difficult to control. It then gives the history of centrifuge proliferation, first by black market transfers, followed by indigenous development, with a focus on important questions like the role of secrecy, tacit knowledge, and the human and industrial resources required for success. The discussion of the potential for states to acquire centrifuges is followed by a discussion of how difficult it would be to prevent centrifuges from being used for weapons purposes. The centrifuge has certain technical characteristics that make both safeguards and counterproliferation difficult or impossible. The chapter concludes with a critical review of several recent proposals for coping with proliferation. While this report does not outline a specific solution, it suggests that none of the supply-side options appear to offer any hope of success. As such, a focus on the demand for nuclear weapons, rather than on the capabilities, is perhaps the only way to mitigate centrifuge proliferation in today's technologically advanced world.

THE EMERGENCE OF SUPPLY-SIDE CONTROLS

Plutonium, not uranium, was the material of choice during the first half of the nuclear age. Plutonium powers nearly all the nuclear weapons of the first four nuclear-weapons states. In time, however, some countries found it convenient to enrich uranium, and its popularity as a bomb fuel has been growing ever since. China (1960), South Africa (1977), and Pakistan (1979) all used highly enriched uranium (HEU) for their first nuclear weapons, and three of the five most recent efforts to acquire nuclear weapons — those of Iraq, Libya, and Iran — had centrifuge-based HEU production as a central focus of the program. North Korea, the most recent nuclear-weapons state, started with a plutonium capability but has since replaced it with a centrifuge capability.[7] Of this group, only Syria has focused more on plutonium, its centrifuge program having been interrupted at an early stage.[8]

The increasing popularity of uranium enrichment as a route to the bomb is due in part to an easing of the technical hurdles involved in its production. In the 1950s, uranium enrichment entailed the construction of large, energy- and resource-intensive plants. A basic nuclear reactor was comparatively simple, and although it would be difficult to hide because reactors produce large quantities of heat, it could be easily justified as part of a peaceful research program. Large research reactors could even be bought from advanced nuclear countries, while enrichment plants had yet to be commercialized.

To counter the growing threat of plutonium, non-proliferation advocates began to minimize its role in the nuclear fuel cycle. Research reactors that produced weapons-grade plutonium as a byproduct of operation were replaced by models that used low-enriched fuel and only produced plutonium very slowly. For electricity production, the light-water power reactor was promoted over other more weapon-friendly designs. For power production, the light-water reactor needs to be refueled only every 1 to 2 years. More frequent refueling is needed to produce the kind of weapons-grade plutonium preferred by weapons designers, and such refueling would easily raise suspicions. Finally, the spent-fuel reprocessing facilities required to extract plutonium from reactor fuel (needed regardless of reactor type) were delegitimized by emphasizing that they were neither economic nor technically essential elements of the civilian nuclear fuel cycle. The United States took the lead in establishing this norm by abandoning its own reprocessing efforts in the 1970s.[9]

But while the plutonium route was getting more difficult, the uranium route was becoming easier. One of the unintended consequences of anti-plutonium policies was that reactors that were poor at producing plutonium also require enriched uranium to operate. During the 1950s and 1960s, this material could only be practically purchased from major weapons states like the United States or the Soviet Union. This gave nuclear-weapons states the ability to regulate to some extent the nuclear activities of nonweapons states. In the early-1960s, all this began to change. A new enrichment technology—the gas centrifuge—was invented. Unlike its predecessor, the centrifuge was a small-scale, affordable technology that was potentially within the reach of a great number of states.[10] By 1970, the centrifuge had become the most economically viable method for enriching uranium, and it remains so today.[11] Whereas the preceding enrichment technology, called gaseous diffusion, had required massive amounts of infrastructure and was therefore successfully developed only by nations with large nuclear programs, centrifuge plants could be built on smaller scales out of simple modules that were individually cheap to build and inexpensive to operate. Countries that depend on the United States or the Soviet Union for the supply of nuclear fuel began to look toward the centrifuge as a way to free those countries from dependency. The Netherlands, Germany, Israel, the United Kingdom, France, China, Australia, Sweden, Italy, India, and Japan all started centrifuge programs for the nominal purpose of self-sufficiency. In 1973, the perils of dependency were dramatized when the United States, fearing that the demand for enriched uranium would outstrip its capacity to supply, briefly closed its order books, reinforcing the perceived importance of self-sufficiency.

Officials in the United Kingdom (UK) and United States immediately recognized that gas centrifuges could be used to make weapons.[12] Further still, they understood that the small footprint and low electricity requirements of a gas-centrifuge plant would make weapons production by means of the centrifuge difficult or impossible to detect. In 1960, Chairman of the U.S. Atomic Energy Commission (AEC) John McCone warned of this problem:

> [D]o not minimize the potential importance of this process. . . . If successfully developed, a production plant using the gas centrifuge method could be simply housed. Its power requirements would be relatively small, and there would be no effects of the operation which would easily disclose the plant. Although the gas centrifuge does not pose an immediate prospect for the production of weapons material, there is no doubt in my mind it will introduce an additional complicating factor in the problems of nuclear arms among nations and our quest for controlled disarmament.[13]

The United States acted immediately to classify centrifuge design information worldwide. Delegations were sent to every research program in the West, all of which complied. Unfortunately, this effort came too late. The United States had already published most of the basic information required to build a centrifuge in a series of technical reports, now widely distributed around the world. Within a decade, nine countries had successful gas-centrifuge programs, despite the U.S. Government's classification efforts.[14]

With time, centrifuges have become easier and easier to build. By the end of the 1990s, additional technical publications and advances in computing and manufacturing had come together to create a situation in which nearly any country—including developing nations—could access the technology and information needed to build a proliferation-scale centrifuge program (whether they could organize themselves well enough was a different matter).[15] Compounding this problem, Pakistan, Iraq, and Iran all built centrifuge programs for nuclear weapons, raising global awareness of the technology and demonstrating its proliferation advantages. Today, many regard the centrifuge as the proliferation technology of choice.[16]

THE DEVELOPMENT AND SPREAD OF CENTRIFUGE TECHNOLOGY

Traditionally, the nonproliferation community looks towards the most recent cases of proliferation for guidance on how to improve the nonproliferation regime. The approach has the advantage of being empirical, but selecting the dependent variable can lead to misinterpretations of causality. It is better to look at a broader history of centrifuge development and the spread of centrifuge technology to understand the true nature of centrifuge proliferation. With over 20 historical cases, there is a considerable basis for drawing new conclusions about the nature of the proliferation problem.

Black Market Networks and State-To-State Transfers.

The spread of the gas centrifuge is widely, but incorrectly, understood to be primarily the work of black market networks. Most famous is A. Q. Khan, a Pakistani metallurgist who worked for a Dutch centrifuge contractor. In 1975, he stole design and supplier information to help Pakistan build a centrifuge capability for its nuclear weapons program. Khan later sold this stolen technology to Libya, Iran, North Korea, and possibly China (then already in possession of basic centrifuges, but interested in learning as much as it could). He offered it to several others, including Iraq and Syria. Pakistan's program and many of its subsequent retransfers were aided by a number of Canadian, Dutch, Swiss, British, and especially German engineers, and some of the German engineers appear to have assisted independently centrifuge programs in Iraq, South Africa, and Brazil.

While black market transfers are not unimportant in the history of centrifuge proliferation, their importance is often exaggerated. China, Pakistan, India, Iran, South Africa, and Brazil all had centrifuge programs prior to the receipt of foreign assistance. These programs were either already successful or would almost certainly have been successful if left to their own autonomous development.[17]

In some cases, assistance may even have been counterproductive. Consider Pakistan, for example. The drawings A. Q. Khan provided were for an advanced and difficult-to-make centrifuge. The machine was immensely complex relative to most entry-level designs, and this diversion probably slowed his country's nuclear progress relative to what could have been accomplished had Pakistan simply worked on its own. According to histories of the program and statements from its former head, as well as the program's chief scientist, the information initially supplied by A. Q. Khan was not even complete: It lacked key manufacturing specifications that would have hinted at the difficulty of making the advanced machine, and it forced the program to make a number of compromises that ultimately eliminated the performance advantages the advanced machine supposedly offered. The result was a machine that performed less well than the machines usually developed by programs of independent development. Iran, Libya, and Iraq also had centrifuge programs prior to their interaction with the black market. The receipt of black market assistance may have improved the level of funding and support from political leadership, but the technical information they received was frequently problematic. Those who turned to A. Q. Khan for assistance also received incomplete, unreliable assistance and were directed to develop the same highly problematic design with which Pakistan started. Only South Africa and Iraq appear to have received foreign assistance that was technically sound and of sufficient timeliness that it advanced the date at which a meaningful centrifuge capability could have been produced. In both of these cases, assistance came from highly experienced German engineers, not A. Q. Khan.[18]

Despite the limited benefit the black market has had for centrifuge proliferation, the potential for high-impact technology transfers remain. The South African and Iraqi cases demonstrate that technically competent foreign consultants can accelerate a program — especially if it consists of hands-on engineering guidance and well-annotated design documents. Another appropriate concern is the transfer of a complete, turnkey centrifuge plant purchased outright from a foreign nation. North Korea has been known to provide this kind of comprehensive assistance for missile programs and is believed to have provided a complete nuclear reactor to Syria on such a basis. The same could easily be done with a centrifuge plant. Among all states currently possessing centrifuge technology, Iran and North Korea are the most probable suppliers of the technology, given their status as nonproliferation pariahs and their relative freedom from international political constraints.[19] The problem of state-to-state and black market transfers thus remains, but it is not the only path of concern.

A Larger History of Independent Development.

The alternative to buying or bartering for centrifuge technology is to develop it indigenously. In fact, most countries with centrifuges acquired them in this way. By studying these programs, we can learn about the resources required to build a proliferation-scale centrifuge capability from scratch and the potential for other states to do the same in the future.

The history of independent development goes back to the late-1950s. The basic Soviet centrifuge, from which all modern designs are derived, was perfected in 1953. Austrian and German scientists captured by the Soviet Army during World War II were used as a source of skilled labor. Starting with an unsuccessful American design, these German prisoners of war (POWs), in collaboration with Soviet scientists, were able to evolve a very successful machine. When the POWs were repatriated in 1956, they carried in their heads the basic principles of the successful design. In 1957, this information spread to three new countries. U.S. intelligence obtained Soviet design information through interviews with one of the POWs; West Germany hired two of the POWs to build centrifuges; and Dutch centrifuge designers met one of the POW engineers at a conference and learned of the basic design concepts in a long discussion. In 1958, the United States commissioned one of the POWs — Gernot Zippe — to come to the United States and replicate the Soviet machine.[20]

Until this point, the basics of modern gas-centrifuge design were not public knowledge. The knowledge was spreading slowly in the expert community, but there was no physical documentation of how the

Soviet centrifuge worked. Then between 1958 and 1960, the reports written by Zippe in fulfillment of his contract with the AEC were released to the public by the U.S. Government.[21] While the AEC considered its own centrifuge research secret at the time, doubts as to the potential of the Soviet design and the inconvenience of classifying reports written by a foreign national were sufficient for the AEC to ignore its own classification guidelines.

The publication of Zippe's reports appears to have fueled a rapid expansion in the number of centrifuge programs around the world. The United States did what it could to classify all further centrifuge research at home and abroad, but new programs nonetheless emerged in Israel (circa 1960), France (1960), China (1961), Australia (1965), Sweden (1971), Italy (1972), India (1972), Japan (1973), and Brazil (1979). Although nominally for peaceful purposes, many, if not most, of these programs were motivated by the understanding that centrifuges would give their countries a latent nuclear-weapons capability; and almost all of them developed the centrifuge using the reports released by the U.S. Government as the basis for their research. Along with this accidental proliferation, the United States also deliberately transferred the technology to the British government in 1960 and informally assisted the Israeli effort by allowing Israeli students to study centrifuge physics with U.S. centrifuge experts during the 1970s and 1980s.

Detailed histories are available for a number of independent programs. They reveal that the effort needed to build the basic, Soviet-style centrifuge is considerably smaller than the effort needed to build the more difficult designs that were provided by A. Q. Khan. The engineers in the early U.S. and British centrifuge programs, for example, had essentially no prior knowledge relevant to centrifuges and, unlike the scientists involved in the Manhattan Project, had only modest educations. Both programs started in 1960 and had access only to basic metalworking equipment, similar to what might be found today in a college machine shop. The technical staff never numbered more than 15 persons. Despite modest resources and the small effort, these programs were able to perfect a centrifuge design suitable for mass production in a little over a year (about 15 months). The Australian program is another interesting case. Notable because it is the slowest program of independent development on record, it took Australia almost 6 years to go from nothing to a working cascade of proliferation-relevant centrifuges. However, the program was also the smallest: It started with three and at no point exceeded six persons.

The record of centrifuge development for 20 historical cases is summarized in Figure 8-1. The average time taken to develop a basic centrifuge so it is ready for mass production across all historical programs with known dates is 25±11 months (about 1 to 3 years, in round terms). Note that these initiatives were mainly from the 1960s and 1970s. A present-day program could also benefit from more modern machine tools, vastly more numerous open-source publications about centrifuge design, desktop computers to aid in design and diagnostics, and the Internet to ease the sourcing of technical information.

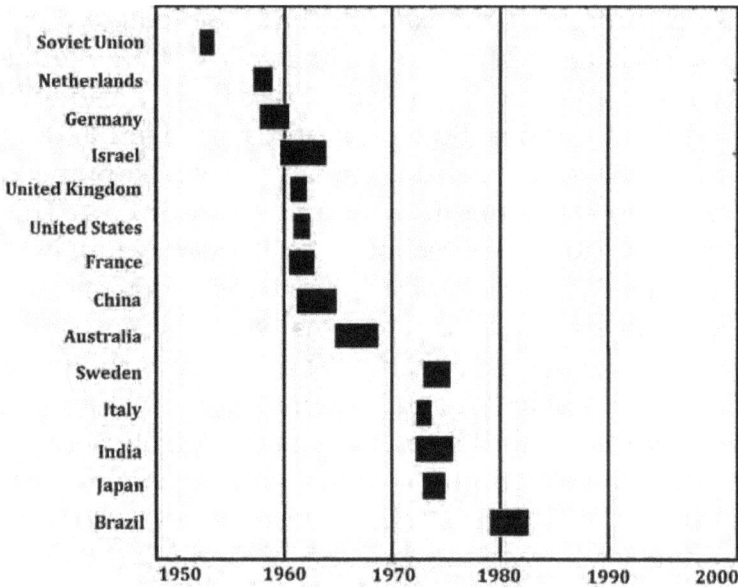

Figure 8-1. Time Required to Develop a
Centrifuge Capability Suitable for Weapons
Production by Programs Operating Free of
Foreign Assistance.

The mass production of centrifuges, along with the
operation of a centrifuge plant, is a larger but technically easier effort than the research and development
(R&D) phase. About 5,000 Soviet-type centrifuges are
needed to produce 25 kilograms (kg) of weapons-
grade (enriched to greater than 90 percent) uranium
per year, the approximate quantity needed for a first
generation implosion-type weapon, or one-half the
amount required for a primitive gun-type weapon. A
program of this scale would be consistent with many
historical weapons programs.[22] Mass production of
the basic Soviet-type centrifuge does not require specialized tooling or skilled labor. The British program,

240

for example, built its first pilot plant by hiring unskilled labor ("milkmen") to make centrifuge parts on
an assembly line. If such an assembly line were able
to produce 20 centrifuges per day, this would be sufficient to produce the 5,000 needed for a proliferation-
sized plant in 1 year. The effort might require 15 to 30
workers. Thus, the core staff sizes required for a basic
centrifuge program are small. A small cadre of half-a-
dozen suitably trained engineers and a slightly larger
force of unskilled but trainable laborers can probably
be organized in nearly any country. Building a centrifuge program may still be outside the capability
of loosely organized terrorist groups, but the task is
within the capability of a small engineering firm with
a few dozen people.

The cost of a program would also be modest. The
first German version of the Soviet-type centrifuge
was built by a firm named DEGUSSA in 1959 and had
about the same performance as Iran's IR-1 centrifuge
(possibly slightly better).[23] This centrifuge was offered
for sale for a small-batch cost of U.S.$235 per centri-
fuge, about U.S.$1,800 per centrifuge in 2012 currency.
Assuming the DEGUSSA price reflected the actual
cost of production, the centrifuge portion of the plant
might be built for less than U.S.$10 million in 2012
currency. The majority of the final costs of the plant
might actually be associated with noncentrifuge costs,
such as building costs, piping, and control systems.

Not only are centrifuges economical and relatively
straightforward to make, but the potential of export
and other technology controls to stop indigenous centrifuge programs is very limited. Under normal circumstances, export controls can only be applied to a
small subset of highly specialized materials and tools.
The simple Soviet-style centrifuge needs few, if any,

of these. The most controlled items are the materials for rotor tubes and the variable-frequency drives used to run the centrifuges. Centrifuge rotors need high strength, lightweight materials. They are export controlled when pre-formed into centrifuge rotor shapes. Unformed material, however, is not controlled, and most countries could acquire the simple tools needed to produce centrifuge rotors from the raw inputs. Even if the raw materials were themselves controlled, most states could probably produce a suitable material domestically. Variable-frequency drives designed for high-speed centrifuges are also controlled, but alternatives exist: The British program, for example, used hi-fi stereo amplifiers to drive its first centrifuges. The basic Soviet machines require no other export-controlled tools or materials.

Counterproliferation.

If it is not possible to prevent the acquisition of centrifuge technology with export controls, might it be possible to reverse proliferation programs using counterproliferation strategies such as diplomacy, interdiction, sabotage, military destruction of facilities, or regime change? The critical prerequisite for all of these strategies is that the proliferation program becomes known to the international community well before the successful acquisition of a weapon. Historically, most nuclear weapons programs have been detected at early stages, but most centrifuge programs have not. For example, Iraq's nuclear weapons ambitions were suspected from the early-1980s because of its visible pursuit of various fuel cycle technologies, but its centrifuge program was missed. The international community did not learn about it until after

242

the 1991 invasion, when Iraq disclosed its existence during an IAEA inspection.

The Soviet Union opened its first gas-centrifuge plant in 1957 and proceeded to build additional plants in the following years as it slowly replaced much of its older diffusion capability with centrifuges. Although the Soviet nuclear weapons program was the target of an intense intelligence effort, the United States routinely assessed that the Soviets did not have centrifuges. The Soviet capability was only learned of after the collapse of the Soviet Union, when Russia told the United States about the existence of these plants, some 34 years after the first large-scale plant opened.

China's centrifuge program started in 1961, 3 years prior to China's first nuclear weapons test. A U.S. national intelligence estimate released after China's 1964 test states with conviction that China did not have centrifuge technology, but internal histories from China's program indicate that the Chinese were working on second-generation centrifuges at the time. In the 1970s, U.S. intelligence received a human source tip-off that China had made weapons-grade uranium with centrifuges, but as the intelligence community had no way to validate this tip, it continued to assess that there was no credible evidence of a centrifuge plant in China.[24] Knowledge of China's centrifuge remained ambiguous for almost 2 decades.

The record of detection is somewhat better for states that were connected to the black market. Newspaper reports indicate that Iran's program must have been detected before January 1991, within 6 years of the program's inception, and may have been detected earlier. According to data published in the 2005 report of *The Commission on the Intelligence Capabilities of the United States Regarding Weapons of Mass Destruction,* however, Libya's program went undetected for approximately 16 years. The report also notes that a "disproportionately large volume" of U.S. intelligence was related to Libya's procurement activities—in other words, its dealings with the A. Q. Khan network—and that little or no information was known about Libya's internal activities. It appears that A. Q. Khan not only slowed down programs by providing incomplete assistance and problematic designs, but he also caused his customers' programs to be discovered because he was being watched. Similarly, IAEA officials have said it was German centrifuge engineer Gotthard Lerch, a member of the Khan network, who led investigators to South Africa's secret centrifuge program.

Independent centrifuge programs rarely have connections to known proliferators. The detection of these programs, therefore, would require either technical signatures that indicate the presence of a centrifuge plant or penetration of the country's political or technical leadership. Regarding technical signatures, centrifuge plants are very easy to hide. Nuclear reactors have visual and thermal signatures that can usually be seen using satellite reconnaissance, and they emit radioactive and chemical tracers that can be detected in the environment. Centrifuge plants have no such signatures, nor do they produce significant electromagnetic emanations that might reveal their existence at significant distances. This may help explain why Russia and China's programs went undetected for so long.

The value of human and signals intelligence is also in question when it comes to finding centrifuge plants. The U.S. intelligence community failed to detect Iraq's and Libya's centrifuge programs in early stages, despite the overt nuclear-weapons ambitions of these countries. North Korea was long suspected of having a centrifuge program because of visits by North Koreans to Pakistan and North Korea's purchase of several centrifuge-related materials, but it was not possible to confirm that this constituted anything more than curiosity. The existence of centrifuge research could not be validated. Even though North Korea has revealed that it does, indeed, have a centrifuge program, there are strong reasons to suspect that other facilities exist, but their locations are still unknown. These cases suggest that penetrating the inner workings of suspected proliferators is not always easily achieved, even with obvious targets. This in turn implies that the opportunities for trying counterproliferation mainly depend on states having connections to known proliferators, and that even when a state is a known proliferator, there may be sufficient ambiguity for it to be difficult to formulate a coherent counterproliferation strategy.

The Counterproliferation Track Record.

Should good fortune lead to detection, it still does not ensure that counterproliferation will work. The early detection of Pakistan's and Iran's program was used to stifle procurement efforts and delay the program, but it did not produce a reversal. Military attacks on nascent nuclear weapons programs have had mixed success. Thus far, Israel appears to have ended Syria's nuclear program when it bombed the Al-Kibar reactor in 2007. This reactor was detected because of its black market connections to North Korea, and the program remained dead after the attack because Syria lacked an indigenous capability to rebuild. Israel tried to do the same with Iraq's Osirak reactor in 1981, but in that case, the bombing appears to have strengthened Iraq's nuclear ambitions and drove the country to pursue the indigenous development of centrifuges, which were not detected. One might conclude that military intervention can work in a case such as Syria, where the state is heavily reliant on a foreign supplier and not comfortable with re-establishing the relationship, but a more fully indigenous capability is apt to be harder to eliminate unless the military attack also kills most of the engineers who retain the knowledge required to build a weapons program.

Diplomatic engagement is yet another way to counter a revealed nuclear weapons program. This has been most successful in cases where significant political pressure could be brought to bear, such as in the nuclear weapons programs of Taiwan and South Korea. However, pressure approaches appear mainly to have worked for states that depend on a strong nonproliferation advocate for their security. Pressure has not been successful, for example, in the cases of Pakistan and Iran.

Centrifuges complicate diplomatic efforts because the technology is a legitimate part of the peaceful nuclear-fuel cycle. This allows a proliferator to declare its capability as entirely peaceful and stare down criticisms, and so slow the rate at which the international community can apply political pressure to reverse the program. In principle, an incentives-based approach might work where negative pressure has not, but the incentives would need to outweigh the perceived benefit of having a nuclear weapon or nuclear-weapons option. In Libya's case, it was sufficient to provide major sanctions relief, minor cooperation in peaceful nuclear activities, and general rapprochement. In 1994, North Korea responded positively to incentives of massive foreign aid and negative security assurances, although the motivations behind North Korea's program were not fundamentally addressed, and the weapons program later re-emerged. The abandonment of a program might require more than simple bilateral incentives. Influential states may need to reshape the security environment of the proliferator to eliminate a security threat that is the essential motivation behind the nuclear weapons program. Such steps can be exceptionally difficult when the proliferator is seen as a violator of international agreements and norms, and thus not worthy of security assurances.

RECONSIDERING NONPROLIFERATION OPTIONS

Centrifuges, because of their technical qualities — including the ease with which they can be made, the difficulty of detecting them, and the lack of visible distinction between weapon oriented and peaceful facilities — are a challenge the like of which existing nonproliferation institutions have never known. A modification of the current approach, which relies heavily on technology controls and detection, is obviously necessary if the nonproliferation regime is to keep up with the changes brought about by the centrifuge.

Technology Controls.

Most nonproliferation institutions created since the signing of the Nuclear Nonproliferation Treaty (NPT) were designed to address state-to-state technology transfers, like those that backed the nuclear weapons programs of China, India, Brazil,[25] and Iraq. The newest of these institutions, such as UN Resolution 1540, the Proliferation Security Initiative, and various national export and financial regimes, respond specifically to black market technology transfers like those that backed the centrifuge programs of Pakistan, Iraq, Libya, North Korea, and Iran. Neither set of institutions attempts to address the problem of indigenous technology development.

However, history shows that there has been no lack of interest in indigenous technology development: Pakistan, Iraq, Iran, Libya, and South Africa all started indigenous centrifuge programs before receiving outside assistance. It was only because outside assistance was later obtained that these programs shifted their mode of operation to one of dependency, which resulted in opportunities for detection and counterproliferation. As nonproliferation champions work to eliminate black market agents, future proliferators might be more likely to stay the course with indigenous programs and, in doing so, keep their programs secret and increase their probability of proliferation success.

Unfortunately, export controls are unable to restrict access to the basic technologies needed for a Soviet-type centrifuge, the type that has formed the basis of nearly every indigenous centrifuge program in history. A more expansive set of export controls, called "catch-all" controls, prohibit the sale of any general-use item if there is a reason to believe it might assist a WMD program. Catch-all controls can restrict the technologies needed for simple Soviet-type centrifuge programs, but they will only be implemented if the program has been detected in advance, and history shows that indigenous centrifuge programs have a remarkable ability to stay secret for decades.[26] Catch-all controls also require that most states agree that the proliferator is seeking weapons because, without consensus on this point, the proliferator's access to international markets will not be effectively blocked. Building this consensus is difficult because centrifuge plants built for peaceful purposes have an inherent weapon-making capability.

Enrichment Regimes: Reducing the Motivations to Build Centrifuge Plants for Peaceful Purposes.

The situation previously described suggests that policymakers might more productively focus their attention on the underlying motivations for building a centrifuge capability. One possible motivation, apart from weapons, is the desire to use centrifuges to make fuel for a civilian power reactor. This entirely peaceful application is not itself problematic, but these peaceful plants give their possessors a weapons making capability that might be used later. More significantly, countries can build centrifuge plants claiming peaceful intent but actually harbor ulterior motivations to acquire a weapons or a weapon-making capability. This has led to a growing interest in restricting the legitimate use of centrifuges—although it is not yet clear whether this is a useful way to restrict proliferation, given the potential for small-scale indigenous production of centrifuges. Nevertheless, policymakers are now considering technology regimes that would establish legal or normative limits on the use of centrifuges for peaceful purposes.

In principle, states with nuclear power are justified in building a national centrifuge plant for the purposes of energy security (provided their reactors use enriched uranium as fuel; CANDU-type reactors do not require enriched fuel). Proposals have been made to suppress this justification by banning national plants and by creating alternative ways of guaranteeing the supply of reactor fuel. The most dramatic include international agreements to establish enrichment-free zones, such as the now-defunct agreement between North and South Korea. A widely used tactic is to implement so-called "flag rights" on raw uranium. Flag rights constitute a bilateral agreement between the buyer and seller of raw uranium. The buyer promises not to enrich it in a domestic facility without the prior approval of the seller. One can also imagine reverse-flag rights in which the buyer of a reactor promises not to put domestically enriched uranium fuel into it without prior consent. Softer proposals allow national enrichment but seek to render it less attractive for those with proliferation intent. These proposals include multilateral enrichment and fuel-supply guarantees, both of which would help ensure that all states have unfettered access to enrichment services, making it more difficult to justify the creation of a national enrichment plant.

In general, all of these enrichment regimes are conceptually similar to the normative regime that helped limit the spread of plutonium reprocessing plants in the 1970s and 1980s. However, these regimes are more formal and also may be more difficult to enforce. When a state chose not to build a reprocessing plant, it lost its ability to establish a large-scale weapons program easily, but it still retained an ability to build a quick-and-dirty reprocessing setup in the event of a crisis. Restrictions on enrichment would be more complete. Without a plant, there is no quick-and-dirty enrichment-based path to the bomb. At the same time, forgoing enrichment has greater consequences for civil nuclear power. Reprocessing of spent fuel essentially made no sense from an economic perspective and thus had arguably little to no role in a civil nuclear power program, whereas national enrichment clearly does. Thus, a ban on enrichment, while more complete in its ability to block some proliferation routes, may be more difficult to justify and sustain. Given both the more severe effect on proliferation potential and the consequences for national energy security, nations may be less willing to sign up to such a regime in the first place.

Multinational Enrichment Plants.

One of the most widely supported enrichment regimes is to require that all new (and possibly existing) enrichment plants be operated as part of a multinational consortium. Multinational ownership does not eliminate a country's ability to pursue centrifuge technology overtly, but it requires that other states be mutually invested in the plant's operation and, in principle, may reinforce the barriers to using it for weapons. It may still be technically possible for the government on whose territory the plant is built to take unilateral control of the facility and produce weapon quantities of HEU in just a few days. In this sense, the host state retains a quick-and-dirty option, and this may satisfy many states that see no immediate need for a weapon but nonetheless want a weapons capability as insurance. The political costs of taking over a multinational plant are likely to be higher than simply violating safeguards and using a national plant, but whether the additional costs are significant compared to the political costs of leaving or violating the NPT in the first place is debatable. If they are not, the multilateral arrangement has not increased the political barriers to proliferation.

Even if the barriers were not significantly enhanced, it could still be argued that a multinational arrangement would facilitate counterproliferation by legitimizing a military strike against the plant, or some equivalent forced shutdown, if it were being used to make weapons without international approval. Co-investors from the multinational consortium presumably would have standing to ask for, or to execute, an attack against the plant once it had been taken over.[27] Unfortunately, a multinational employee base would also provide a hostage opportunity that could be used to deter an attack on the facility, or defeat any autodestruct system.

Uncertainty about the ability of multinationals to increase the barriers to proliferation deserves to be taken seriously, because a multinational requirement might have unintended consequences that exacerbate the proliferation problem. For example, the legitimization of multinational plants might facilitate acquisition of a capability, or worse, overt centrifuge research that could be used to build a parallel centrifuge program in secret. The mantle of a multinational consortium might also help a state buy high-performance centrifuges from a commercial vendor, thereby enhancing its breakout capability beyond what would have been available in a go-it-alone approach. In other words, the world might wind up with more threshold states rather than fewer.

Fuel Banks, Guarantees, and Lifetime Supply Contracts.

Other types of technology regimes seek to deter the construction of enrichment facilities altogether or ban them by default except when meeting specific criteria. These include fuel banks and other guarantees of supply. These regimes will tend to be a more significant barrier to proliferation because states participating in them would not, in general, have a recognized reason to build centrifuge plants. On the other hand, the potential for these regimes to capture proliferation aspirants is in question.

One such proposal is to implement a legal or normative ban on national enrichment, balancing the loss of freedom with an improved assurance of supply backed by a fuel bank or other kind guarantee. The difficulty with these proposals is in their appeal. It is not clear if a state exists that would value a contrived fuel-security mechanism enough to trade the option of building a national plant for the benefit of that mechanism. The reason is that the marketplace traditionally has provided all the assurance that the state needs by default, and the additional benefit of the mechanism is seen to be marginal at best. States for which the market might not provide significant assurance because they are ultimately subject to some political manipulation probably are no more assured by an international fuel-supply assurance mechanism, which is also likely to be subject to political manipulation, perhaps even more so than the market. As such, the number of proliferation aspirants that the arrangement would capture successfully might prove to be vanishingly small.

Cabals, Flag Rights, and Agreements of Cooperation.

The no-reprocessing regime of the 1970s successfully stayed the completion of several reprocessing plants, yet the regime was completely voluntary. Participating states were explicit that they were not willing to give up their right to reprocess spent fuel.[28] In the enrichment case, states may be less likely to buy into the normative regime because the lack of economic incentives and the centrality of enrichment for fuel-supply security are too important to forgo on a voluntary basis. A more coercive approach might be needed to formalize the regime. Flag rights, in which suppliers of uranium or nuclear reactors ask recipients to forswear enrichment, are at the leading edge of the coercive approach. A stronger sort of coercion could be had if nuclear suppliers colluded to withhold all nuclear fuel and civil technology from states that are not otherwise willing to give up their unconstrained right to enrich. There is a collective-action problem in building this kind of supplier cabal, as there is a strong incentive for a supplier government to be the last holdout in the creation of the regime, and thereby benefit its domestic nuclear industry by providing access to a broader market. However, these problems are not insurmountable and have been overcome in the past, such as in the creation of the Nuclear Suppliers Group. One such approach has been outlined in the International Nuclear Fuel Cycle Association proposal and it, or a similar arrangement, may be worth pursuing.[29]

Problems with Technology Regimes.

All technology-control regimes aimed at stopping the spread of legitimate enrichment plants prevent states from building totally self-sufficient nuclear power programs, with the result that a state legitimately could reject the regime on the basis of economics and energy-security grounds alone.[30] If a large number of states do this up front, the regime has little chance of success. The coercive regimes resolve this problem by creating a difficult choice for client states, but the inequity in these regimes may begin to erode support for the broader NPT regime, which is already plagued by problems of inequity.

Even if a regime could be implemented, it is necessary to ask to what extent banning legitimate enrichment activities helps to prevent nuclear proliferation. In the reprocessing case, the normative ban prevented large-scale nuclear programs, but states retained an unattractive quick-and-dirty option. Use of that option still required the diversion of plutonium-bearing reactor fuel to a makeshift or clandestine reprocessing plant. The diversion of fuel would almost certainly be detected by safeguards, so any breakout attempt would come with the large political cost of violating the NPT regime. Thus, proliferation via the plutonium route was attractive only in an emergency situation, and even then only feasible on a small scale. By contrast, the historical record suggests that indigenous centrifuge programs can be built and kept secret for years, even decades at a time. Secrecy has even been effective in countries like Iraq and Libya that were known to have nuclear-weapons ambitions and presumably were under intense scrutiny. The technology regime would compound the political costs associated with detection, but the probability of detection is low. One must ask if the extra political risk of getting caught violating the technology regime — computed as the probability of detection multiplied by the political cost — is substantial relative to the existing political cost of overtly violating the NPT with a national facility. If the extra risk is small relative to an NPT violation, then the technology-control regime has not added much. Furthermore, this gain then needs to be weighed against the potential negative consequences of legitimizing proliferation of fuel-cycle facilities or exacerbating the already contentious inequity in the NPT regime. It is obviously impossible to quantify these effects and make an actual computation of the relative costs and benefits, but this article has argued that detection rates are probably small and that support for a technology-control regime will be tepid given the legitimate purpose of centrifuges. Both of these arguments lead to the conclusion that the benefit of attempting to control states' legitimate access to centrifuge technology is probably small, and may even be negative.

REDUCING THE MOTIVATION TO BUILD NUCLEAR WEAPONS

Instead of focusing on export controls that limit access to technology, which this article has argued are inoperative in the case of indigenous centrifuge development, or the motivations behind building legitimate facilities, which appear to be of uncertain value, it may be better to direct nonproliferation efforts to reducing the motivation for nuclear weapons.

Classically, it is argued that states tend to pursue nuclear weapons because they feel a security threat justifies the need or because nuclear weapons are seen as a symbol of great-power status or as tools of coercion.[31] Nonproliferation in the centrifuge age may require that nonproliferation advocates better prepare themselves to address these motivations face on. More attention may need to be given to the security situation of states that feel their existence is threatened by more powerful states and which thus seek nuclear weapons as a kind of existential guarantee. Nuclear-armed states that are not at peace with their neighbors (e.g., Israel) may need to reconsider the value of their own nuclear armaments if they prefer to maintain a conventional rather than nuclear standoff. Established nuclear powers may need to accelerate progress towards the reduction, and ultimately the complete elimination, of their arsenals if they are to deny weapons the symbol of great-power status. Finally, champions of the nonproliferation regime may need to be prepared to offer security guarantees of various sorts when a potential proliferator emerges on the international stage. All these require major changes in the way states conduct their foreign policy. They are unlikely to happen easily, but they are increasingly important in an age when nearly any state can make a proliferation-scale centrifuge program covertly, using only indigenous resources.

ENDNOTES - CHAPTER 8

1. George W. Bush, "President Announces New Measures to Counter the Threat of WMD," Speech at National Defense University, Washington, DC, February 14, 2004, available from *www.fas.org/irp/news/2004/02/wh021104.html*.

2. Mohamed ElBaradei, "Nobel Lecture," Lecture at the Nobel Peace Prize Ceremony, Oslo, Norway, December 10, 2005.

3. Houston G. Wood, Alexander Glaser, and R. Scott Kemp, "The Gas Centrifuge and Nuclear-Weapon Proliferation," *Physics Today*, Vol. 61, No. 9, September 2008.

4. Chaim Braun and Christopher F. Chyba, "Proliferation Rings: New Challenges to the Nuclear Nonproliferation Regime," *International Security*, Vol. 29, No. 2, October 2004, pp. 5–49.

5. For example: Alexander Montgomery, "Ringing in Proliferation: How to Dismantle an Atomic Bomb Network," *International Security*, Vol. 30, No. 2, 2005, pp. 153-187; David Albright, *Peddling Peril: How the Secret Nuclear Trade Arms America's Enemies*, 1st Ed. New York: Free Press, 2010, Chap. 12.

6. R. Scott Kemp, "Nonproliferation Strategy in the Centrifuge Age," Ph.D. Dissertation, Princeton University, Princeton, NJ, 2010.

7. Siegfried S. Hecker, "What I Found in Yongbyon and Why It Matters," *APS News*, March 2011.

8. *Implementation of the NPT Safeguards Agreement in the Syrian Arab Republic*, GOV/2011/30, Vienna, Austria: International Atomic Energy Agency (IAEA), May 24, 2011.

9. Frank von Hippel, *Managing Spent Fuel in the United States: The Illogic of Reprocessing*, Research Report No. 3, International Panel on Fissile Materials, January 2007.

10. The technology, deployed on a large scale in the Soviet Union in 1957, was not perfected in the West until 1960. Information sufficient to start a well-directed centrifuge plant was published in Gernot Zippe, *The Development of Short Bowl Ultracentrifuges*, Progress Report No. ORO-216, Charlottesville, VA: Division of Engineering Physics, Research Laboratories for the Engineering Sciences, University of Virginia, November 6, 1959.

11. For the history of the effort to commercialize the centrifuge, see R. B. Kehoe, *The Enriching Troika: A History of Urenco to the Year 2000*, Marlow, UK: Urenco Limited, 2002.

12. This was not immediately evident, given the difficulties experienced by the Manhattan Project of enriching uranium to high levels with gaseous diffusion. A U.S. research program showed it to be trivial circa 1963; see Union Carbide Nuclear Company *Proposal for the Development of the Gas Centrifuge Process of Isotope Separation*. KA-621, Papers of Ralph A. Lowry, Oak Ridge National Laboratory, Oak Ridge, TN: Union Carbide Nuclear Company, July 11, 1960.

13. U.S. Atomic Energy Commission, *Major Activities in the Atomic Energy Programs, January–December 1960* Washington DC: U.S. Government Printing Office, 1961, Appendix 20, p. 500.

14. Australia, Brazil, China, France, Germany, India, Iran, Iraq, Israel, Italy, Japan, Libya, the Netherlands, North Korea,

Pakistan, South Africa, the Soviet Union, Sweden, the United Kingdom (UK), and the United States.

15. For the most part, this change was due to general progress in mechanical and aerospace engineering. Of the documents originating from centrifuge programs, perhaps the most important were Zippe, *The Development of Short Bowl Ultracentrifuges*, final report No. ORO-315, Charlottesville, VA: Research Laboratories for the Engineering Sciences, University of Virginia, July 1960; S. Whitley, "The Uranium Ultracentrifuge," *Physics in Technology*, Vol. 10, No. 1. January 1979. pp. 26–33; S. Whitley, "Review of the Gas Centrifuge Until 1962," *Reviews of Modern Physics*, Vol. 56, No. 1, January 1984, pp. 41–97; H. G. Wood and J. B. Morton, "Onsager's Pancake Approximation for the Fluid Dynamics of a Gas Centrifuge," *Journal of Fluid Mechanics*, Vol. 101, 1980 pp. 1–31; F. Doneddu, P. Roblin, and H. G. Wood, "Optimization Studies for Gas Centrifuges," *Separation Science and Technology*, Vol. 35, No. 8, 2000, pp. 1207–1221.

16. Kurt M. Campbell, Robert J. Einhorn, and Mitchell B. Reiss, *The Nuclear Tipping Point: Why States Reconsider Their Nuclear Choices,*. Washington, DC: Brookings Institution Press, 2004, p. 339.

17. South Africa claims to have had an early centrifuge program, but no information has been published in the public domain to back this claim. However, South Africa did have a successful uranium-enrichment program based on a different technology, called the vortex or 'stationary-wall centrifuge' process. This process was used to make South Africa's HEU-based nuclear weapons.

18. These German engineers were at times also associated with Khan, but they also operated separately from Khan, providing better assistance. Libya is an outlier here. Its program was never successful, either prior to or after A. Q. Khan's assistance, largely because the country could not organize a stable development program.

19. Sheena Chestnut, "Illicit Activity and Proliferation: North Korean Smuggling Networks," *International Security*, Vol. 32, No. 1, July 2007, pp. 80–111.

20. Gernot Zippe, "Unclassified Spots on History of Modern Gascentrifuges," *Workshop on Gases in Strong Rotation*, 1983.

21. Gernot Zippe, Jesse W. Beams, and A. Robert Kuhlthau, *The Development of Short Bowl Ultracentrifuges*, Progress Report No. ORO-210, Charlottesville, VA: Ordnance Research Laboratory, University of Virginia, December 1, 1958; Gernot Zippe, *The Development of Short Bowl Ultracentrifuges*, Progress Report No. ORO-202, Charlottesville, VA: Ordnance Research Laboratory, University of Virginia, July 1, 1959; Zippe, *The Development of Short Bowl Ultracentrifuges*, Progress Report No. ORO-216; Zippe, *The Development of Short Bowl Ultracentrifuges*, Progress Report No. ORO-315. On the spread of the report, see Kemp, "Nonproliferation Strategy in the Centrifuge Age."

22. Many countries start with programs of this scale: It is estimated that Pakistan started with a capability of one bomb every 2 years, around 1983. North Korea, during its plutonium program, maintained a capability of less than one weapon per year. Iraq's nuclear weapons program, though interrupted by war, had plans to build about one bomb every 2 and 1/2 years. South Africa's nuclear weapons program produced at a rate less than one bomb every 2 years.

23. The IR-1 was supposed to operate at about 2.5 kg-Separative Work Unit (SWU)/year, but in reality, it operates at only 0.6-1.0 kg-SWU/year. The DEGUSSA machine was claimed to have had a performance of about 1.02 kg-SWU/year. There was an error in the calculation of this figure. The actual performance was probably closer to 0.9 kg-SWU/year. See Zippe, *The Development of Short Bowl Ultracentrifuges*, p. 87.

24. U.S. Central Intelligence Agency, "Nuclear Energy," *Weekly Surveyor*, January 12, 1970.

25. According to Myron Kratzer (then Science and Technology counselor and in 1975, as Senior Deputy Assistant Secretary for Nuclear Energy at the U.S. Department of State), Brazil's program was the major impetus behind the Nuclear Suppliers Group. The Indian nuclear test of 1974 served mainly to finalize an already active discussion. See Myron Kratzer, interview with the author, April 19, 2011.

26. Catch-all controls, while useful in principle, are highly imperfect. They are more easily bypassed through the creation of front companies intended to deceive exporters of the true end use. The detection of front companies is an intelligence function and is also highly imperfect. Compliance is also problematic because manufactures of normally uncontrolled items may not be aware that they needed an export license for particular buyers. Furthermore, not all states implement catch-all controls.

27. Such a right could be further codified in the original terms of cooperation. However, legal barriers have not deterred some states, notably Israel, from attacking facilities that belonged wholly to Iraq and Syria. Still, other states than these may feel more apprehension about military attacks.

28. On the opinions of states, see the proceedings of the International Fuel Cycle Evaluation (INFCE), October 19–21, 1977, Washington DC, as published by the IAEA. (I assume by proceedings, he means part or all of the 8 volumes plus 1 summary volume published by the IAEA in 1980, which covered all 134 INFCE working group sessions from 1977-79 and the final plenary conference held in Vienna, Austria, in February 1980.)

29. See Christopher E. Paine and Thomas B. Cochran, "Nuclear Islands: International Leasing of Nuclear Fuel Cycle Sites to Provide Enduring Assurance of Peaceful Uses," *The Nonproliferation Review*, Vol. 17, No. 3, July 2010.

30. Even if a national centrifuge program were not competitive with the international market, a small centrifuge program might still be reasonable and economic insurance against a possible fuel-supply cutoff. To demonstrate this, consider that estimates for the real levelized cost of nuclear power per kilowatt-hour (kWh) can range from $0.09 in optimistic, forward-looking case studies to values in excess of $0.20/kWh (2008 dollars) for first-of-a-kind construction in regulation-heavy regions. Assume, for the sake of calculation, a median value of $0.15/kWh. Nearly all of that cost comes from the capital charge and staffing costs; less than $0.01 is the cost of fuel. At 85 percent capacity factor, the plant operates for about 7,500 hours/year. Thus, the real levelized cost for a nonoperating full-size 1 gigawatt/year (GWe/yr) plant

is about $1.1 billion/year, approximately the cost of a modern centrifuge plant able to support approximately 30 GWe of nuclear power annually. For a primer on the economics of nuclear power, including estimates of the real levelized cost, see Massachusetts Institute of Technology, *The Future of Nuclear Power*, July 2003. The estimated cost of a modern enrichment plant is based on the projected cost of the Urenco/LES plant in Eunice, NM; see Michael Knapik, "LES Hopes for Fresh Start in New Mexico," *Nuclear Fuel*, September 1, 2003.

31. Another common explanation is that nuclear weapons can be used to satisfy domestic constituencies, but presumably, constituencies that are large and well supported want nuclear weapons for one of the listed reasons.

PART III:

HOW WELL CAN WE SAFEGUARD
THE PEACEFUL ATOM?

CHAPTER 9

HOW WELL WILL THE INTERNATIONAL ATOMIC ENERGY AGENCY BE ABLE TO SAFEGUARD MORE NUCLEAR MATERIALS IN MORE STATES?

Patrick S. Roberts

The International Atomic Energy Agency (IAEA) will confront new proliferation risks if its safeguards system must operate in a world with more nuclear facilities in more and riskier places. The usual suggestions for upgrading IAEA safeguards focus on increasing the agency's resources and improving technology. Yet improved technology and more resources for inspections will not help unless the agency can develop standards to gauge the strength of the safeguards system, and unless countries confront the problem of what to do after the IAEA detects a violation.

Nuclear power will likely spread to new countries and new kinds of facilities in the coming years. Even a modest expansion of nuclear power will require more safeguards inspections, which at the very least raises budgeting problems for the agency. Even if money were no obstacle, however, it is not clear that the agency could simply scale up its operations to meet new demands. Scaling up is more complicated than making successful models from the past even larger. Consider one famous example, the RMS *Titanic*. The largest passenger steamship in the world was considered to be the pinnacle of engineering until it sank after striking an iceberg on its maiden voyage.

The IAEA may not be the *Titanic*, but the metaphor raises the question of how one would know whether the agency is headed for a disaster. After all, the Titanic's engineers never expected their creation to become synonymous with hubris and disaster. To evaluate whether IAEA safeguards can function without a breakdown while monitoring and inspecting more kinds of nuclear facilities in more countries, the agency needs clear and transparent standards. The IAEA has used timeliness detection goals, which are based on calculations about whether the agency can detect the diversion of a significant quantity of nuclear material within the minimum time needed to make a bomb.[1] Timeliness detection goals have the advantage of being clear, but they involve calculations about hundreds of facilities that are very low risk, and the agency does not have the resources to meet these goals for all facilities. As a result, the IAEA is moving toward state-based declarations to evaluate safeguards risk and performance, but these standards are still being developed, and it will be a challenge to apply them to diverse countries in a fair and equitable way.

So far, there is scant evidence for the agency actually having prevented diversion in a timely manner. In the four most prominent cases in recent memory of illicit nuclear activity, the IAEA appeared to formally meet its timeliness detection goals while countries pursued illegal nuclear activities. The agency detected violations in only one of these cases, North Korea. The agency's safeguards division sees promising new tools in new technologies, training, and legal authority in the Additional Protocol (AP), but adopting these tools is sometimes slowed by goal conflict within the agency. The IAEA exists to provide technical assistance in developing nuclear power, prevent military diversions, and enhance safety, but it is not clear which of these should take priority.

Even if the agency is able to improve its detection capabilities through better technology, training, and standards, the safeguards system cannot be relied upon to react quickly enough to a diversion of nuclear materials toward a military program. Finally, there is no consensus on how to handle countries that may be in violation of IAEA and Non-Proliferation Treaty (NPT) agreements. To fulfill the goals of nonproliferation, the IAEA must be able to detect violators, and the world political community, along with the agency, must be able to enact a sufficient penalty for violation.

If the agency does not change what it and the United Nations (UN) Security Council will do after a violation is discovered, then the system of preventing the proliferation of illicit nuclear materials is at risk of collapse. Iran could withdrawal from the NPT without penalty, and other states, including Syria, could continue to deny the IAEA access to suspect locations. Meanwhile, the IAEA's credibility is at risk because it is expected to verify agreements in countries that it can at best only monitor but not truly safeguard.

INCREASING DEMAND FOR NUCLEAR POWER IN PROLIFERATION-RISKY REGIONS

Developing economies demand new energy sources, while North America and Europe are showing a greater resistance to the costs and potential consequences of nuclear power. Therefore, new nuclear reactors will likely be built in regions where the risks of proliferation are the highest.

While there is great uncertainty surrounding the price of energy in the future, it is clear that demand for energy will grow in the coming decades because of modernization and population growth. The International Energy Agency, the chief international organization for monitoring energy demands, predicted growth in energy demand in each of its three scenarios from 2008 to 2035. Growth in demand is the only reliable prediction, however. The executive summary for the 2012 *World Energy Outlook* begins with the statement, "The energy world faces unprecedented uncertainty."[2]

Investors do not know the price of fuels in the future or the precise risk of delays in plant construction. Nuclear power requires a high initial investment in the facility and in supporting technical staff, but long-term operational costs are relatively low. (From the perspective of investors, however, it can be difficult to disaggregate initial investments from long-term ones.) New nuclear plant construction often faces unpredictable delays and cost overruns that would decrease returns on investment to equity investors.[3] Furthermore, there are limits to the world's capacity to increase nuclear plant production and technical nuclear training; for example Japan Steel Works is the sole maker of certain reactor parts.[4]

Despite the uncertainty, more than 45 countries are considering embarking upon nuclear power programs, according to a May 2012 report from the World Nuclear Association.[5] It is unlikely that all of these countries will develop nuclear power soon, but some of them likely will. After Fukushima, Japan, nuclear power is likely to hold steady or slightly recede in Europe and the United States, although nuclear power plant construction has been part of climate change and energy legislation discussions in the United States.[6] There are 443 nuclear plants in operation around the world, and 64 new plants are under construction. Demand, however, appears to be shifting from Europe and North

America to Asia and the Middle East, where many countries have expressed interest in building nuclear plants to meet their energy needs.[7] More than half of the reactors currently under construction are in Asia, one-fourth in Eastern Europe, and some in the Middle East. Many of these countries are investigating new kinds of nuclear plants that pose safeguards challenges. IAEA Director General Yukiya Amano expects "more than 20 new states, including many developing countries, [to] bring their first nuclear power plants online within 2 decades."[8]

Fast-paced growth in emerging economies, led by Brazil, China, India, and Indonesia, will drive the bulk of demand.[9] India remains committed to nuclear power even after the Fukushima disaster. "We are determined that our expanded nuclear power programme will follow the highest standards of nuclear safety and security," Indian Prime Minister Manmohan Singh said at a summit on nuclear security in 2012.[10]

Because investment in nuclear power provides such an unpredictable return on investment, it is likely that plants will be built in countries whose political leadership has decided that nuclear power is a strategic goal rather than an economic investment. In 2009, Citigroup concluded that:

> . . . it is extremely unlikely that private sector developers will be willing or able to take on the Construction, Power Price, and Operational risks of new nuclear stations. The returns would need to be underpinned by the government and the risks shared with the taxpayer / consumer. Minimum power prices (perhaps through capacity payments), support for financing, and government backed off-take agreements may all be needed to make new nuclear viable.[11]

The uncertainty in nuclear power investment and the high fixed costs make it more likely that certain kinds of countries will find it worthwhile to invest in nuclear power: large countries with ample resources; countries looking to develop power for political prestige; or countries that factor military uses into the cost of nuclear power investment. It is also more likely that countries that do not factor the high and uncertain cost of waste disposal into the investment calculation will pursue nuclear power more than those that do factor in disposal. All of these considerations make it more likely that countries attentive to military considerations and located in regions at risk of proliferation will invest in nuclear power, rather than countries in regions that are less at risk for proliferation, or countries that rely more on market pricing for energy.

WILL MORE INSPECTIONS BE NEEDED?

If there will be an expansion of nuclear power into new countries and new types of facilities, the IAEA will need to perform more inspections. In 2010, the IAEA carried out 1,750 inspections, 423 design information verifications, and 142 complementary accesses.[12] By 2030, the IAEA expects global nuclear electrical generating capacity to grow from between 40 and 120 percent. It also expects between 10 and 25 new countries to attain nuclear power.[13] Facilities in these countries, or expanded facilities in countries already with nuclear power, will require more inspections. This, at the very least, raises budgeting questions because IAEA member states are reluctant to approve budget increases. Former Deputy Director General of the IAEA and head of the Department of Safeguards Olli Heinonen concluded in 2010 that, "We must do more with less without compromising the necessary safeguards assurances. Smarter and better verification techniques and technologies should be explored."[14]

In 2012, the IAEA had 1,125 facilities under safeguards, and with the expansion of nuclear activity into new states, it may have more. New facilities require more inspections than existing ones where the agency can rely more on accounting procedures for verification. New facilities will require more work because the agency must ensure that they are built according to standards, and it must work with states to develop a method for making initial and then annual declarations of nuclear material. The agency's budgetary growth is limited, however, by UN policies to maintain zero real growth in budgets.[15] An expansion of nuclear power around the world will provide a bigger nuclear haystack of sites to inspect but no more than the current number of inspections given budgetary constraints.

From 1987 to 2010, the amount of significant quantities under IAEA safeguards increased by five times, but the number of inspection days remained roughly constant.[16] A significant quantity is a standardized measure of nuclear material defined as "the approximate amount of nuclear material for which the possibility of manufacturing a nuclear explosive device cannot be excluded."[17] Inspection days refer to the number of days inspectors worked. Therefore, inspectors inspect more material than ever before using the same amount of human resources.

The IAEA's Safeguards Division, one of several such divisions, carries out a number of tasks meant to provide member states with a reasonable assurance that states are accounting for the nuclear material in their possession and that they are not diverting material or facilities for illicit use. The agency's safeguards are based on assessments of the correctness and completeness of a state's declarations of nuclear material and nuclear-related activities. Verification measures include on-site inspections, visits, and ongoing monitoring and evaluation. Basically, two sets of measures are carried out in accordance with the type of safeguards agreements in force with a state. First, the agency verifies state-issued reports of declared nuclear materials. Nuclear accountants check the record books to monitor consistency, and inspectors perform "material accountancy" by physically installing and observing tamper-proof seals on storage vessels and cameras in sensitive areas. These inspections verify nondiversion, essentially keeping tabs on nuclear material that states use for nuclear power to make sure that none goes missing by being diverted for weapons uses. Second, as of 2012, 115 of 181 countries with IAEA safeguards agreements concluded an agreement with the IAEA known as the Additional Protocol (AP).[18] (Another 23 have signed the protocol but not brought it into effect). This is a legal agreement that permits the agency to perform more intensive supervision to not only verify nondiversion, but also to provide some evidence about the absence of undeclared nuclear material in a state.

Inspections come in several types. For new states and states making changes in their programs, the agency conducts **ad hoc inspections** to verify initial reports of nuclear material or report on changes to the report. Beyond that, the agency conducts **routine inspections** according to a predefined schedule and, occasionally, with very short notice. In either case, the agency technically enters with the permission of the state and, in some cases, there are delays as states attempt to negotiate arrival and travel schedules with their hosts. The agency has the right to conduct routine inspections, but without the Additional Protocol or other agreements in place, it is limited to locations with a declared nuclear facility or declared sites that handle nuclear material. (There is a possibility that a state has nuclear material in undeclared or secret sites about which the agency could gather information but not perform on-site inspections). The agency has the authority to conduct **special inspections** if it believes that the information provided by the state is not adequate for the agency to fulfill its safeguards responsibilities, but this power has rarely been used. Member states are uncomfortable with the agency entering a state without the state's permission because such inspections would appear to infringe upon national sovereignty. In addition, the agency conducts **safeguards visits** to verify that the design, construction, and decommissioning of facilities are conducted according to standards and the information contained in countries' official reports.

ARE SAFEGUARDS INFINITELY EXPANDABLE?

The IAEA has developed a range of safeguards tools, from accounting procedures that date back to the agency's earliest days, to basic tools such as tamper-proof seals and cameras, to more advanced methods for environmental sampling to measure traces of nuclear materials. These tools may provide some efficiency in performing safeguards in new places, particularly environmental sampling, which can be conducted at a distance. Most likely, however, each of these tools and each type of inspection will have to be repeated in new countries and new facilities with the expansion of nuclear power.

This increase raises a question: Are IAEA safeguards infinitely expandable, or do they reach a breaking point at some level of capacity? It is not clear how one would know whether safeguards might reach a breaking point. Some systems are relatively easily scaled up. They can operate at higher levels of capacity, using a larger version of the same structure. Among computer programs, contemporary peer-to-peer fire sharing systems without a central node are easily scaled up because increasing the number of peers does not increase demand on the system beyond the overall carrying capacity.[19] (This is an improvement over the first online peer-to-peer music file sharing systems that fell victim to severe bottlenecks).[20]

In other cases, scaling up is much more complicated than making successful models even larger. In one famous example, the RMS *Titanic*, its scaled-up design failed for a number of reasons, including a lack of sufficient lifeboats. After the accident, new regulations required that there be enough lifeboats on board to carry everyone on the ship. Ocean liners kept communication systems open 24 hours a day (since no one heard the *Titanic*'s distress call), and ships were much more careful about spotting and avoiding icebergs.

Though the *Titanic* became a paradigm for the hubris of engineers, it was not destined to fail on its maiden voyage. It was a victim of a faulty design **and** bad luck. Had the ship not struck an iceberg and sank, people would have concluded that the celebrated scaled-up ocean liner design was a success. Shipbuilders would have built more ships modeled after it, and likely even larger ships.[21] Designers might have added even more luxuries and reduced the number of lifeboats. Scaling up could have continued for years, maybe decades, until there was a catastrophic failure.

How do we know that the IAEA's safeguards system is not ripe for catastrophic failure? Safeguards could be like a *Titanic* that has been lucky enough not to strike an iceberg.

THE PROBLEM OF STANDARDS

To evaluate whether IAEA safeguards can function adequately with more kinds of nuclear facilities in more countries — or even whether they function adequately now — the agency needs clear and transparent standards. Standards would clarify how well the agency is meeting the obligations found in its statute, which came into force on July 29, 1957, and has been subsequently amended. Its safeguards obligations in particular have been interpreted with almost ecclesiastical attention to nuance. The IAEA's safeguards mission is to verify, "Through its inspection system that States comply with their commitments, under the Non-Proliferation Treaty and other non-proliferation agreements, to use nuclear material and facilities only for peaceful purposes." Until the discovery that Iraq had a clandestine weapons program, however, the statement was interpreted to read, "to use **declared** nuclear material and facilities. . . ."[22] In other words, the agency read its mandate as to verify only the facilities that member states agreed to have inspected, leaving out other research, industrial, and military sites.[23] The agency took seriously the idea that member states that consented to agency inspections might attempt to conceal facilities only after the discovery of the Iraq program. The agency's mission in a post-Iraq world is more complicated if it cannot assume that countries that consent to inspections are acting in good faith and are attempting to fulfill the obligations of the NPT.

275

One way to measure how the agency is doing would be to chart its progress relative to clear and transparent standards. The agency has developed quantitative measures of performance such as "timeliness detection" and "risk of early detection," but it is difficult to find data on these. Even how much money the agency spends on particular country inspections remains inaccessible to anyone who is not agency staff or a qualified member state representative. Timeliness detection goals refer to measures specific to facility and material types of how quickly inspectors should be able to identify that a diversion has occurred, and these goals are used to establish the frequency of inspections.[24] Instead of evaluating and measuring primarily at the facility level, the IAEA is now moving toward a system in which it evaluates state declarations for their consistency and thoroughness. At present, state criteria remain rudimentary, and measuring state criteria rather than facilities risks making verification an even more political issue than it is presently. If the agency can develop relatively neutral criteria for evaluating state declaration, however, the state-level criteria could prove useful, especially when combined with other measures such as timeliness detection.

Standards have their virtues. It is difficult to gauge progress and improvement without measureable standards to show where performance is lacking. It is not clear, however, that the standards for IAEA safeguards are accurate performance measures. Unlike the *Titanic* example, what counts as a failure for the IAEA is ambiguous. A nuclear explosion or nuclear war is as much or more a political failure than an administrative one that can be blamed on the IAEA.

However useful the data gathered during safeguards evaluations, their interpretations are not self-evident. The agency performs careful analysis of state declarations about nuclear material possession and transfers, and sometimes the agency finds that these reports are incomplete. But state-based reports alone do not provide clear standards because their completeness and accuracy depends on factors other than whether a state is engaging in illegal activities. For example, rich countries with well-developed bureaucracies are better able to produce complete reports, as are countries that have produced such reports in the past and merely make routine updates. Countries with less developed bureaucracies may produce incomplete reports out of inexperience and lower capacity, not an intent to deceive.

Timeliness detection goals do provide a clear standard for having met a goal. But it is not clear that these are used as standards any longer, since the agency is moving toward using state-based declarations. Furthermore, the agency's move away from facility-based timeliness detection goals was in part a result of problems with those goals. It was not clear how accurately the goals were being measured, and even so, the agency rarely met the goals and lacked the resources to perform significantly more or longer inspections.[25]

Safeguards are designed to detect the diversion of a significant quantity of nuclear material within a conversion time, which is the minimum time needed to build a bomb using diverted materials. Timeliness detection goals and inspection schedules are created according to the relevant conversion time. (The conversion time is defined as 7-10 days for plutonium or highly enriched uranium [HEU], 1-3 months for plutonium in spent fuel, and about 1 year for low enriched uranium [LEU]). The presence of safeguards and timeliness detection goals suggests that the IAEA aims to detect military diversions before they result in bombs, and even to prevent an attempt at diversion from occurring undetected.

The agency's performance in preventing diversion in a timely manner is mixed. In the four most prominent cases in recent memory of illicit nuclear activity, the IAEA appeared to formally meet its timeliness detection goals (though detailed such evaluations are not available to the public).[26] Iraq, North Korea, Iran, and Libya all engaged in illegal nuclear activities, and it is generally agreed that all embarked on some stage of an illegal nuclear program. Nevertheless, the IAEA detected violations in only one case, and even there, the evidence is mixed. In Iraq, the United States discovered the country's nuclear program just before the first Gulf War. After the war began and the agency gained access, IAEA inspectors learned that Iraq secretly enriched uranium and carried out reprocessing experiments in buildings at Tuwaitha, which is not covered by IAEA inspections agreements.

In Iran, an opposition group, The National Council of Resistance of Iran, provided initial evidence of illicit activities there. In Libya, the Libyan government announced its nuclear weapons program, though some government intelligence agencies may have known about it previously. In North Korea, however, IAEA inspectors operating under additional legal authority did uncover diversions and illicit activity, and the inspections were supported by U.S. satellite imagery.[27] The IAEA has discovered other cases of materials unaccounted for and various discrepancies in nuclear accounting, but there are no other known cases of the IAEA detecting diversions of nuclear material to an illicit nuclear program. The recent history of the agency's safeguards program leads to a disturbing conclusion. The agency's traditional safeguards of regular inspections and material accountancy failed to detect and deter all known cases of member states' illicit nuclear programs. The North Korean case, however, shows that when given additional legal mandates, technical support, and political support, the agency can verify nuclear programs and make discoveries that qualify as meeting a performance goal — detecting illicit nuclear activity.

Many of the proponents of reform for the IAEA advocate increasing the number of safeguards agreements the IAEA has with member states and expanding the legal and technical tools available to the agency by increasing the number of states covered by the AP. Former Deputy Director General of the IAEA and Head of the Department of Safeguards Pierre Goldschmidt writes that the "Department of Safeguards doesn't have the legal authority it needs to fulfill its mandate and to provide the assurances the international community is expecting from its verification activities."[28] Goldschmidt advocates giving the IAEA greater authority and better technology so that it might more quickly detect violations and with greater certainty.

AFTER DETECTION, WHAT?

Standards can help gauge the agency's performance, but the real problems for dealing with more facilities in more countries and countries at higher risk of proliferation are structural. The problem with current standards is that the IAEA inspection system cannot be relied upon to react quickly enough to a diversion even once the agency detects a violation. To quote a classic article on arms control and disarmament, "After detection, what?"[29]

Much of the debate over safeguards focuses on how to detect violations. The agency developed verification procedures and measures for effectiveness such as timeliness detection goals, and it is investing in promising new detection measures such as environmental and remote sampling. New investments may increase the agency's ability to detect violations and even to identify nuclear material in unauthorized locations. Technical questions about how to better detect violations dominate discussion about safeguards, as the conference agenda of a recent meeting in Vienna, Austria, shows.[30] Debates over how to improve safeguards usually result in requests for larger budgets, more and better training for inspectors and analysts, and better technology.[31]

The IAEA presents the effort to detect violators as its central safeguards mission, and in public, the agency portrays more effective inspections as the key to detection. In 2002, IAEA Director General Mohamed El Baradei said:

> Inspections by an impartial, credible third party have been a cornerstone of international nuclear arms control agreements for decades. Where the intent exists to develop a clandestine nuclear weapons programme, inspections serve effectively as a means of both detection and deterrence.[32]

Detecting violations is not enough to achieve deterrence, however. Just as important are the consequences of a violation once it has been detected. While it is important that the IAEA be technically capable of detecting a violation, the agency's Board of Governors, the UN, and world governments must be able to react quickly and effectively once a violation has been discovered. Focusing solely on the IAEA's technical capacity and resources risks neglecting the important political challenges to dealing with a violation.[33] In private and in expert-level discussions, safeguards officials worry that the agency lacks the capacity and authority to address cases of proliferation that are outside its mandate. In a July 2011 speech, IAEA Deputy Director General for Safeguards Herman Nackaerts said that:

> The [safeguards] system was manifestly failing in its primary objective, namely, to detect activities that **did** raise potential compliance issues and proliferation concerns—such as those undertaken, for instance, in Iraq, Libya, Syria and Iran.

The reason Nackaerts thought that the system was "manifestly failing" was that "major proliferation challenges have arisen in States with limited nuclear fuel cycle facilities, and involved previously exempted or undeclared nuclear material."[34] If proliferators were outside the IAEA's authority, according to this logic, then increasing the agency's capacity and legal authority would be the solution.

Technical and legal improvements within the agency will not be sufficient, however. A violator will not be deterred by the IAEA's technical ability to detect a violation alone. The violator will be deterred by a calculation that the consequences of a violation will be too great to risk detection. Even if the agency's standards and technical ability to detect a violation improve, a nation or an entity considering proliferation will not be deterred if it thinks that it can ignore, forestall, or withstand the consequences of detection. The IAEA realized in Iraq and elsewhere that a country could attempt to conceal violations or hide from inspectors. Similarly, a country could attempt to escape the consequences of detection through political strategies. To fulfill the goals of nonproliferation, the IAEA must be able to detect violators, and the world political community, along with the agency, must be able to enact a sufficient penalty for violation.

If a country's violation is clear, then the IAEA has an easier job, and it can more easily refer the matter to the UN and world community. But in most cases, violations are not clear. The evidence is mixed, or the violation is clear to those in the know, but it is discovered through secret intelligence provided by a member state, and it cannot be scrutinized or it does not have the same level of credibility as a violation discovered by IAEA inspections.

The IAEA's inspection system is based on inspecting and auditing declared sites. If a country wants to hide a small nuclear program, it can probably escape detection from the agency, as happened with Libya's nuclear program. Beginning in the 1990s, Libya had traded in uranium and other illicit material without IAEA detection.[35] Former IAEA Director General Mohamed ElBaradei conceded that "the system cannot detect easily concealable small items. Any verification system cannot do that."[36]

The IAEA often relies on state intelligence agencies to provide satellite and other sensitive data. If intelligence is provided in secret, however, the accused state can question the veracity of the information and the motivations of its sources. In 2011, state intelligence agencies provided documents to the IAEA showing that Iran had sought and found foreign help to learn the steps necessary to build a nuclear weapon. The intelligence reports showed how, among other things, a former Soviet nuclear scientist provided sensitive information to Iran while under contract in Tehran in the 1990s. Because the information was provided in secret without details about the sources and probably by countries hostile to Iran, Iran questioned the credibility of the information. A former nuclear official in Iran, Ali Akbar Salehi, described the controversy following the information as "100 percent political" and explained that the IAEA is "under pressure from foreign powers."[37] Iran's ability to question secret intelligence about its nuclear program delayed international action to impose consequences.

Even if the IAEA does find noncompliance with the NPT and safeguards agreements and evidence of trafficking in illicit materials, a state could blame non-state actors — legitimately or illegitimately — for the trades. Export controls and safeguards do not easily cover nonstate actors. By blaming nonstate actors, a state could avoid sanctions and other consequences of illicit activity.

In one famous example, A. Q. Kahn was the mastermind behind an illicit nuclear enterprise that sold secrets to Iran, North Korea, and Libya, but the Pakistani government claimed that he was acting without their approval and that Pakistan should not be held responsible. Pakistani authorities apprehended Kahn in 2003, and Pakistani President Pervez Musharraf pardoned him in exchange for a confession. Some investigations speculated that Kahn cooperated with the Pakistani government, while others found that he cooperated with corrupt parts of the government but escaped detection by the rest of the Pakistani authorities.[38]

ADDITIONAL PROTOCOL AS A DETERRENT OR A BURDEN?

The IAEA's safeguards structure assumes that once a violation is detected, the violator will repent, or world opinion will impose consequences on the violator. Sometimes this happens. In 2004, South Korea revealed that it had failed to report nuclear material used in experiments to enrich uranium as recently as 2000. South Korea also revealed other undeclared materials used in enrichment and other experiments as far back as 1979. The revelations came as part of South Korea's declarations in ratifying the AP, which expands the IAEA's authority to investigate both declared and undeclared nuclear facilities. The protocol also requires that countries declare more of their nuclear activities than is required under traditional safeguards. In November 2004, the IAEA Board of Governors noted "serious concern" about Korea's unreported activities, but the board did not refer South Korea to the UN Security Council, though referral is within the agency's rights.[39] The IAEA is required to report findings of a country's noncompliance with safeguards agreements to the Security Council, but what rises to the level of noncompliance that merits reporting is open to interpretation.

In the South Korean case, the system worked as many people hoped it would. The AP served as a deterrent of sorts, leading the country to at least report previously unreported material, sites, and activities and possibly leading the country to stop activity that might have continued otherwise, and to ensure greater accountability in its nuclear programs. (South Korea claimed that nuclear scientists conducted these experiments without telling high-level political officials.)

While South Korea is a possible example of the AP's deterrent ability, the AP may lead to additional burdens on the agency in the future. Some experts claim that adoption and implementation of the AP by additional countries will improve the effectiveness of safeguards. The AP is a legal document that enables the IAEA not only to verify the nondiversion of declared nuclear material but also to provide assurances as to the absence of undeclared nuclear material and activities in a state. In short, the AP gives the IAEA increased access to sites and information in a state. For the AP to be effective, and for safeguards generally, timeliness is everything. Safeguards are designed to detect the diversion of a significant quantity of nuclear material (defined as enough to make one crude bomb) within a conversion time, which is the minimum time needed to build a bomb using diverted materials. The expansion of the AP to new states, combined with the need for timeliness and budget constraints, produces a worrisome new safeguards equation. Member states expect the AP not to cost more than current safeguards, but an increased number of facilities under the protocol multiplied by the cost of increased data analysis and environmental sampling per facility that comes with the AP, divided by a constant safeguards budget, equals a reduced number and intensity of inspections. The AP will likely lead (and has led) to countries covered by the AP being subject to fewer inspections, even as the IAEA obtains greater authority to provide assurances about declaring undeclared sites.

Despite the potential for the AP to increase expectations of the IAEA beyond what the agency can accomplish, the adoption and implementation of the AP has been an article of faith among proponents of the

IAEA's safeguards program. In 2005, Director General Mohamed ElBaradei wrote that:

> I believe that, for the Agency to be able to fulfill its verification responsibilities in a credible manner, the Additional Protocol must become the standard for all countries that are party to the Treaty on the Non-Proliferation of Nuclear Weapons.[40]

Similar calls for wider adherence to APs came from the UN General Assembly, by member states at the 2000 and 2010 NPT Review Conferences, and by states at IAEA General Conferences. Yet if the IAEA expands verification activities under the AP without increasing its budget at a greater rate than in the past, it will have to seek new efficiencies, likely by reducing the number and intensity of inspections or decreasing inspections in countries not considered to be proliferation risks, a practice that opens the Agency to charges of unfairly applying its standards.[41] Reducing inspections by increasing randomization may make sense from a cost-benefit standpoint, but it opens a window for diversion, especially if the inspections do not occur within the timeliness window.[42]

The AP confers a preferred traveler status to a country, allowing the IAEA to reduce inspections once it has reached a finding of "nondiversion." A clever state, terrorist, or criminal (perhaps without state knowledge) may seek to attain the activation of the AP and then engage in diversion activities. Given the intensive background work required to implement the AP and inspect new nuclear facilities, the IAEA will likely reduce inspections for its "preferred travelers" because of resource constraints. Despite its advantages, the AP provides an increased possibility for diversion.

With more nuclear facilities in more countries, whether under the AP or not, the IAEA faces a tough decision about whether to lower the false alarm rate and potentially increase the possibility for diversion. The IAEA has an acceptable false alarm rate of approximately 5 percent.[43] In other words, the agency tolerates the inspectorate alleging that materials are unaccounted for when, in fact, there exists a good explanation about 5 percent of the time. The 5 percent rate, multiplied by an increasing number of inspections, leads to two options. First, the agency could pursue a politically unacceptable high number of false alarm reports and confront member states, who will demand that the agency lower the false alarm rate. Governments do not like the agency to give them negative publicity, give excessive attention to sensitive nuclear and industrial processes that are secret, or insult national pride. Second, the agency could allow more nuclear material to remain unaccounted and make a security case to member states and the Board of Governors for the IAEA to increase its permissible false alarm rate. It is not clear which of these will happen, but the agency's leadership will have to make a choice. While technical constraints shape the process of material accountancy, the greatest challenges are political.

THE LIMITS OF LEGAL TOOLS

Former IAEA officials agree that the legal framework of the IAEA needs revision. The NPT prohibits non-nuclear weapons states from making weapons, but it allows states to get extremely close to making weapons. The treaty has been interpreted as providing an inalienable right to peaceful use, which overlaps with much of what is needed for military purposes.[44] Much of the delay in imposing penalties on Iran for its nuclear activities comes from disagreement about whether and to what degree Iran is in violation of the NPT. Iran maintains that it is fulfilling its obligations under the treaty and that hostile powers are unfairly seeking to deny its right to peaceful use of nuclear materials.[45]

The current safeguards legal framework relies on member states' voluntary cooperation with the inspection process. IAEA inspectors are more like door-to-door salesmen asking for permission to enter a home than they are like police investigators demanding access. The IAEA also lacks the legal tools to gather information in case of noncompliance, and the international system has no regular sanction available if a country forbids or delays IAEA access, and no pre-established penalty if a country withdraws from the NPT. Without enforcement, detection risks lead to nothing more than empty threats. Alternatively, detection could lead to ad hoc penalties imposed by the UN Security Council or by coalitions of interested countries that risk being criticized as unfair. The agency will be in a better position if governments can agree on penalties for violation that apply to all countries and are agreed upon before any particular country is in violation. Without reform of the legal processes governing IAEA inspections, the IAEA will have more of the same—more evasion, as in the cases of Iran and Syria, and perhaps many more cases if nuclear power expands around the world.

Former safeguards officials agree that the IAEA and the UN need to have more authority for inspections and a standard procedure for dealing with evidence of diversion. The international system also needs agreement on what to do about countries that leave the NPT. Without standard measures for what counts as a violation and standard procedures for dealing with it, the IAEA risks a loss of credibility as a neutral arbiter when a crisis occurs and the agency does not have an impartial procedure to address these crises. For the IAEA to fulfill its obligation under the statute—and to maintain credibility with the world, fairly or unfairly—the agency must be able to detect violations of safeguards agreements and the NPT and be clear about the consequences if countries do not co-operate in addressing violations.

PROSPECTS FOR REFORM

Goal conflict within the IAEA will not make resolving these challenges easy. The agency has at least three major goal conflicts. First, member states are unclear about whether the agency should privilege promoting nuclear power, preventing military diversions, or maximizing safety. Article II of the Agency's statute provides for the first two missions simultaneously:

> The Agency shall seek to accelerate and enlarge the contribution of atomic energy to peace, health and prosperity throughout the world. It shall ensure, so far as it is able, that assistance provided by it or at its request or under its supervision or control is not used in such a way as to further any military purpose.[46]

The goals of promoting peaceful use and preventing diversion are in conflict if nuclear power expansion makes identifying diversion more difficult because inspectors look for the "needles" of diversion in a bigger nuclear haystack, with more nuclear facilities, materials, transfers, and knowledge around the world. Furthermore, if, after Fukushima, the agency devotes more of its budget to safety and less to other aims, then the agency's missions are locked in a zero sum game.

Second, the agency's responsibilities for inspection and verification conflict with member states' desire for control and secrecy in nuclear energy. The agency wants unfettered access to facilities for safety and safeguards inspections and verification, but member states want their companies to protect proprietary information and want IAEA inspectors to have as little interference as possible with industrial routines. As the nuclear facilities under safeguards expand beyond light-water reactors, new safeguards demands may make reconciling these conflicts more difficult. Third, the agency's conflicts over budget growth will force a choice as the number of states and kinds of facilities under safeguards expands. The agency can choose to constrain budget growth to near 0 percent a year and keep staff and financial burdens on member states to a minimum, or it can expand budget obligations on member states and expand its responsibilities and capacities. Each of these three goal conflicts will force the agency and its member states to make choices at higher levels of operations.

The potential for goal conflicts prompts a question: Can the IAEA reasonably be expected to safeguard much larger nuclear programs throughout the world, including in many more countries than at present? Scaling up the agency's operations, human capital, and technology at the same rate as has been done in the past is not sufficient, but the IAEA's Secretariat can develop workarounds to improve performance. The agency can engage in selective scaling up, purchasing new surveillance equipment, better laboratory facilities, and training for inspectors. These efforts alone, however, are not sufficient, and the agency is not likely to dramatically increase its budget through the current financing structure. To reduce goal conflict and improve performance, the agency could reorganize according to mission, separating elements of the agency that promote nuclear power from those that conduct safeguards. The IAEA already has separate divisions, but the separation could go further, perhaps providing separate reports to the Board of Governors and to the director general, leaving it to the director general to find the appropriate balance among organizational goals and not to bureaucrats further down the organizational chart. This might resemble the competing intelligence analysis given to the President of the United States after intelligence reorganization following September 2001.[47]

Simply grafting a new branch onto an organization may not prove to be sufficient separation among divisions. To incorporate additional perspectives in the decisionmaking process, the IAEA could seek new sources of funding based not just on membership but on use.[48] The current safeguards funding structure has a free rider problem in that those who benefit from nuclear power do not always pay more for safeguards, and those who do not benefit sometimes pay an amount disproportionate to their use. The excessive payers are especially reluctant to increase safeguards funding. Nevertheless, countries that benefit more from nuclear power could ante up for safeguards. Industry, too, could contribute to safeguards based on use either through member states or through other bodies. Some IAEA officials balk at the idea of anyone other than member states contributing to the agency's work since it is a creature of member states. Yet nongovernmental organizations (NGOs) have made contributions to safeguards before, and Taiwan pays a safeguards fee even though it is not a member state.

While the agency can seek new sources of funding to meet new challenges, the agency could also be honest about its limits. If governments and NGOs were more aware of the technical and political limits of the agency's capacity, then they may not be so quick to give the agency new missions or to spread nuclear materials around the globe, both of which would make the agency's job easier than it would be otherwise. Former Secretary General Mohamed ElBaradei's memoir presents an expansive portrayal of the agency's powers, which may undermine the agency's authority in the long term if it is shown to be perpetually underperforming compared to expectations.[49] For instance, ElBaradei asserts that the agency can, with the AP, declare that a state has no undeclared material. But proving the negative is impossible. The UN Monitoring, Verification and Inspection Commission conducted inspections in Iraq and did not find weapons of mass destruction (WMD), but it could not prevent the March 2003 invasion of Iraq, which was justified on the grounds that Iraq had WMD.

Better transparency about performance goals could help the agency be more clear about its limits. Agencies resist establishing and publicizing data for a host of reasons: outcomes are difficult to measure; inequalities breed envy; and publicity could spawn a media blame-game. Nevertheless, performance metrics could help improve performance or secure more resources, even if the agency falls short. If the agency perpetually underperforms, it could make a case that member states expect far too much. The agency has developed quantitative measures of performance such as "timeliness detection" and "risk of early detection," but it is difficult to find data on these measures. Traditionally, on-site verification efforts concentrated on the states with the largest nuclear programs, not on the programs that necessarily posed the greatest proliferation risks. Under traditional safeguards, "60 percent of the agency's verification efforts was expended in just three states," according to IAEA Deputy Director General for Safeguards Herman Nackaerts.[50] Thus, the more materials and sites a state declared, the more it was inspected, independent of any analysis of the proliferation risk in a state or the state's history of cooperation. The IAEA is moving toward a system in which it evaluates state declarations for their consistency and thoroughness, and away from evaluating and measuring facilities. If the agency can develop relatively neutral criteria for evaluating state declaration, however, the state-level criteria could prove useful, especially when combined with other measures, such as timeliness detection.[51]

Perhaps because of the agency's exceptional professionalism among international organizations, and because of successes in the cases of North and South Korea, some politicians and the IAEA's own leaders overstate the capabilities of the agency. Yet the safeguards system is fragile, and, without reform, more countries will advance to the brink of building weapons, even as they maintain their legal obligations under the NPT's right to peaceful use.

Some reformers advocate improving the IAEA's technical abilities and giving more discretion to the agency's Secretariat.[52] Not doing so, they argue, would risk further politicizing issues in the UN Security Council, where countries would pander to their most powerful constituencies. The UN Security Council lacks neutral dispute resolution procedures unlike, for example, the World Trade Organization. Meanwhile, the IAEA has more legal authority than it uses. The agency's statute allows for it to suspend technical cooperation to problematic member states who value the agency's expertise, and the agency could ask for the international Court of Justice or other litigation bodies to intervene in disputes. In short, working around the UN Security Council gives the agency the best chance at deterring violators.

Another school of thought advocates reform in cooperation with the UN Security Council and the UN as the best path forward for the IAEA. The UN Security Council and UN are the most effective bodies for aggregating world opinion and for legitimate enforcement action, the argument goes, and their participation is essential in an equitable system of nonproliferation that imposes consequences for violations. Not all consequences should be punitive, however. Studies of regulation enforcement show that different kinds of violators respond to different kinds of consequences.[53] For example, regulators should be reasonable toward more cooperative organizations, harsh with chronic evaders, and conciliatory with repentant organizations.

Mohamed ElBaradei has been a leading proponent of comprehensive reform, recommending that the IAEA and the UN Security Council work together to, "effectively deter, detect, and respond to possible proliferation cheats."[54] Former Head of the Department of Safeguards Pierre Goldschmidt warns that if reform is not adopted, "we will see more of the same and we should not be surprised if one day Iran withdraws from the NPT and other states like Syria continue to deny the IAEA access to suspect locations."[55] In other words, he predicts a breakdown of the nonproliferation system. Goldschmidt offers a series of thoughtful international-level reforms. These include:

> imposing penalties for withdrawal from the NPT; announcing that safeguards agreements and the IAEA's power of special inspection will outlast withdrawal; giving the agency new tools for monitoring the nuclear trade; and being clear and objective about when a violation occurs and the steps a country must take to address the violation; and, in some cases, imposing penalties for violations that would ultimately improve the agency's effectiveness and credibility.[56]

When asked about these proposals, experts in the nonproliferation community provided few solid arguments against them except that they will be difficult to achieve. Enacting these reforms will require that enough interested governments take the matter to the international community, specifically the UN and the UN Security Council. In the meantime, the IAEA can do its part by sounding the alarm for reform and calling attention to the limits to what it can safeguard, especially if nuclear power expands.

ENDNOTES - CHAPTER 9

1. Safeguards Glossary, 2001 Edition," International Atomic Energy Agency, Vienna, Austria, p. 23, available from *www-pub.iaea.org/MTCD/publications/PDF/nvs-3-cd/PDF/NVS3_prn.pdf*. Inspection intervals are based on the time required to divert a significant quantity for non-peaceful uses, but a significant quantity could be more than enough to make a bomb. The IAEA safeguards glossary defines significant quantity as: "the approximate amount of nuclear material for which the possibility of manufacturing a nuclear explosive device cannot be excluded."

2. "World Energy Outlook for 2010," *International Energy Agency*, Paris, France, 2010, p. 1.

3. Citigroup Global Markets finds recent delays in construction in several new nuclear power plants reduce the likelihood that nuclear power will be considered a viable investment by the market. Citigroup finds that nuclear power investments are economically viable only when governments intervene in the market to provide explicit assurances of a minimum return on investment. Peter Atherton *et al.*, "New Nuclear: The Economics Say No," *Citigroup*, November 2009, p. 4-6, available from *npolicy.org/article_file/New_Nuclear-The_Economics_Say_No.pdf*.

4. *Ibid*, p. 8.

5. "Emerging Nuclear Energy Countries," *World Nuclear Association*, London, UK, May 2012, available from *www.world-nuclear.org/info/inf102.html*.

6. Ryan Lizza, "As the World Burns," *New Yorker*, October 11, 2010.

7. Heather Timmons, "Emerging Economies Move Ahead With Nuclear Plans," *The New York Times*, March 14, 2011; Trevor Findlay, "The Future of Nuclear Energy to 2030 and its Implications for Safety, Security and Nonproliferation: Overview," *The Centre for International Governance Innovation*, Waterloo, Ontario, Canada, 2010; Jaeyeon Woo, "Seoul's U.A.E. Deal Caps Big Sales Push," *The Wall Street Journal*, December 29, 2009.

8. Yukiya Amano, "International Cooperation Vital for Nuclear Renaissance," *Le Monde*, March 7, 2010.

9. "World Energy Outlook 2011," *International Energy Agency*, Paris, France, p. 2.

10. "India Needs Nuclear Energy, Says PM Manmohan Singh," *BBC News India*, March 27, 2012, available from *www.bbc.co.uk/news/world-asia-india-17520589*.

11. Atherton *et al.*, p. 3.

12. "Safeguards Statement for 2010," *International Atomic Energy Agency*, Vienna, Austria, available from *www.iaea.org/OurWork/SV/Safeguards/es/es2010.html*.

13. Herman Nackaerts, "A Changing Nuclear Landscape: Preparing for Future Verification Challenges," *International Forum on Peaceful Use of Nuclear Energy and Nuclear Non-Proliferation*, Vienna, Austria, February 2, 2011, available from *www.iaea.org/newscenter/statements/ddgs/2011/nackaerts020211.html*. The IAEA's 2009 projections high growth scenario forecast that nuclear power production could double by 2030, and its low-growth scenario has nuclear power production increasing by about 40 percent. See IAEA, *Energy Electricity and Nuclear Power Estimates for the Period up to 2030,* Vienna, Austria, 2009.

14. Olli Heinonen, "Belfer Center Release," Cambridge, MA: Harvard University, October 12, 2010, available from *www.iaea.org/Publications/Reports/Anrep2009/table_a5.pdf*.

15. "The Agency's Programme and Budget, 2010-11," *International Atomic Energy Agency,* Vienna, Austria, August 2009, available from *www.iaea.org/About/Policy/GC/GC53/GC53Documents/English/gc53-5_en.pdf*.

16. "Annual Reports, 1987-2010," *International Atomic Energy Agency,* Vienna, Austria.

17. "Safeguards Glossary, 2001 Edition," *International Atomic Energy Agency,* Vienna, Austria, p. 23, available from *www-pub.iaea.org/MTCD/publications/PDF/nvs-3-cd/PDF/NVS3_prn.pdf*.

18. Mark Hibbs, "The Unspectacular Future of the IAEA Additional Protocol," *Carnegie Proliferation Analysis*, April 26, 2012.

19. Lucia D'Acunto, Tamas Vinko, and Johan Pouwelse, "Do BitTorrent-like VoD Systems Scale under Flash-Crowds?" IEEE Communications Society P2P 2010 Proceedings Paper, 2010.

20. Yingwu Zhu, Xiaoyu Yang, Yiming Hu, "Making Search Efficient on Gnutella-like P2P Systems," Proceedings of the 19th IEEE International Parallel and Distributed Processing Symposium, 2005; A. Rowstron and P. Druschel, "Pastry: Scalable, decentralized object location, and routing for large-scale peer-to-peer systems," *Proceedings of the 18th IFIP/ACM International Conference on Distributed System Platforms (Middleware)*, Heidelberg, Germany, November 2001, pp. 329–350.

21. Henry Petroski, *Success through Failure: The Paradox of Design,* Princeton, NJ: Princeton University Press, 2006, pp. 95-96.

22. "IAEA Mission Statement," available from *www.iaea.org/ About/mission.html.* On the relationship of declared to undeclared sites, see IAEA, *The IAEA's Safeguards System, Ready for the 21st Century,* Vienna, Austria, undated, available from *www.iaea.org/ Publications/Booklets/Safeguards2/part5.html.*

23. This authority is in accord with non-nuclear weapons states who are members of the NPT under the Model Safeguards Agreement pursuant to IAEA INFCIRC 153. For more on how the agency interpreted its safeguards mission in the past, see Herman Nackaerts, "IAEA Safeguards Cooperation as the Key to Change," talk to the INMM 52nd Annual Meeting, July 18, 2011, available from *www.iaea.org/safeguards/documents/IAEA_Safeguards_Cooperation _as_the_Key_to_Change.pdf.*

24. "Safeguards Glossary," available from *www-pub.iaea.org/ MTCD/publications/PDF/nvs-3-cd/PDF/NVS3_prn.pdf.*

25. See *Falling Behind: International Scrutiny of the Peaceful Atom,* Henry Sokolski, ed., Carlisle, PA: Strategic Studies Institute, U.S. Army War College, 2008.

26. Performance according to timeliness detection goals is reported in the Safeguards Implementation Report. Annual and facilities-specific data is not available to the public, but occasionally the agency reveals performance data as an illustration in various reports. In the United States, relevant bureaucrats and congressional committees have access to this data.

27. For a chronology of the IAEA's discoveries in North Korea, see the agency's fact sheet available from *www.iaea.org/news-center/focus/iaeadprk/fact_sheet_may2003.shtml.*

28. Pierre Goldschmidt, "Looking Beyond Iran and North Korea for Safeguarding the Foundations of Nuclear Nonproliferation," see Chap. 11 in this volume, p. 1.

29. Fred Charles Iklé, "After Detection: What?" *Foreign Affairs*, Vol. 39, No. 2, January 1961, pp. 208-220.

30. See the *Symposium on International Safeguards: Preparing for Future Verification Challenges*, Vienna, Austria, November 1-5, 2010, available from *www-pub.iaea.org/iaeameetings/38095/ Symposium-on-International-Safeguards-Preparing-for-Future-Verification-Challenges.*

31. Pierre Goldschmidt, "Identifying the Right Skills and Expertise for the Challenges of the 21st Century: Where to Find Them? How to Retain Them?" IAEA Safeguards Symposium, Vienna, Austria, November 1, 2010.

32. *IAEA Safeguards: Stemming the Spread of Nuclear Weapons*, Vienna, Austria, 2002, p. 1, available from *www.iaea.org/ Publications/Factsheets/English/S1_Safeguards.pdf.*

33. The IAEA often makes statements such as, "With wider access, broader information and better use of technology, the Agency's capability to detect and deter undeclared nuclear material or activities is significantly improved." While this is probably true, better technology and more information is only half the battle. What to do after detection is the other half. See "IAEA Safeguards: Stemming the Spread of Nuclear Weapons," *International Atomic Energy Agency Information Series Division of Public Information Factsheet,* 2002.

34. Herman Nackaerts, "IAEA Safeguards: Cooperation as the Key to Change," Keynote Address to the INMM 52nd Annual Meeting, July 18, 2011.

35. Borzou Daragahi, "Details Told of Libya's Nuclear Bid," *The Los Angeles Times*, September 13, 2008.

36. Elbaradei, "Coming Clean Background," *PBS NewHour*, December 30, 2003.

37. Joby Warrick, "IAEA Says Foreign Expertise Has Brought Iran to Threshold of Nuclear Capability," *The Washington Post*, November 6, 2011.

38. Douglas Frantz and Catherine Collins, *The Nuclear Jihadist: The True Story of the Man Who Sold the World's Most Dangerous Secrets . . . And How We Could Have Stopped Him,* New York: Twelve Books, 2007; Adrian Levy and Catherine Scott-Clark, *Deception: Pakistan, the United States and the Global Nuclear Weapons Conspiracy,* New York: Walker & Company, 2005.

39. Paul Kerr, "IAEA: Seoul's Nuclear Sins in the Past," *Arms Control Today,* December 2004.

40. IAEA, *Non-Proliferation of Nuclear Weapons and Nuclear Security,* Vienna, Austria, May 2005, p. 2, available from *www.iaea. org/Publications/Booklets/nuke.pdf.*

41. Dapo Odulaja, "Broader Use of Statistical Techniques in the Design of Advanced Safeguards Approaches," *2010 Safeguards Symposium: Preparing for Future Verification Challenges,* Vienna, Austria, November 2010, p. 3.

42. It is difficult, if not impossible, to obtain actual data on the relationship between the probability that random inspections will take place within the timeframe of detection goals. The IAEA does make some theoretical models for random inspections and timeliness goals available, but these models do not necessarily reflect practice.

43. U.S. Congress, Office of Technology Assessment, *Nuclear Safeguards and the International Atomic Energy Agency,* OTA-ISS-615, Washington, DC: U.S. Government Printing Office, June 1995, p. 45; Marvin M. Miller, "Are IAEA Safeguards on Plutonium Bulk-Handling Facilities Effective?" Washington, DC: Nuclear Control Institute, August 1990, available from *www.nci.org/k-m/ mmsgrds.htm.*

44. The first clause of article IV of the NPT reads: "Nothing in this Treaty shall be interpreted as affecting the inalienable right of all the Parties to the Treaty to develop research, production and use of nuclear energy for peaceful purposes without discrimination and in conformity with Articles I and II of this Treaty."

45. Peter Crall, "Iran's Nuclear Program: An Interview with Iranian Ambassador to the IAEA Ali Asghar Soltanieh," *Arms Control Today*, October 2011.

46. IAEA, Article II.

47. Richard A. Posner, *Uncertain Shield: The U.S. Intelligence System in the Throes of Reform*, Lanham, MD: Rowman & Littlefield Publishing, 2006, pp. 55-86; John Diamond and Judy Keen, "Bush's Daily Intel Briefing Revamped," *USA Today*, August 25, 2005, p. A1; Douglas Jehl, "Intelligence Briefing for Bush is Overhauled," *The New York Times*, July 20, 2005, p. A18.

48. Tom Shea and Henry Sokolski have suggested this idea. See *Falling Behind*, pp. 13, 36.

49. Mohamed ElBaradei, *The Age of Deception: Nuclear Diplomacy in Treacherous Times*, New York: Metropolitan Books, 2011. For example, the agency said that it is not yet "in a position to conclude that there are no undeclared nuclear materials or activities in Iran." This assumes that the agency could potentially be in such a position. See the "Report on the Implementation of Safeguards in the Islamic Republic of Iran," GOV/2006/15, *International Atomic Energy Agency*, February 27, 2006.

50. Herman Nackaerts, "IAEA Safeguards Cooperation as the Key to Change," talk to the INMM 52nd Annual Meeting, July 18, 2011, available from *www.iaea.org/safeguards/documents/IAEA_Safeguards_Cooperation_as_the_Key_to_Change.pdf*.

51. In "IAEA Safeguards Cooperation as the Key to Change," Nackaerts describes the new safeguards thusly:

> This means moving away from such a heavy reliance on routine quantitative measurements and the mechanistic application of generic criteria. Instead, it requires taking into account a wide range of factors—qualitative and quantitative, reaching an informed judgment based upon a detailed analysis and evaluation of all the information available to the Agency, and then deciding to act accordingly. To do

this, our focus needs to be on each State as a whole, rather than solely on the nuclear material and particular facilities within that State.

52. Sebastian Harnish, "Minilateral Cooperation and Transatlantic Coalition-Building: The E3/EU-3 Iran Initiative," *European Security*, Vol. 16, No. 1, March 2007, pp. 1-27.

53. John T. Scholz, "Voluntary Compliance and Regulatory Enforcement," *Law & Policy*, Vol. 6, Issue 4, October 1984, pp. 385-404.

54. Mohamed ElBaradei, "A Recipe for Survival," *The New York Times*, February 16, 2009, available from *www.nytimes.com/2009/02/16/opinion/16iht-edelbaradei.1.20216338.html?_r=1&pagewanted=all*.

55. Pierre Goldschmidt, E-mail Interview with author, December 4, 2011.

56. "Looking Beyond Iran and North Korea for Safeguarding the Foundations of Nuclear Nonproliferation."

CHAPTER 10

INTERNATIONAL ATOMIC ENERGY AGENCY
INSPECTIONS IN PERSPECTIVE

Olli Heinonen

The nuclear nonproliferation regime continues to face a broad array of challenges. It is easy to see why new solutions are needed. The world is undergoing rapid changes on many fronts—including technologically. The Nuclear Nonproliferation Treaty (NPT) entered into force 40 years ago. It should not surprise us that the solutions of 1970 are not a perfect fit to the challenges of the 21st century. In particular, during the last 2 decades, we have seen three major developments related to nuclear proliferation: (1) the increased dissemination of nuclear technology and nuclear "know-how"; (2) a renewed drive on the part of (a few) state as well as nonstate actors seeking to acquire nuclear weapons; and (3) the emergence of clandestine nuclear procurement networks. The following reality is before us: Either we continue to overcome the vulnerabilities or accept a nominal international nuclear safeguards system. Some improvements to the system do require political will and agreement, but there are also technical fixes that can be implemented by the International Atomic Energy Agency (IAEA) in capitalizing on its verifications objectives and without additional external endowed authorities.

Which, then, are the areas that the IAEA should focus on in providing for a more effective safeguards regime? Looking back, proliferation cases of the last 2 decades have shared several common features. Proliferation countries used clandestine facilities and exploited loopholes within the IAEA verifications system itself, such as acquiring nuclear material like uranium from ore or yellow cake where no actual IAEA verification is applied. Much of the clandestine work also took place at undeclared/unreported facilities. The logic is simple—the IAEA verification system is good enough to detect diversion of nuclear material but still falls short of any foolproof method to detect clandestine facilities or, to a certain extent, a misuse of declared facilities. Efforts have been made to address these deficiencies. The IAEA verification system was overhauled 15 years ago by providing the inspectorate with more access rights, namely introducing the Additional Protocol and an early provision of design information. A state-level approach system was also implemented, where the state's nuclear program is assessed annually using a holistic method. Equally vital to the IAEA's verification scheme was the inculcation of an investigative inspection culture, a cornerstone to complement better access and information provided for under a reformed verification regime.

At the same time, while approaching safeguards from a more critical approach, the IAEA state level concept system also has, as its byproduct, a reduction of actual on-site inspection activities by complementing it with enhanced information analysis. This means that the physical presence of inspectors, the eyes and arms of the international community, at the nuclear installations is less frequent but occurs at a fairly predictable rate. By and large, this is a progressive step for the Agency to provide for cost effectiveness while stepping up and maintaining safeguards assurances. For the state level approach to function optimally, all the ingredients must be in place to ensure broader conclusions are drawn on a state's nuclear activities and cooperation from states themselves.

Concerns are certainly heightened when a state uses sensitive technologies such as uranium enrichment or reprocessing. Article III of the Nuclear Non-proliferation Treaty (NPT) lays down as the objective to prevent diversion of nuclear energy to nuclear weapons. This objective is wider than just the detection of diversion of nuclear material. Is the Agency's verifications system able to see those indications?

To better understand the modifications and fixes needed, it is important to look at the current verification system. One is in the area of nuclear "transparency." Another is in the area of expanding unannounced and special inspections. A third is in the area of monitoring to prevent a "breakout" scenario.

UNANNOUNCED INSPECTIONS

Routine inspections—the most common type of inspections—are carried out according to a defined schedule. The Agency's right to carry out routine inspections under comprehensive safeguards agreements is limited to those locations within a declared nuclear facility, or other locations containing nuclear material, through which nuclear material is expected to flow (strategic points). The safeguards agreement also has a provision that allows for unannounced inspections. Unannounced inspections have been the core of safeguards at enrichment plant sites but are rarely used elsewhere.

Unannounced inspections in reality are not quite as its name implies but are carried out with a short advance notice, typically about 2 hours, so that appropriate security arrangements can be completed and the state inspectors can be present. These requirements arise from the subsidiary arrangements concluded be-

305

tween the IAEA and inspected state. Such short notice inspections have been carried out at the light water reactors in South Korea, where they form part of a quality assurance component of remote monitoring that is being applied. In Japan, such inspections have been used at its fuel fabrications plants to monitor nuclear material flow. At the uranium enrichment plants worldwide, this scheme has been used to ensure that the enrichment cascades are not modified or used to produce high-enriched uranium. More recently, the short notice inspections have also been modified to cover nuclear material flows at the enrichment plants.

Currently, the IAEA annually conducts a total of few hundred person-days of unannounced inspections, which are only a fraction of the total annual 8,000 person-days of inspection. The unpredictability of such (unannounced) inspections is meant to enhance the effectiveness of the inspection regime. They are also incorporated with the Agency's integrated as well as traditional safeguards approaches. Many other steps also go into providing assurances, *inter alia*, periodical confirmation of nuclear material balances, covering of credible scenarios for diversion of nuclear material and misuse of facilities, and detection of signs of turning nuclear material for weapons purposes before it actually takes place. The implementation parameters between traditional and integrated safeguards approaches may differ from each other, but in both cases, the verification system has to ensure that the detection probabilities remain high and timely.

SPECIAL INSPECTIONS

The IAEA also has in its toolbox the authority to conduct "special inspections." The term "special inspection" has an unnecessarily negative connotation that resulted from the circumstances under which it was discussed in the past. Cases linked to the call for special inspections to be evoked included North Korea, Iran, and Syria. States, indeed, would not suffer any disadvantage from the use of special inspections. On the contrary, the possibility of dispelling doubts through such inspections should be an advantage to all except those states who have violated their commitments. A debate in the Board on the use of special inspections will not, under current circumstances, bring a solution to the problem of its use. Rather, the IAEA Secretariat should be constantly encouraged to use all its inspection rights, including that of special inspections as an option. The proliferation cases mentioned previously also reveal that the IAEA deputy general's position and interpretation is instrumental in the willingness to evoke special inspection procedures. To date, it appears that the bar of its actual use remains unnecessarily high.

Nuclear Transparency.

In the early-1990s, during IAEA Board discussions on strengthening the Agency's safeguards, former Deputy General Hans Blix advocated "transparency visits" to clarify questions and ambiguities on several states' nuclear programs. The IAEA conducted several such visits to North Korea, South Korea, Iran, Taiwan, and South Africa. Such visits were also carried out at a later date to remove some ambiguities in Japan and elsewhere. The outcomes of the various transparency visits were of mixed success. In the case of North Korea and Iran, the alarm bell had not rung on the clandestine activities of their nuclear programs.

Instead, transparency visits to Iran in the 1990s provided a false sense of security, both within the IAEA as well as to the larger international community. The IAEA provided assurances through press statements following each of the various transparency visits made to Iran, without inspectors taking environmental samples (with one exception) or conducting more technical investigations. This significance cannot be understated since inspectors **did** come up with proof of undeclared nuclear material.

Short of an implemented Additional Protocol that provides the legal basis for more intrusive inspections, transparency visits also remained at the behest of the "goodwill" of the inspected state. Moreover, the limitations of transparency visits were not well understood by the Board and the public. As such, it was unclear whether assurances made by the IAEA about a lack of evidence of external reports of undeclared nuclear activities undertaken in Iran at sites visited by the IAEA were derived as a result of rigorous safeguards carried out. The danger of providing a sense of complacency had telling consequences on implementing safeguards in Iran years down the road, as we are now witnessing.

The Secretariat should be clear that transparency implies openness, communication, and accountability. Occasional visits by diplomats or invitations to IAEA officials do not replace Iran's Nonproliferation Treaty, the United Nations (UN) Security Council, and IAEA obligations. Since Iran continues to remain in deficit in fulfilling those international requests, it is in equal deficit in its transparency with regard to its actions and nuclear activities.

MONITORING TO PREVENT NUCLEAR BREAKOUT SCENARIOS

In proliferation cases such as Iran where there remains risks involved for a breakout scenario with the presence of enrichment and unresolved military dimensions on its nuclear program, the case for effective nuclear verifications is at the same time both more important and more challenging. For the case of Iran, an Additional Protocol-plus type of agreement implemented in a cooperative manner by Iran would be required. There are also important distinctions to be made between Iran and other countries that conduct enrichment or have a developed nuclear fuel cycle. Japan, Germany, and the Netherlands, for example, are presented as "latent or virtual nuclear weapon states." There are four differences in these countries compared to Iran: there are no indications of nuclear weapon related research and development; these states are in compliance with the terms of their safeguards agreements; they are applying additional protocols; and they do not appear to have uranium enrichment excess to their needs. Prevailing conditions in these countries may change over time. With proper safeguards in place, detection should be easier to detect at an early stage.

This brings up a related point concerning the IAEA's reporting practices. The vital nature of noticing and reporting early signs in any state that could involve potential diversion or cause for suspicion of safeguards failures cannot be overstressed. Early corrective measures can be useful. In more serious cases of potential subversion of peaceful nuclear use, early notification to the IAEA Board, including on safeguards implementation problems, can provide for more timely ratification or ramifications. To this end, the IAEA *Annual Safeguards Implementation Report* would also benefit from more detailed findings and conclusions drawn and written in a more targeted approach to fulfilling such assurances.

CONCLUSION

As measures are developed to continue to strengthen the IAEA verifications regime, it is equally important to ensure its thorough implementation and maximum effectiveness. A robust IAEA inspections regime contributes directly to preventing, detecting, and deterring untoward nuclear activities. This is no easy task. To this end, critical self-reflection, recognition of weaknesses in the inspections system, and seeking the best possible solutions are necessary to provide the meaningful assurances to uphold nuclear nonproliferation.

ENDNOTES — CHAPTER 10

1. International Atomic Energy Agency, Press Release 92/11, February 14, 1992.

CHAPTER 11

LOOKING BEYOND IRAN AND NORTH KOREA FOR SAFEGUARDING THE FOUNDATIONS OF NUCLEAR NONPROLIFERATION

Pierre Goldschmidt

Those who do not remember the past are condemned to repeat it.

George Santanaya[1]

International Atomic Energy Agency (IAEA) safeguards are both the principal means of verifying a state's compliance with international nuclear obligations, as well as detecting the potential transgression of these obligations. In the coming years, the IAEA will be asked to safeguard an increasing number of nuclear facilities, including new types of facilities (such as laser enrichment and pyroprocessing plants, floating nuclear power plants, and nuclear propelled submarines) and decommissioned ones. It will need additional funds to procure new types of and more effective equipment,[2] and expertise to carry out these additional responsibilities.[3]

But the real issue does not stem from resource constraints. Even with greater human and financial resources, there is nothing more the Agency would have done in fulfilling its verification mandate in Iran and North Korea. The real constraint was identified by current IAEA Deputy Director General for Safeguards Herman Nackaerts in a July 2011 speech:

Experience has shown that proliferation risk is not only associated with the amount of declared nuclear material that a State possesses or the number and type of declared facilities. Indeed, the major proliferation challenges have arisen in States with limited nuclear fuel cycle facilities, and involved previously exempted or undeclared nuclear material. . . . [The safeguards] system was manifestly failing in its primary objective, namely, to detect activities that did raise potential compliance issues and proliferation concerns—such as those undertaken, for instance, in Iraq, Libya, Syria and Iran.[4]

There are two main reasons the safeguards system has been "manifestly failing." First, the Department of Safeguards does not have the legal authority it needs to fulfill its mandate and to provide the assurances the international community is expecting from its verification activities. Second, the Department lacks the necessary cooperation and transparency from Member States of the IAEA. Redressing both deficiencies would significantly strengthen the role of IAEA safeguards in preventing further proliferation.

LIMITED LEGAL AUTHORITIES

Under the Article III.A.5 of the IAEA Statute, safeguards are:

designed to ensure that special fissionable and other materials, services, equipment, facilities, and information . . . under [Agency] supervision or control are not used in such a way as to further any military purpose.[5]

To reach that objective, Article XII.A.6 provides that the Agency will have the right and responsibility:

to send into the territory of the recipient State inspectors . . . who shall have access at all times to all places and data and to any person who by reason of his occupation deals with materials, equipment, or facilities which are required by this Statute to be safeguarded, as necessary . . . to determine whether there is compliance with the undertaking against use in furtherance of any military purpose.[6]

This excellent and forward looking mandate was agreed to more than half a century ago. Unfortunately, in practice, the commitments accepted by Non-Nuclear-Weapon States (NNWSs) under Comprehensive Safeguards Agreements (CSA)[7] and even the Additional Protocol (AP)[8] are much more limited.[9]

Under a CSA (with or without an AP), a state has the right to construct a uranium enrichment facility and to produce not only low-enriched uranium (LEU), but also highly-enriched uranium (HEU), or to extract plutonium from spent nuclear fuel, as long as these activities and material are declared and placed under IAEA safeguards. This right holds even if there is no clear economic justification for undertaking these activities. However, in such a case, it seems legitimate for the international community to wonder, in light of Article III.A.5 of the IAEA statute, whether such legal activities are undertaken in furtherance of any military purpose.

It is likely that in the future, should they decide to do so, an increasing number of NNWSs will acquire the necessary scientific, technical, and industrial capability to manufacture nuclear weapons. To increase the likelihood that those states will be deterred from making such a decision — most likely under maximum secrecy, since it would be a clear violation of Article II of the Nuclear Nonproliferation Treaty (NPT) — it is necessary that the international community be informed of any indications of nuclear weapons activities as soon as possible. Maximum IAEA scrutiny in such states should therefore be a priority.

Some possible indicators that would raise suspicion about a military nuclear program include:

- The state has denied or unjustifiably delayed access to locations by IAEA inspectors and/or is not fully cooperating with the Agency;
- There is a domestic enrichment or reprocessing facility in a state that has no AP in force;
- The state is producing and stockpiling uranium enriched beyond 5 percent uranium 235 (U-235);
- The state's military establishment is directly or indirectly involved in "peaceful" nuclear-related activities (including procurement);
- The state has previously been found in breach or in noncompliance with its safeguards agreement;
- There has been a nuclear weapons program in the past;
- The state has publicly threatened to withdraw from the NPT;
- There are serious indications that the state is acquiring or developing the non-nuclear components of a nuclear device;[10]
- The state is developing or otherwise acquiring ballistic missiles or other means of delivering nuclear warheads; and,
- There is evidence that national scientists are undertaking research on nuclear explosions or related disciplines suitable to nuclear weapons development.

These individual activities may not be illegal,[11] but a combination of many of them in the same state should be a matter of concern and a reason for the IAEA to increase its verification activities in and scrutiny of that state. If the Agency is unable to do so because the state is not fully cooperating, the secretariat should explicitly report these findings to the IAEA Board of Governors, at least in the publicly available background statement of the annual Safeguards Implementation Report (SIR).

IMPROVING COOPERATION AND TRANSPARENCY

All states that have been called out by the IAEA secretariat for failing to report nuclear material and activities in accordance with their safeguards obligations were implementing a State System of Accounting for and Control of nuclear material (SSAC), which was not fully independent of nuclear operators and state authorities, and did not provide unrestricted access and cooperation to IAEA inspectors. This has been the case in Iraq, North Korea, Iran, Libya, the Republic of Korea, Egypt, and Syria. It is therefore not surprising to note that under "Areas of Difficulty in Safeguards Implementation," the SIR for the year 2010 reports that:

> The performance of State and regional authorities and the effectiveness of SSACs and RSACs [Regional Systems of Accounting and Control] have a significant impact upon the effectiveness and efficiency of safeguards implementation. In 2010, in some States SSACs still did not exist. Moreover, not all existing State and regional authorities have the necessary authority, independence from operators, resources and technical

capabilities to administer the requirements of safeguards agreements and additional protocols. In particular, some States do not impose and verify proper nuclear material accountancy and control systems at nuclear facilities and LOFs [locations outside facilities] to ensure the required accuracy and precision of the data transmitted to the Agency.[12]

The 2008 SIR, for instance, stated:

> The Agency was informed in 2004 by Egypt's SSAC, the Atomic Energy Authority (AEA), that it did not have the authority necessary for it to exercise effective control of all nuclear material and activities in the State. A Presidential Decree was issued in May 2006 to redefine the AEA's authority. Ministerial Decrees were issued in October 2006 for the practical implementation of the Presidential Decree. The AEA then undertook a State-wide investigation of its nuclear material holdings, during which additional, previously unreported, nuclear material was identified, including several depleted uranium items for which Egypt subsequently provided accounting reports.[13]

The Egyptian Atomic Energy Agency's incomplete authority is an explanation, but not an excuse, for the lack of effective control of all nuclear material and activities in the State.

This example demonstrates once more the necessity for the IAEA Board of Governors to request the Secretariat to provide an evaluation of the effectiveness and necessary independence of SSACs, starting with those states that have previously been found to be in breach of their safeguards obligations.[14] It is as important to guarantee this independence and effectiveness (in particular in States with no AP in force) as it is to assess those of national safety authorities.[15]

In this regard, one wonders whether an objective evaluation of the Brazilian-Argentine Agency for Accounting and Control of Nuclear Materials would conclude that this organization is sufficiently independent from the operators of nuclear facilities and from the Brazilian and Argentinean authorities, and whether it fully and satisfactorily cooperates with the Agency. This last question is particularly relevant given that the 2010 SIR notes that short notice random inspections, which are critical to verifying material flows in conversion and fuel fabrication plans, are still under discussion and not yet being implemented in Argentina and Brazil.[16]

Although it is not public, it is rumored that the 2010 SIR also mentions that three states restricted Agency access, two states did not report material that should have been reported, and three states did not permit environmental sampling. These are very important shortcomings and, for the sake of transparency, as well as effectiveness, the Secretariat should name these states.

STRENGTHENING FOUNDATIONS OF NONPROLIFERATION

The objective of IAEA safeguards is to help prevent proliferation by deterring states from seeking nuclear weapons due to the risk of early discovery of a nuclear weapons program. For deterrence to be effective, states must be convinced that any deliberate noncompliance has a high probability of being detected early and that a noncompliant state that does not cooperate fully and proactively with the IAEA to resolve the problems will inevitably face serious consequences. Further, the Agency should be seen as exercising its existing legal authority to the fullest. In particular, whenever justified by the circumstances, it should promptly make use of its right to conduct special inspections at suspicious undeclared locations when states are otherwise denying access.[17]

Recently, the obligation of states to provide early design information about new facilities and the Agency's right to verify it have been challenged by Iran's refusal to comply with its safeguards obligations. The IAEA Director General should make it clear in a document to the Board of Governors that, when and where such refusals occur, they will be recognized for what they are: noncompliance. The Agency should not be complacent toward states that are violating their obligations.

However, the weakest link in the nonproliferation regime today is not the performance of the IAEA Department of Safeguards, but that of the international community in responding to noncompliance. Before the next crisis occurs, generic procedures for responding to noncompliance should be discussed and agreed upon. With a "veil of ignorance" about which states might be involved in the future, such discussions should be easier and less acrimonious than in the heat of a specific crisis. Moreover, agreement upon a set of standard responses to be applied evenhandedly to any state found in noncompliance — regardless of who its allies might be — would significantly enhance the credibility of the nonproliferation regime.

Against this background, a necessary first step is for the IAEA to acknowledge where it has acted inconsistently in the past. In particular, the Board of Governors should adopt a resolution recognizing that failures and breaches committed by South Korea and Egypt in 2004 and 2005 respectively, constituted cases of noncompliance with their safeguards agreements. This resolution, without seeking any punitive measure against either state, would correct damaging precedents by reasserting the impartiality and universality of procedures for reporting noncompliance as envisioned in the IAEA Statute.

For its part, the United Nations (UN) Security Council should adopt legally-binding generic resolutions that would set out a "roadmap" for responding to noncompliance. Experience demonstrates that in investigating safeguards violations in a state not fully and proactively cooperating with the Agency, the IAEA needs, for some limited period of time, enhanced legally binding authority to conduct effective inspections in that state. Such authority extending beyond that provided by the AP can only be granted by a Chapter VII UN Security Council resolution.[18]

Furthermore, considering the precedent of North Korea's 2003 withdrawal from the NPT, it would be wise to plan for the possibility of another state withdrawing as well. As a deterrent, it is essential that the UN Security Council adopts a Chapter VII resolution declaring that the withdrawal of a noncompliant state from the NPT is a threat to international peace and security. In order to secure the irreversibility of safeguards on nuclear material and sensitive fuel-cycle facilities even if a state withdraws from the NPT, the Board of Governors should urge all states with enrichment or reprocessing facilities to conclude "back-up" safeguards agreements that would not terminate in case of NPT withdrawal.[19] Such a facility-specific safeguards agreement would be subsumed to the state's CSA without any additional cost to either the state or the IAEA. Countries like Germany, the Netherlands, Japan, Brazil, and Argentina should lead by example.

The current difficulties in resolving the problems the IAEA is facing in Iran, North Korea, and Syria demonstrate the necessity to act now to ensure that, when the Agency confronts the next proliferation crisis, it has the tools, authority, and political support to avoid repeating history.

If adopted, concrete measures such as those recommended would significantly strengthen the nonproliferation regime and make a real difference in protecting against nuclear proliferation. It depends now on the political will of key governments to make this a reality before the next crisis occurs.

ENDNOTES - CHAPTER 11

1. George Santayana (1863-1952) was a Spanish-American philosopher, essayist, poet, and novelist.

2. For instance, improved surveillance systems, seals, and containment verification equipment, portable and resident non-destructive assay equipment, and new types of equipment to increase the IAEA capability to detect possible undeclared nuclear-related activities.

3. Including well-qualified and trained safeguards inspectors with analytical skills, as well as expertise and resources to carry out disarmament verification activities at the request of member states.

4. Herman Nackaerts, "IAEA Safeguards: Cooperation is the Key to Change," presentation for the Institute of Nuclear Materials Management, July 18, 2011, available from *www.iaea.org/safeguards/documents/IAEA_Safeguards_Cooperation_as_the_Key_to_Change.pdf.*

5. *Statute of the IAEA,* October 6, 1956, available from *www.iaea.org/About/statute.html.*

6. *Ibid.*

7. See IAEA, *The Structure and Content of Agreements between the Agency and States Required In Connection With the Treaty on the Non-Proliferation of Nuclear Weapons*, INFCIRC/153 (Corrected), June 1972, available from *www.iaea.org/Publications/Documents/Infcircs/Others/infcirc153.pdf*.

8. *Model Protocol Additional to the Agreement(s) between State(s) and the International Atomic Energy Agency for the Application of Safeguards*, INFCIRC/540 (Corrected), September 1997, available from *www.iaea.org/Publications/Documents/Infcircs/1997/infcirc540c.pdf*.

9. The main areas of limitation are relating to access to information, to persons, to locations, and to data and documents. A detailed analysis of these limitations and the way they should be corrected can be found in Pierre Goldschmidt, "IAEA Safeguards: Dealing preventively with noncompliance," Washington, DC: Harvard University Belfer Center and Carnegie Endowment for International Peace, July 12, 2008, available from *www.carnegieendowment.org/files/Goldschmidt_Dealing_Preventively_7-12-08.pdf*.

10. As long as no nuclear material is used, a state is entitled, without having to report to the Agency, to study and test the effect of shock waves on non-nuclear materials; to develop high explosives for high-precision applications such as shaped charges; to undertake theoretical studies of the effect of nuclear explosions; or to develop or procure neutron sources (e.g., for applications such as oil well logging) that can also be used as initiators in nuclear weapons. The NPT prohibits manufacture by NNWS of nuclear explosive devices. It seems generally accepted that this includes the production of components that would only have relevance to a nuclear explosive device.

11. Except the first one.

12. IAEA, *Safeguards Statement for 2010*, para. 43, available from *www.iaea.org/safeguards/documents/es2010.pdf*.

13. *Safeguards Statement for 2008*, para. 45, available from *www.iaea.org/safeguards/documents/es2008.pdf*.

14. Pierre Goldschmidt, "Concrete Steps to Improve the Nonproliferation Regime,"April 5, 2009, available from *www. carnegieendowment.org/2009/04/05/concrete-steps-to-improve-nonproliferation-regime/4mp*.

15. In 2007, the IAEA undertook an Integrated Regulatory Review Service of the Japanese Nuclear and Industrial Safety Agency (NISA). Among the 10 main recommendations made by the IAEA, almost 4 years before the Fukushima accident, were: "NISA should more clearly define its expectations with respect to reporting of minor inspection findings and events, in order to screen them for early identification before they become a problem"; and another one which also sounds particularly familiar to safeguards inspectors:

> NISA should ensure that its inspectors have the authority to carry out inspections at the site at any time, on a continual basis. This would ensure that inspectors have unfettered access to the site, to interview people, and to request the review of documents at any time rather than just at prescribed inspection times as in the law. This applies to both the construction and the operational inspection programmes.

IAEA, *Integrated Regulatory Review Service (IRRS) to Japan*, IAEA-NSNI-IRRS-2007/01, June 25-30, 2007, available from *www. nsr.go.jp/archive/nisa/genshiryoku/files/report.pdf*.

16. *Safeguards Statement for 2010*, para. 46.

17. Delaying for months, or even years, a request for special inspection would, in most cases, be self-defeating because the state concerned would have plenty of time to remove any incriminating evidence.

18. A draft of such a resolution and the precise description of the necessary additional IAEA verification rights can be found in Annex I of Goldschmidt, "Concrete Steps to Improve the Nonproliferation Regime."

19. A facility-specific IAEA INFCIRC/66-type safeguards agreement, contrary to a Comprehensive Safeguards Agreement, does not lapse when an NNWS withdraws from the NPT.

PART IV:

IGNORING NUCLEAR WEAPONS
PROLIFERATION INTELLIGENCE

CHAPTER 12

CASTING A BLIND EYE:
KISSINGER AND NIXON FINESSE
ISRAEL'S BOMB[1]

Victor Gilinsky

It is now widely accepted that 1969 marked a turning point in U.S. policy regarding Israeli nuclear weapons. A "stopping point" may be a better description. The pivotal moment appears to have come in a private, unrecorded September 1969 meeting between Richard Nixon and Golda Meir: She is supposed to have admitted having the bomb, and Nixon is supposed to have promised that, as long as Israel kept its bomb under wraps, the United States would not ask questions about it. Up to that point, the United States had been urging Israel to join the Nonproliferation Treaty (NPT).[2] After the 1969 meeting, as General Yitzhak Rabin (the Israeli Ambassador at the time) put it, the subject "dropped off the agenda." In fact, the entire subject of Israeli nuclear weapons dropped off the U.S. foreign policy agenda.

This history is still important today because the subject is still off the U.S. agenda. In fact, the U.S. Government is still committed to keeping Israel off the international nonproliferation agenda.[3] But the pretense of ignorance about Israeli bombs does not wash anymore. President Barack Obama looked foolish, or worse, when he said he did not want to "speculate" whether any countries in the Middle East had nuclear weapons.[4] The evident double standard undermines efforts to control the spread of nuclear weapons worldwide.[5]

It is useful, therefore, to try to understand the 1969 origins of the current approach toward Israeli nuclear weapons and to inquire about the continuing validity of U.S. promises at the time. We have more material to work with, since a few years ago the Nixon Library released many Nixon-era White House documents related to Israeli nuclear weapons, including recommendations to the President from his national security advisor, Henry Kissinger. The released documents — some of them formerly Top Secret — provide a fascinating glimpse into the White House policy reviews before the critical meeting with Meir.

The story has now been told in some detail, most recently by Avner Cohen, who used the 1969 Nixon-Meir meeting as the point of departure for his critique of Israel's policy of "opacity," or total secrecy about its bomb.[6] What strikes me about this, and other accounts of the 1969 U.S. policy shift, is that, however interesting they are, these accounts are focused mainly on the Israeli side of the interaction. From my own brief look at the documents, there is rather more to the story of interest from the U.S. point of view.

Let me sketch some points that strike me about: (1) the Kissinger-directed White House policy analyses and recommendations; (2) Nixon's own handling of the Israeli nuclear issue; and, (3) the current weight of Nixon's promises to Meir, including any promise to shield Israel from the NPT.

NIXON SUBMITS THE NON-NUCLEAR PROLIFERATION TREATY FOR APPROVAL

It was President Nixon, by the way, who ratified U.S. membership in the NPT after President Lyndon Johnson had negotiated it and signed it. Nixon had no particular attachment to the Treaty—it does not even rate a mention in his memoirs—and neither did Kissinger.[7] Still, Nixon submitted the NPT to the Senate soon after he entered office and received its approval in March 1969. Apparently, Nixon was persuaded the United States did not thereby give up any freedom of action. In any case, he had no intention of pressing other countries to adhere to it.[8] However little Nixon thought of the NPT, other senior officials did take it seriously, and the ratified Treaty formed part of the backdrop to dealing with Israel's rapidly evolving nuclear weapons project. Since Israel was not one of the NPT-authorized five nuclear powers, the confrontation with Israel was to be the first test of the universality of the new Treaty.

DECISION ON *PHANTOM II* AIRCRAFT LEFT FROM THE JOHNSON ADMINISTRATION

The immediate nuclear-related Israeli question Kissinger had to address actually had to do with conventional arms—whether to permit delivery of 50 F-4 *Phantom* aircraft that Israel had bought in the last days of the Johnson administration. The F-4 was the top fighter-bomber in the world, and the Israelis wanted it badly. The outgoing administration had written into the F-4 contract the possibility of delivery cancellation if it appeared Israel was getting nuclear weapons.

The Defense and State Departments had wanted, as a condition of the F-4 sale, an explicit Israeli pledge not to build nuclear weapons.[9] Israel offered instead its standard declaration that it would "not be the first country to introduce nuclear weapons into the Middle East."[10] The U.S. interpretation of this was that not "to introduce" nuclear weapons meant not to obtain them.

But Rabin would not agree, nor would he provide an alternative definition. When Defense Assistant Secretary Paul Warnke, who was handling the plane sale, asked, "What do you mean by 'introduce'?" Rabin responded with, "What do you mean by 'nuclear weapon'?"[11] The discussion went round and round until finally Rabin allowed—and this stuck as the Israeli interpretation—that an **unadvertised and untested** nuclear device would not be a nuclear weapon. This made explicit that Israel's declaration did not exclude physical possession of nuclear weapons.

Warnke would not yield on the F-4 sale, so Rabin found ways to get around the Department of Defense (DoD).[12] Seventy senators signed a letter to the President supporting the sale. Arthur Goldberg and others spoke directly to President Johnson, who then ordered the DoD to approve the F-4 sale without conditions.[13] Despite this order, Defense Secretary Clark Clifford permitted Warnke to say in his approval letter to Rabin that the United States retained the option to withhold delivery if Israel was not complying with its pledge not to introduce nuclear weapons—as the United States understood it.[14] Since the planes were not yet built, the final decision on their delivery was left to the incoming Nixon administration.

KISSINGER LAUNCHES POLICY REVIEW ON ISRAELI NUCLEAR WEAPONS

To make the new administration's decision more difficult, intelligence indicated the Israeli nuclear weapons project was advancing rapidly and possibly had already succeeded in producing bombs. (U.S. experts had been visiting Dimona more or less annually since the early-1960s, supposedly to ensure

328

the work there stayed "peaceful," but the Israelis had easily hoodwinked them.)[15] Israel was also producing Jericho missiles, which, because of their low accuracy, could only have been intended for carrying nuclear warheads. Additionally, as Kissinger later informed the President, there was "circumstantial evidence that some fissionable material available for Israel's weapons development was illegally obtained from the United States by about 1965."[16]

It was against this background that Kissinger ran a White House study (National Security Study Memorandum [NSSM] 40) in mid-1969, responding to the issue of Israeli nuclear weapons. The principal participants were the DoD, the Department of State, the Joint Chiefs of Staff, and the Central Intelligence Agency (CIA). They all agreed that Israeli acquisition of nuclear weapons raised the prospect of a more dangerous Middle East and undermined efforts to control proliferation worldwide. They also agreed that a major U.S. effort to stop the Israelis was justified. But they did not agree on what that meant.

In truth, it was too late to stop the manufacture of Israel's first bombs. Any possibility of keeping Israel from going any further depended entirely on the United States—on which Israel depended for advanced weapons—making this a firm condition of the weapons supply. But as the Johnson administration history showed, this condition would not be easy to make stick in the U.S. domestic political environment.

The DoD and the Joint Chiefs, as they did under the previous administration, advocated withholding delivery of the F-4 *Phantom* jets to gain an Israeli commitment not to build nuclear weapons or nuclear missiles, or at least not to deploy them. The State Department, on the other hand, wanted to avoid a confrontation with Israel, in part to preserve political capital for Arab-Israeli peace negotiations. It advocated keeping weapon sales and nuclear issues on separate tracks and proposed a series of well-meaning but ineffectual steps to deal with the nuclear issue.[17] The State Department rationalized that there was still time for negotiations over the issue, that the Israelis had still not completed nuclear weapons, and that, in fact, they really only wanted a nuclear option and might stop on their own. If the Israelis did not stop, the State Department advised, we should at least "make a record for ourselves" of having tried to stop them.

In the hope of facilitating Israeli adherence to the NPT, the State Department offered the view that reasonable interpretation of the NPT's Article III would draw the difference between **maintaining** and **exercising** the option to manufacture nuclear explosives. In other words, State was saying that so long as a country had not taken the last step in nuclear weapon manufacture, it could be judged to be in conformance with the Treaty.

In his recommendation to the President on possible Israeli adherence to the NPT, Kissinger went even further in watering down the meaning of the Treaty. He wrote:

> The entire group agreed that, at a minimum, we want Israel to sign the NPT. This is not because signing will make any difference in Israel's actual nuclear program because Israel could produce warheads clandestinely. Israel's signature would, however, give us . . . a way of opening the discussion. It would also publicly commit Israel not to acquire nuclear weapons.

Kissinger apparently believed that the Israelis might actually sign the NPT—a course they pretended to be evaluating—with the thought of still keeping clandestine bombs. And he was willing to go along with that arrangement.

In the end, the touchstone of U.S. seriousness about stopping Israel's nuclear weapons program was still a willingness to tie delivery of the F-4 *Phantoms* to the nuclear issue. This Kissinger did not propose to do—it seems, on the basis of Nixon's guidance—although he kept the door open to doing so at a later stage. He concluded that holding the planes back would unleash a fierce political response against the administration from Israel's domestic supporters, and that this was too high a price for the administration to pay to uphold the principle of nonproliferation.[18] Without the leverage of the fighter aircraft deal, however, there was no chance of gaining Israeli agreement on the nuclear issue. The only option left was to see what could be salvaged in terms of appearances.

In writing to the President about what the United States really wanted, Kissinger subtly shifted the ground away from trying to stop the Israelis from accepting their nuclear weapons but trying to: (1) avoid the appearance of U.S. complicity in Israel becoming a nuclear power; and, (2) keep Israel's bomb from leading to Arab pressure on the Soviets to match it.[19] "While we might ideally like to halt Israeli possession," Kissinger wrote, "what we really want at a minimum may be just to keep Israeli possession from becoming an established international fact." In other words, if no one knew that Israel had bombs, that was almost as good as if the bombs did not exist—and it was a lot cheaper in political capital.

To make this work, both the United States and the Soviet Union had to pretend total ignorance of the situation. In the case of the U.S. Government, with its difficulty in keeping secrets, it would be best if the government really was ignorant of the truth and so should stop asking questions. The Israelis had to go along with this by keeping their bomb under wraps, but of course, they were going to do so anyway. In short, after all the high-level White House analyses of what to do about Israeli nuclear weapons, the recommended option was for the U.S. Government to stick its head in the sand.

Kissinger and the top U.S. diplomats still pursued Israeli adherence to the NPT, just as had their predecessors in the Johnson administration, and continued fencing with Rabin over the meaning of "introduce" in the Israeli nuclear mantra—again, without resolution. The fact was that, by August 1969, the first of the F-4s were already getting delivered to the Israelis. They did not have to give in on anything.

NIXON DECIDES

Since we have Kissinger's memoranda and his formal recommendations, it is tempting to see in them the intellectual lineage of the President's decision. There is, however, a tendency to exaggerate the importance of the written bureaucratic record—and the work of advisors altogether. High-level decisions often move on other tracks. In the end, it appears that Nixon did in his private meeting with Meir on the nuclear issue— the meeting on that day covered other important topics—what he would have done anyway, quite apart from any advice he got. He gave the Israelis a pass on their nuclear weapons program primarily because he wanted them on his side in what he saw as his worldwide struggle with the Soviets. He did not care about the NPT and ignored Kissinger's (seemingly genuine) recommendation to pursue an Israeli signature.[20] Nixon seems to have decided the United States would not pursue the question of Israeli nuclear weapons, would not press Israel to join the NPT, and would end the by-then farcical U.S. "visits" to Dimona.[21]

It would also have been natural for Nixon to want to keep the entire arrangement secret, for one thing, to avoid charges of complicity in Israel's nuclear program. Similarly, Meir agreed to keep, or acquiesced in keeping, the existence of her weapons secret, which she had every incentive to do, anyhow.[22]

Nixon had already set his course in favor of providing Israel with advanced weapons during the 1968 presidential campaign. He said:

> The United States has a firm and unwavering commitment to the national existence of Israel . . . as long as the threat of Arab attack remains direct and imminent . . . the balance must be tipped in Israel's favor.[23]

In speaking to a Jewish group, Nixon explicitly promised that, if elected, he would send the 50 *Phantoms*, and he told Rabin the same in a private meeting.[24]

A March 1970 memorandum written by the President to Kissinger provides further insight into Nixon's thinking underlying the 1969 Nixon-Meir deal.[25] Nixon wrote the memorandum after his decision in early-March 1970 to delay delivery of a later batch of F-4 *Phantoms* provoked a storm of protest from Israel's U.S. supporters.[26] He had held up the planes because, with an eye on possible Soviet reaction, he did not want to tip the military balance in the Middle East too far in favor of Israel. His willingness to hold up delivery of the F-4s is interesting in itself. This is the same act that Kissinger earlier judged as too risky politically for reasons related to nuclear proliferation or the NPT. But Nixon was prepared to make it for reasons he thought were important.[27]

In the March 1970 memorandum, Nixon told Kissinger that, in further talks with Meir and Rabin, Kissinger needed to "lay it on the line." Nixon said the key to his own pro-Israel stance was opposition to Soviet expansion. He was counting on Israel to stand with the United States. The Israelis had to understand that their "only reliable friends are the hawks in this country," not the liberals. RN (as Nixon referred to himself) "does not want to see Israel go down the drain and makes an absolute commitment that he will see to it that Israel always has 'an edge'." Nixon pointed out that he did not get many Jewish votes in New York, Pennsylvania, California, or Illinois—the implication of which was pretty clear.[28] At the same time, he said, his "silent majority" voters would expect Israel to oppose Soviet expansion everywhere. He also stated they:

> will not stand for a double standard . . . it is a question of all or none. This is cold turkey, and it is time that our friends in Israel understood this. . . . **Unless they understand it and act as if they understood it beginning now they are down the tubes.**[29]

Nixon was irked that U.S. Jews were hawks when it came to Israel but doves on Vietnam, and he obviously wanted the Israelis to help straighten out his domestic political opponents. But what mattered to Nixon most was that Israel stand fast with him against Soviet expansion. That is primarily what the 1969 Nixon-Meir deal concerned.

WHAT U.S. OBLIGATIONS REMAIN FROM THE DEAL?

That 1969 deal still casts a shadow over U.S.-Israeli relations. There are reports that in 2009, President Obama provided Prime Minister Netanyahu with a letter that was said to "reaffirm" the 1969 agreement in writing.[30] In light of this, it is worthwhile to reconsider the assumptions of the original 1969 deal and to ask to what extent they are still valid today.[31]

In their dealings with both the Johnson and Nixon administrations, the Israelis accepted that not being "the first to introduce nuclear weapons into the Middle East" meant keeping their weapon's existence secret and not performing nuclear tests. By Kissinger's account, Nixon emphasized these conditions to Meir as the "primary concern."[32] Despite this, the Israelis conducted a nuclear test in 1979 in the oceans below South Africa.[33] More importantly, everyone now knows about the existence of Israel's nuclear weapons. There is no longer even any ambiguity.

There were a number of reasons the United States worried in the past about public knowledge of Israel's nuclear weapons: One was that the Soviets might then have had to help the Arab countries in some way that increased the risk of a U.S. confrontation with the Soviets. But now the Soviets are gone. Another reason was the fear that public knowledge of the Israeli nuclear weapons program would undermine the NPT, especially in the Middle East, by forcing Arab governments to respond with nuclear programs of their own. Now everyone outside Israel already knows and talks freely about Israeli nuclear weapons. Still another reason was the concern that knowledge about the Israeli weapons might expose the United States to charges of complicity in the Israeli nuclear program. But it is precisely the current policy of pretended ignorance about Israel's weapons that makes the United States look foolish, hypocritical, and complicit to boot.

In the end, it is up to the Israelis to decide how they want to deal with **their** half of the 1969 deal—whether to stick with "opacity." But it is up to the United States to decide how to deal with **our** half—whether to continue the U.S. Government's taboo on discussing Israel's nuclear weapons. Whatever reasons there may be to continue to do so, they do not include obligations flowing from the 1969 Nixon-Meir deal.

ENDNOTES - CHAPTER 12

1. This chapter was previously published in Henry Sokolski, ed., *The Next Arms Race*, Carlisle, PA: Strategic Studies Institute, U.S. Army War College, 2012.

2. As early as 1960, President Dwight Eisenhower met with his top Cabinet officials and military leaders to discuss the problems raised by information that the Israelis, in Secretary of State Christian Herter's words, were "operating a plutonium production plant." Defense Secretary Thomas Gates said, "Our information is that the plant is not for peaceful uses." The President made clear that the issue went beyond the Middle East. He said, "We are now faced with the question of what to do as further countries become atomic producers." He told the group the United States needed to tell the Israelis that we wanted the IAEA to inspect the plant "as a matter of course." See General A. J. Goodpastor, Memorandum regarding a December 9, 1960, conference with the President, January 12, 1961.

3. Consider, for example, the Obama administration's hostile reaction to the proposal, coming out of the 2010 NPT Review Conference, for a 2012 meeting to discuss a nuclear-free Middle East, a goal the United States claims to support. Although the U.S. delegation had voted for the entire document—presumably to avoid an embarrassing conference failure—the Obama White

House immediately thereafter attacked the language of the meeting proposal.

4. Helen Thomas, at the President's first televised news conference, February 9, 2009.

5. Israel was not the only country whose nuclear weapons program was eased by ad hoc considerations that overwhelmed U.S. support for the NPT. This also happened in U.S. interactions with India and Pakistan over their nuclear programs, and, at one point, even with North Korea. In fact, U.S. policy toward India's nuclear program is surely a close second to that toward Israel's nuclear program in its glaring inconsistencies with stated nonproliferation policy. Israel was, however, the first country to face down U.S. nonproliferation policy — immediately after the signing of the Treaty — which created a precedent for U.S. acquiescence in NPT holdouts that was later exploited by other countries.

6. Avner Cohen, *The Worst Kept Secret: Israel's Bargain with the Bomb*, New York: Columbia University Press, 2010. See also the article by William Burr and Avner Cohen in the May 2006 issue of *Bulletin of the Atomic Scientists*: "As long as Israel kept the bomb in the basement — which meant keeping the program under full secrecy, making no test, declaration, or any other visible act of displaying capability or otherwise transforming its status — the United States could live with Israel's 'non-introduction' pledge. . . ."

7. Nixon's memoirs contain no index entry for the NPT and apparently no reference to the Treaty in the book. See Richard Nixon, *RN: The Memoirs of Richard Nixon*, New York, Grosset & Dunlop, 1978. Kissinger's memoirs make two glancing references, one to the 1968 signing by Johnson, and the second to German concerns about discriminatory treatment under the Treaty. See Henry Kissinger, *White House Years*, Boston, MA: Little, Brown and Co., 1979.

8. After deciding to back the Treaty, Nixon instructed U.S. diplomats not to push it too hard, and especially not to lean on the Germans. See also Robert Dallek, *Nixon and Kissinger: Partners in Power*, New York: HarperCollins 2007, p. 136:

In [early] 1969, Nixon . . . urged the Senate to approve a nuclear nonproliferation treaty (NPT) signed by Johnson. Nixon's commitment to an NPT carried no political or economic costs. His internal directive supporting ratification emphasized that adherence to the treaty neither created new commitments abroad nor broadened existing ones. Nor would the treaty cause any international difficulties for the United States, since Nixon had no intention to pressure other countries to follow America's lead.

9. An indication of this comes through in a 1966 cable from Secretary of State Dean Rusk to the U.S. Ambassador in Israel. Rusk described his conversation with the Israeli Ambassador, who repeated what was by then the formulaic "[Israel] would not be first to introduce nuclear weapons in the Near East." Rusk told him, "Nothing would be more disastrous" for Israel than to get nuclear weapons, and he urged the Israelis to accept international inspection. Rusk noted, "If Israel is holding open the nuclear option, it should forget U.S. support. We would not be with you. . . ." Telegram from the Department of State to the Embassy in Israel, Washington, DC, July 28, 1966.

10. A formulation usually attributed to Shimon Peres, who improvised it in response to an unexpected question from President John Kennedy.

11. Memorandum of Conversation, "Negotiations with Israel — F-4 and Advanced Weapons, November 12, 1968," approved by Paul Warnke.

12. In his memoirs, Rabin comments on getting involved in U.S. campaign politics:

Sensitive souls may find the notion of setting a Democratic president against his Republican successor distasteful. If so, they will only be demonstrating their ignorance of the ways and means of American politics. It is not enough to say that in pursuing his country's welfare an ambassador to Washington is entitled to take advantage of the ongoing rivalry between the two parties. The fact is that for his efforts to bear fruit, he is obliged to do so; and any ambassador who is either unwilling or unable to maneuver through America's political landscape to advance his country's interests would do well to return home.

338

Yitzhak Rabin, *The Rabin Memoirs*, Berkeley, CA: University of California Press, 1979, p. 142.

13. It is hard to know what Johnson really thought about Israel getting the bomb. He seemed to genuinely care about the NPT and getting Israel to sign it. But there seems to have been another side, too, as indicated by a story told by Arnold Kramish. In 1967, Kramish had somehow gotten an invitation to visit Dimona. Before leaving, he called U.S. Ambassador Walworth Barbour in Tel-Aviv. "Oh, no," Barbour shouted. "If you learn anything about Dimona, I'd have to tell the President, and then he would have to do something, and he doesn't want to." T. C. Reed and Danny Stillman, *The Nuclear Express: a Political Examination of the Bomb and its Proliferation*, Minneapolis, MN: Zenith Press, 2009, p. 119.

14. Letter, Paul Warnke to Yitzhak Rabin, November 27, 1968. This arguably still conformed to Johnson's instructions, in the sense that the Israelis were not asked to agree beforehand with the U.S. interpretation.

15. The arrangement was first worked out during the Kennedy administration, but it soon deteriorated. In an account provided by former Ambassador Barbour:

> ... We had considerable difficulty making arrangements for periodic visits which was a window-dressing exercise. The Israelis tried to be as forthcoming, or to appear as forthcoming as possible, at the same time without revealing anything to us. This wining them and dining them and taking them down there with, under great secrecy, sometimes even meeting them at the airport when they arrived, and taking them off the plane, and over around the back, and then clearing them through customs with Russian names and so forth [Laughs], it was all a very unrealistic exercise which went on for many, many years and then finally just petered out when even the United States realized it wasn't getting anywhere. And it became ridiculous.

See May 22, 1981, interview, Kennedy Library Oral History Project. One is left with the impression that the State Department, which coordinated the "visits" (specifically, not inspections), was not especially keen on having the experts learn anything. They apparently did not receive intelligence briefings.

16. This refers to the suspicion that Israel stole highly enriched uranium from the Nuclear Materials and Equipment Corporation (NUMEC) fuel plant in Apollo, PA, whose owners had close Israeli ties. A 1965 inventory found that a loss of about 100 kilograms (kg) could not be explained after accounting for all possible industrial loss pathways. By the time of Kissinger's memorandum, a further 150-kg remained missing. By then, the CIA had concluded that the material ended up in Israel's bomb program. In the early part of the Nixon administration, all the top national security officials, including Kissinger and the President himself, were involved in one way or another in the NUMEC case. See Victor Gilinsky and Roger Mattson, "Revisiting the NUMEC Affair," *Bulletin of the Atomic Scientists*, March/April 2010, p. 61.

17. National Security Study Memorandum No. 40, "Israeli Nuclear Weapons Program—Issues and Courses of Action," Rodger P. Davies State/NEA to Dr. Kissinger, undated but evidently mid-1969, formerly Top Secret ("sanitized").

18. Dallek, *Nixon and Kissinger*, p. 176: "The White House considered tying arms shipments to Israeli promises not to go nuclear, but concerns about domestic political opposition deterred it from making the connection." Kissinger barely mentioned the concern about opposition from domestic Jewish groups, even though that was obviously a major factor. This omission is not surprising, since Nixon had earlier instructed his national security staff not to mention domestic political considerations, so as to maintain an illusory separation. See Richard Reeves, *President Nixon Alone in the White House*, New York: Simon & Shuster, 2001, p. 42, which describes a February 22, 1969, Nixon memo to Rogers and Kissinger regarding Middle East papers from State and the National Security Council: "In the future, I want no references to domestic political considerations to be included in any papers. . . ." It is a reminder to be cautious in relying on the written record. One is dealing with people who operate on several levels and who use their writings for multiple purposes.

19. Henry Kissinger, Memorandum for the President, "Israeli Nuclear Program." The copy in the Nixon Library is undated, but it refers to a Tab A dated July 19, 1969. Emphasis added.

(1) Israel's secret possession of nuclear weapons would increase the potential danger in the Middle East, and **we do not desire complicity** in it. (2) In this case, **public knowledge is almost as dangerous as possession itself**. This is what might spark a Soviet nuclear guarantee for the Arabs, tighten the Soviet hold on the Arabs, and increase the danger of our involvement. **Indeed, the Soviets might have an incentive not to know.**

20. Kissinger did not attend Nixon's private meeting with Meir and, hard as it is to believe, he seems not to have immediately taken in the change in policy. In an October 8, 1969, memorandum to the President, he reports on, among other things, Rabin's answer regarding the prospects for Israeli NPT adherence that the next Israeli government will decide after the upcoming elections. Kissinger commented: "This formulation strikes me as unacceptably weak. It seems to me that signature of the NPT with its loopholes and escape clause would not jeopardize Israel's potential nuclear capability or diminish Arab recognition of its conventional military superiority." He recommended that Nixon press Meir to make a "vigorous personal effort" to gain Cabinet support for an Israeli signature and ratification. This was 2 weeks after the Nixon-Meir private meeting. Perhaps the meeting left the NPT issue up in the air, with Nixon leaving it to Meir to decide.

21. The last "visit" took place in July 1969. The Israelis rushed the U.S. team, as usual. Meir refused a later U.S. request from U.S. Ambassador Barbour for an extra day-long visit. As much as the Israelis controlled the visits, they involved a lot of preparation, and there was always the chance of a slip-up that revealed too much. In reality, the Israelis did not have much to worry about—the Americans apparently never sent anyone who knew Hebrew, and they were used to getting the runaround.

22. Kissinger seems to allude to this in his memoirs: "It would be too much to claim that Mrs. Meir agreed; more accurate to say she acquiesced in a formulation whose meaning **only the future would reveal**." Kissinger, *White House Years*, p. 371, emphasis added. Nixon does not mention the September 1969 meeting in his memoirs. Meir was obviously the cleverest of the lot. Of course, it is possible that she may have been reluctant to agree not to test warheads.

23. Statement by Richard Nixon, *The New York Times*, September 9, 1968.

24. Rabin, *The Rabin Memoirs*, pp. 131, 133.

25. Memorandum for Henry Kissinger from the President, March 17, 1970. In his memoirs, Nixon quotes at length from this memorandum, so it seems to reflect his considered views.

26. Nixon quotes at length from it in his memoirs and describes the background as follows:

> At the beginning of March I decided to postpone our delivery of Phantom jets to Israel. I had heard that the Soviets had come under renewed pressures from their Arab clients to surpass the new American deliveries to Israel, and I hoped that since Israel was already in a strong military position, I could slow down the arms race without tipping the fragile military balance in the region. I also believed that American influence in the Middle East increasingly depended on our renewing diplomatic relationships with Egypt and Syria, and this decision would help promote that goal. . . . One of the main problems I faced in this regard was the unyielding and shortsighted pro-Israeli attitude in large and influential segments of the American Jewish community, Congress, the media, and in intellectual and cultural circles. . . . There was a wave of criticism in the media and in Congress when my decision to postpone the Phantom deliveries was announced. . . . I was annoyed that a number of the senators who were urging that we send more military aid to save Israel were opposing our efforts to save South Vietnam from Communist domination. I dictated a memorandum to Kissinger describing my feelings. . . .

27. Ultimately, of course, the Israelis got the planes. Another angle on the plane delivery decision is presented in a recent biography of John Mitchell, Nixon's Attorney General:

> Max Fisher, the late Jewish industrialist, philanthropist, and pro-Israel lobbyist, remembered pleading with Kissinger in 1970 to speed up American delivery of a few dozen *Phantom* fighter jets for which Israel had paid, but, owing to pressure

from Arab states, never received. Completion of the deal would mark a decisive shift in American policy towards Israel: from neutrality to the guarantee of military supremacy Nixon had advocated as a candidate. . . . Who could convince the President? 'Go see John Mitchell,' Kissinger said . . . Fisher did as he was told—and got what he wanted.

See James Rosen, *The Strong Man: John Mitchell and the Secrets of Watergate*, New York: Doubleday, 2008, p. 127.

28. Although earlier in the memorandum, he says he is not motivated by the "Jewish vote."

29. *Ibid.*

30. See Eli Lake: "Exclusive: Obama Agrees to Keep Israel's Nukes Secret," *The Washington Times*, October 2, 2009: "President Obama has reaffirmed a 4-decade-old secret understanding that has allowed Israel to keep a nuclear arsenal without opening it to international inspections. . . ."

31. In any case, the United States is not obligated to observe an informal private agreement of which there is no written record.

32. Kissinger wrote to Nixon in an October 7, 1969, memorandum: "During your private conversation with Golda Meir, you emphasized that our primary concern was that the Israelis make no visible introduction of nuclear weapons or undertake a nuclear test program."

33. President Jimmy Carter's Science Advisor Frank Press commissioned a panel of academic scientists who devised an ingenious alternative scientific explanation about how the satellite might have been fooled. But every expert intelligence body in the government regarded the satellite signal as a valid indication of a test. Incidentally, such a test was also a violation of the Limited Test Ban Treaty, to which Israel is a party. Reed and Stillman, *The Nuclear Express*, p. 180.

CHAPTER 13

THE 1979 SOUTH ATLANTIC FLASH: THE CASE FOR AN ISRAELI NUCLEAR TEST

Leonard Weiss

INTRODUCTION: THE VELA SATELLITE

In the wake of the 1963 Limited Test Ban Treaty, the United States launched a series of satellites under the name Vela[1] (Vela is a constellation in the southern hemisphere sometimes called "the sails" because of its configuration). The Vela satellites were designed to monitor compliance with the treaty by detecting clandestine nuclear tests either in space or in the atmosphere. The first such satellite was launched in 1963 and the last in 1969. They operated by measuring x-rays, neutrons, and gamma rays, and, in the case of the more advanced units, emissions of light using two photodiode sensors called bhangmeters (a name derived from the Hindi word for cannabis). These satellites had a nominal life of 7 years, after which the burden of detection was to be shifted to a new series of satellites under the Defense Support Program (DSP) with infrared detectors designed to detect missile launches as well as nuclear tests. The Vela satellites, however, kept operating long past the end of their nominal design life; one of them, designated Vela 6911, detected an event on September 22, 1979, that has become a subject of intense interest ever since.

THE MYSTERIOUS FLASH

What Vela 6911 detected was a light pattern that had the characteristic "double hump" shape associated with a nuclear explosion.[2] As a function of time, the observed light pattern of a nuclear test rises to an initial peak of luminosity with a subsequent decline due to the fireball being obscured by the shock wave (a thin layer of highly compressed air). As the shock wave cools, it becomes less opaque and the fireball is then increasingly visible, with luminosity rising to a second peak before declining monotonically (see Figures 13-1 and 13-2).[3]

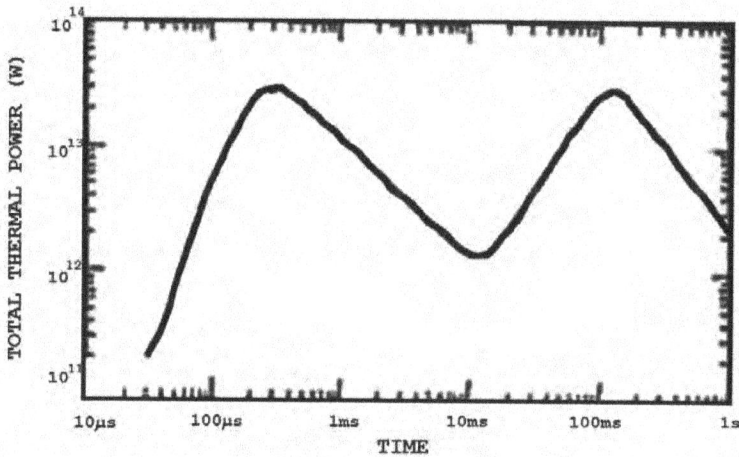

Figure 13-1. Light Pattern for a 19-kiloton Nuclear Test.

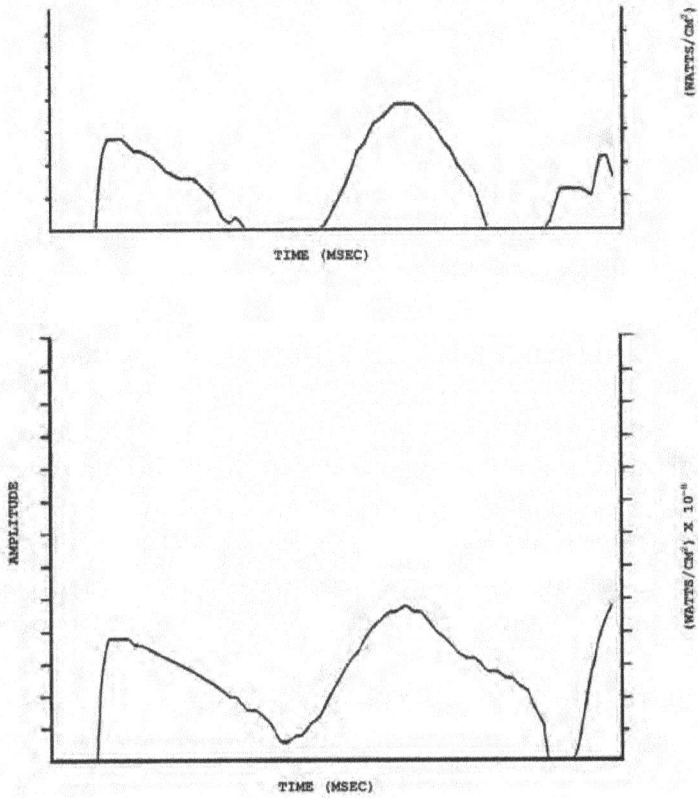

**Figure 13-2. Light Pattern Detected by
Two Bhangmeters of a Vela Satellite for a Known
Nuclear Test
(Signals Above a Fixed Threshold Are Shown).**

Ordinarily, both bhangmeters on the satellite
would have recorded exactly the same signal with an
amplitude or phase difference, depending on the spatial orientation of the satellite with respect to the point
of origin of the blast. However, one of the bhangmeters, possibly because of a malfunction, did not repro-

duce precisely the record of the other.[4] This has been a key element in the argument of the increasingly small community of interested parties who believe that no test took place.

In any case, the U.S. Government acted quickly and began searching for data from sources other than the Vela that could corroborate the event as a nuclear test. This included data from the bhangmeters on the DSP satellites and from the Ionospheric Observatory at Arecibo, Puerto Rico, which might detect an ionospheric wave resulting from an atmospheric test. Aircraft were dispatched to try to obtain evidence of radioactive debris in the atmosphere in the vicinity of what was calculated to be the site of the event. In addition, the Naval Research Laboratory (NRL), which had played an important part in establishing a nuclear test detection system early in the Cold War era, prepared to analyze any data that would be collected by naval ships dispatched to try to collect radiological evidence in the ocean; NRL's task included collecting and analyzing hydro-acoustic and ocean wave data that might also provide evidence of a nuclear test.[5]

The results of these efforts were mixed; the DSP satellites recorded no flash[6] and no radioactive debris was found, but a researcher at Arecibo recorded an ionospheric wave traveling in an anomalous direction that could have been the result of a nuclear test.[7] The NRL analysis of its hydro-acoustic and wave data took time to prepare and, in the end, convinced its scientific director that a nuclear test had taken place.[8] However, the data and analysis are still classified.[9] The lack of an immediate and definitive corroboration that a nuclear event had taken place led to rampant speculation about the event. The initial assessment of the National Security Council (NSC) in October 1979 was that the intelligence community had "high confidence" that the event was a nuclear test.[10] A later NSC report altered this conclusion to one of "a position of agnosticism."[11]

A PROBLEM FOR THE CARTER ADMINISTRATION: WHO DID IT?

In the meantime, the Jimmy Carter administration had to think about the political ramifications of a test, if indeed one had taken place. One problem was that a clandestine test not definitively labeled as such meant that the system for detection could be claimed to be insufficiently reliable, calling into question the ability to detect any Soviet cheating on the Limited Test Ban Treaty, and therefore undermining the value of the second Strategic Arms Limitation Talks (SALT) II Treaty that had been signed in June 1979 and was awaiting a Senate vote on ratification. Carter had made nonproliferation and disarmament a key element of his presidency and was expected to run for reelection in 1980, touting his successes in that arena. A Soviet clandestine test was unlikely, but if the "mysterious flash" was not a Soviet test, who else would have and could have done it?

Initial speculation centered on South Africa[12] because of the calculated geographic location of the event and the knowledge that South Africa was developing nuclear weapons. In addition, a *Washington Post* story revealed that U.S. intelligence had tracked a secret South African alert to some of its naval forces a few days prior to the Vela event and an associated movement of some of its ships in the calculated vicinity at the ostensible time of the event.[13] A January 1980 intelligence report sent to the Arms Control and

Disarmament Agency said South Africa was the most likely perpetrator. But the South African program was actually insufficiently advanced at that point to conduct a small clandestine test, a conclusion that was verified later by the International Atomic Energy Agency (IAEA), among others.[14]

Attention then turned to Israel, which presented the Carter administration with additional political concerns. The Camp David Accords between Israel and Egypt had been brokered earlier that year by President Carter and were going to be an important element of Carter's reelection campaign. Assistant Secretary of State Hodding Carter described the State Department attitude as one of "sheer panic" upon receipt of the news of the Vela incident and that Israel might be involved.[15] The State Department had taken a hard line toward Pakistan in 1977 and 1979, cutting off economic and military assistance as a result of Pakistan's nuclear enrichment and reprocessing imports, which had violated the Symington and Glenn amendments to the Foreign Assistance Act, even though Pakistan was still years away from the ability to test a nuclear device. Under the circumstances, the U.S. Government would be hard pressed to ignore an evident Israeli test, especially since Israel had signed the Limited Test Ban Treaty. To do so would have negative repercussions in the Arab world and possibly blunt progress toward peace in the Middle East, but to take any punitive action against Israel would upset the Jewish diaspora in the United States, an important constituency for Carter and the Democratic Party.

THE RUINA PANEL

To relieve the political pressure created by the Vela event, the Carter administration seized upon the discrepancy between the Vela bhangmeters and speculation that the meters could have recorded a combination of natural phenomena (e.g., lightning plus a meteor strike) that might mimic a nuclear test to parry the growing opinion in intelligence circles that the Vela event was a nuclear test.

The White House asked Frank Press, the President's science advisor and Director of the Office of Science and Technology Policy, to convene a panel of scientific experts to review the available data and determine whether the "double flash" was the result of a nuclear test, a natural phenomenological event, or a satellite malfunction. An MIT electrical engineering professor and long-time consultant to the government on defense matters named Jack Ruina was made chairman of the panel, which included scientific luminaries Luis Alvarez, Richard Garwin, Wolfgang Panofsky, Richard Muller, Alan Peterson, William Donn, Riccardo Giacconi, and F. William Sarles.

The panel was specifically tasked to ignore all political questions concerning the event such as who might be in a position to conduct such a test if it was nuclear.[16] CBS News reported that the administration withheld intelligence data from the Ruina panel, showing that Israel and South Africa were cooperating on the development of missiles that could carry nuclear warheads.[17] This guaranteed that Israel would not be mentioned in the report if the conclusion was that a nuclear test had occurred. Thus, while the Carter administration did not create false intelligence data to reach a desired conclusion, it hoped to create an alternative explanation of the data at hand that could enable it to ignore or counter the conclusion of most of the government's intelligence analysts.

One possibility was the effect of sunlight glinting off the debris of a micrometeoroid that had struck the Vela satellite. Studies had been performed by Mission Research Corporation (MRC) and Sandia National Laboratory suggesting several meteoroid shape and trajectory models that could explain the waveform observed by the Vela bhangmeters. In addition, there was considerable data from an experiment on the spacecraft *Pioneer 10* that might shed light on what kind of optical signals might be detected from meteoroid collisions. In December 1979, Stanford Research International (SRI) was tasked with assessing the probability that the Vela signal was caused by a sunlight-meteoroid interaction, and examined both the *Pioneer 10* data and whether the circumstances postulated in the MRC and Sandia models would actually come about, taking account of the number of sensor observations over the life of the bhangmeters. The SRI report concluded that the *Pioneer 10* data contained insufficient information to make a definitive judgment about the Vela signal's origin but that the aforementioned models would require more than one meteorite strike with a particular set of characteristics to result in the Vela signal of September 22, 1979, and that the probability of this happening was of the order of one in 100 billion.[18] Their calculation was reviewed and affirmed in the context of other data in a 1980 Defense Intelligence Agency (DIA) study.[19]

THE RUINA PANEL'S REPORT

The Ruina panel's report was classified and officially presented on May 23, 1980. An unclassified version was released on September 23, 1980.[20] The report focused on the differences in the measurements obtained by the two bhangmeters and concluded that the signal was probably not that of a nuclear explosion, though it could have been. The panel offered an alternative explanation of the signal, suggesting the possibility that it could have come from sunlight glinting off the debris of a micrometeoroid that had struck the Vela satellite. As already indicated, the probability of a micrometeoroid causing the bhangmeter signals of September 22, 1979, was estimated as one in 100 billion. A personal explanation of the Ruina panel's conclusion was provided by Luis Alvarez in his 1987 memoir,[21] in which he states that he asked DIA to provide a selection of the Vela records that indicated events that were nuclear explosions or were unclear as to their origin but had some signal characteristics associated with a nuclear explosive event.

The latter were called "zoo animals" or "zoo events" in reference to the "zoo-ons" that physicists like Alvarez called the unexplainable tracks in a bubble chamber experiment. In his memoir, Alvarez seems to claim that only one bhangmeter recorded the September 22 "flash" and, on that basis, suggests that the flash was a zoo event. But the panel's report and other accounts of the flash refer to differences in the two bhangmeters recorded intensities rather than a complete nondetection. In a private conversation I once had with Richard Garwin, he spoke merely of "phase differences" between the recorded signals of the bhangmeters, not a failure to detect. More recently, the light signals seen by Vela 6911 on September 22, 1979, have become publicly available (see Figure 13-3), showing detection by both bhangmeters. What Alvarez was probably referring to was not the bhangmeters but a third optical sensor that was used normally to locate the geographic origin of an event but was no longer operating on Vela 6911. A paper by Carey Sublette[22] in the Nuclear Weapon Archive lays out other flaws in Alvarez's defense of the Ruina panel's report, which had concluded that the Vela signal more likely represented a zoo event than a nuclear explosion.

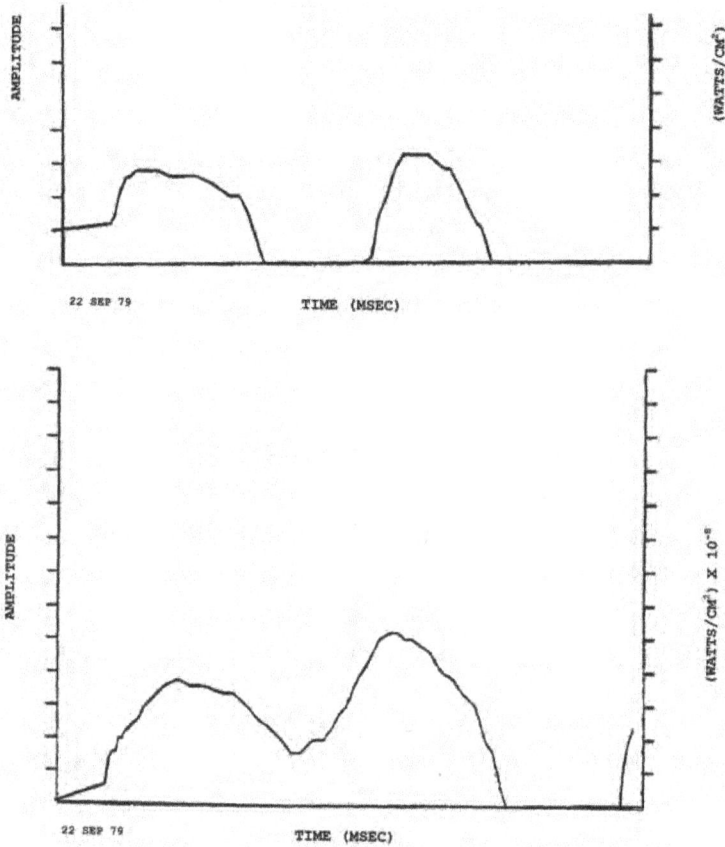

Figure 13-3. Bhangmeter Light Patterns for Event Detected by a Vela Satellite on September 22, 1979.

THE NRL REPORT

It is interesting to compare the U.S. Government's treatment of the Ruina panel's report with other classified documents that suggested more definitively that the Vela event was a nuclear test. In the late-fall of 1981, I interviewed Alan Berman, the former scientific director of the NRL, who had retired from the laboratory and was then the Director of the Marine Laboratory of the University of Miami. I had known Berman for more than a decade as a result of my part-time consulting and research position at NRL. Berman was unanimously viewed at the NRL as a superb scientist and administrator who would never color a scientific data-based conclusion because of political or ideological considerations. My interview with him took place about 18 months after a 300-page NRL report that laid out the laboratory's analysis of the hydro acoustic and other data collected following the Vela event had been completed in the summer of 1980. According to one account, the report concluded that the event was most likely a nuclear test and was accompanied by a large underwater signal resembling signals given by previous nuclear explosions conducted by France in the Pacific in the 1970s.[23]

Berman had said that pulses of underwater sound detected by Navy sensors at two locations following the blast were the strongest corroborative evidence that a nuclear explosion had taken place. Regarding that evidence, he said this evidence was "strong enough to make the case in its own right."[24] The Navy sensors showed that the explosion's signal was reflected off the Antarctic shelf and the reflection was also detected, allowing a calculated estimate of the event's location in the vicinity of Prince Edward and Marion Islands.

The White House ignored the NRL report and referenced only the Ruina panel's report whenever publicly queried. Berman had vociferously objected when the Ruina panel's report was released prior to the completion of the NRL report, and he was still furious when I interviewed him in his office. On two other occasions in late-1980 following the delivery of the NRL report, he had contacted the White House with new information indicating additional support for the conclusion that a nuclear test had taken place, and offering to undertake a broader analysis of the information. But his offer was ignored or rebuffed.[25] One of these contacts was by means of a letter to John Marcum, then a senior advisor to the White House on technology and arms control.[26] Marcum was one of the officials helping the administration deflect attention from the growing consensus in the intelligence community that the Vela signal was nuclear in origin.

FURTHER EVIDENCE OF A NUCLEAR TEST: A PERSONAL MEMOIR

Based on what I learned in a number of briefings, I reached the conclusion that the September 22 event was a nuclear test, and I was not shy in offering that opinion during discussions within the government on nonproliferation issues. But I said nothing publicly. The first news story about the Vela detection occurred on October 25, 1979, when John Scali, then working for ABC News, broke the story of the flash after being briefed by contacts at the Pentagon. But Scali did not claim that the event was a nuclear test. Others, however, did.

One of the most outspoken proponents of the notion that a nuclear test had taken place was Major General George J. Keegan, former head of Air Force Intelligence. Keegan had a long military career before retiring in January 1977. He received much notoriety for claiming that the Soviet Union had achieved a breakthrough in the development of directed energy weapons, specifically in the area of particle beam weapons, and that this constituted a serious shift in the balance of strategic power between the two superpowers. Although both President Carter and Defense Secretary Harold Brown issued public statements refuting Keegan's claim, the administration responded to political pressure from Congress on the issue and significantly expanded the American directed energy program. Later it became clear that Keegan had misidentified a nuclear rocket facility in the Union of Soviet Socialist Republics (USSR) as a particle beam facility.[27] Keegan took a significant hit to his reputation over this error, and he became *persona non grata* within the Carter administration, whose personnel began referring to his claims as "Keegan's Follies." Thus, when Keegan publicly stated his opinion that the Vela event was a nuclear test, the Carter administration lost no time in pointing out how wrong he was in the past on the directed energy weapons issue.

This was brought home to me personally when, at a nonproliferation briefing given by Carter administration personnel, I was taken aside and told that if I persisted in stating my belief that a nuclear test had taken place on September 22, my reputation would take a hit, and I would suffer the same fate as Keegan. Nonetheless, in my role as Staff Director of the Senate Subcommittee on Energy and Nuclear Proliferation, I continued to make numerous requests to see the classified data from Vela 6911, but without success. I felt I was being stonewalled. All this simply reinforced my belief that the Vela event was a nuclear test and that the Ruina panel was engaged in an exercise designed by the White House to give it the ability to point to an alternative scenario—albeit one that had a low probability of occurrence.

But any small doubt I might still have harbored about the origin of the double flash was erased by an event that took place in the office of Senator John H. Glenn of Ohio on March 6, 1981. At the time, I was working as Glenn's chief advisor on nonproliferation issues as well as in my formal position on the Senate subcommittee, of which Senator Glenn was the Ranking Member (he lost the chairmanship when the Republicans took over the Senate in the wake of the 1980 election in which Ronald Reagan beat Jimmy Carter). I received a call to my office that morning from a well-known CBS News reporter named Robert Pierpoint. Pierpoint said that CBS was doing a story on the "mysterious flash," and that he had heard that I had some "interesting" opinions about it. He asked if I would be willing to say those things on camera for possible broadcast on the CBS Evening News show anchored by Walter Cronkite. Perhaps naively, I agreed and gave Pierpoint permission to bring a camera crew to my office, which he did a few hours later. While they were setting up their equipment, the phone rang, and my secretary announced that Senator Glenn was on the phone. The first thing he said to me after I said hello was to tell me that a phone call had been made to his office by the White House, and that (much to my astonishment) the White House had heard that I was going to give an on-camera interview about the Vela event. He asked if that was true, and I said that not only was it true, but the camera crew was in my office as we spoke. Senator Glenn responded by saying that the White House was very upset and that I needed to come to his office immediately to discuss this. I excused myself and told Pierpoint I needed to talk to Glenn for a few minutes.

It took about 3 minutes to walk to Glenn's office, and when I entered his inner office, he was there with his press secretary and erstwhile campaign manager, Steve Avakian. They looked grim. Glenn began by telling me again how upset the White House was about the proposed interview, and he asked me what I intended to say. When I said that "I intend to say that the 'mysterious flash' was a nuclear test," he responded sharply, "No! You can't say that!" He then reiterated how upset the White House was and how damaging the political fallout could be if I went ahead. Glenn said the White House told him that my interview could result in a serious foreign policy problem for the United States. Then he uttered a cryptic comment about how his political enemies would "make hay" over this were I to cause a problem.

Needless to say, I was stunned by all this. I had given interviews before on other issues and had never before been given an order to say or not say something. But I was not about to risk losing my job, so I said I would go back to my office and call off the interview. At this, Avakian jumped in and, with Glenn's evident approval, said "No! You have to go ahead with the interview but you can't say there was a nuclear test!" As I started walking out, I asked who had made the call to Glenn. They said it was John Marcum, the same person whom Alan Berman had written to in an attempt to get the White House to pay attention to the NRL report and the laboratory's capabilities in analyzing any new data. Marcum was now representing the Reagan administration in trying to scuttle unwanted comments and conclusions about the Vela event. Clearly, concerns about Carter's presidential fortunes in September 1979 were not the only reason for White House panic over the "flash." It was now a bipartisan panic, and that meant to me not only that the "flash" was a nuclear test but also that Israel was the likely perpetrator.

I left Glenn's office with my head swimming. How was I going to do an interview on the Vela event without lying and without saying explicitly that I believed it was a nuclear test? I decided the least I could do was to indicate my disdain for the alternative scenario contained in the report of the Ruina panel. I said that "I was surprised at the zeal which some people were bringing to the question of proving that this was not a nuclear event," and used the White House locution that, "If this was a nuclear event, it would present a serious political problem for the United States." I concluded by saying, "I don't think it is possible to lay this event to rest with a report that indicates that a group of people feel that the probability of it not being a nuclear event is perhaps more than half, and on that basis, we all should forget about it and go to sleep."

The comment about the event being a political problem for the United States was code for the problems that would be created by naming Israel as the culprit. I was upset that I had to resort to verbal subterfuge to get my point across, but I was relieved that Pierpoint did not accuse me of bait-and-switch. In fact, the interview was broadcast that night and was the last segment of Cronkite's farewell broadcast as anchor before he personally signed off. But my experience that day in the Glenn office and the representations made of the panicky White House phone calls were the last bits of evidence for me, if any were needed, that Vela 6911 had recorded a nuclear test, and the most likely perpetrator was Israel, probably with South African support. To underscore the unique nature of my interaction with Glenn in this case, I worked for him for another 20 years, gave many interviews, and never was told again what I could or could not say.

360

It was perhaps a coincidence that about 3 weeks after the CBS broadcast, I was finally allowed to see the Vela satellite data I had been seeking for months. I examined the graphed "flash" data along with the group of "zoo events" referred to by Luis Alvarez. Perhaps I should not have been surprised at that point, but, notwithstanding the phase differences between the bhangmeters on Vela 6911, the plot of the data showed the two humps of the classic curve associated with the light intensity from a nuclear explosion (See Figure 13-3). Moreover, there was not a single "zoo animal" that came close to the classic shape in duration and amplitude (see Figures 13-4A and 13-4B). Finding an alternative explanation other than a nuclear test for the "flash" of September 22, 1979, required some serious stretching of the mind by the individuals on the Ruina panel.

FURTHER EVIDENCE SUPPORTING THE CONCLUSION THAT THE "FLASH" WAS AN ISRAELI TEST

In 1991, Seymour Hersh published *The Samson Option*, which described the history of the Israeli nuclear weapons program up to that time. Hersh reported that former Israeli government officials told him that Vela 6911 recorded an Israeli test of a low-yield nuclear artillery shell, and that the test was the third in a series conducted over the Indian Ocean. Hersh wrote that the test was preceded by a visit to the site by two Israeli ships and that elements of the South African Navy were observers. He also describes the panic among White House and State Department officials upon learning of the Vela event. But Hersh ascribes the panic mainly to the Carter administration's

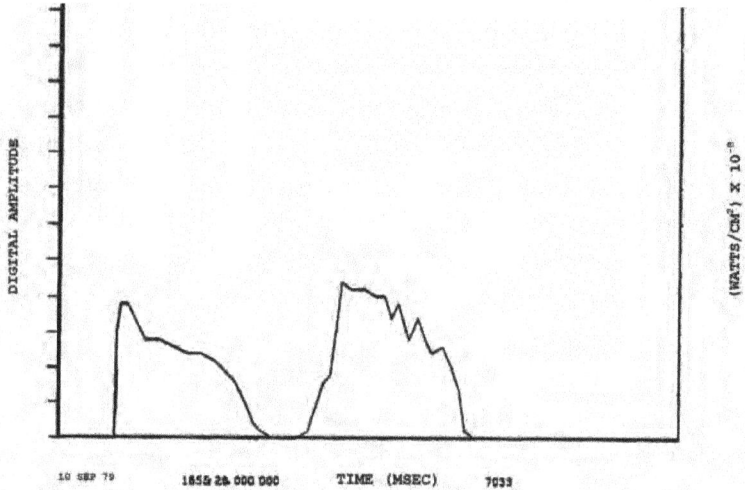

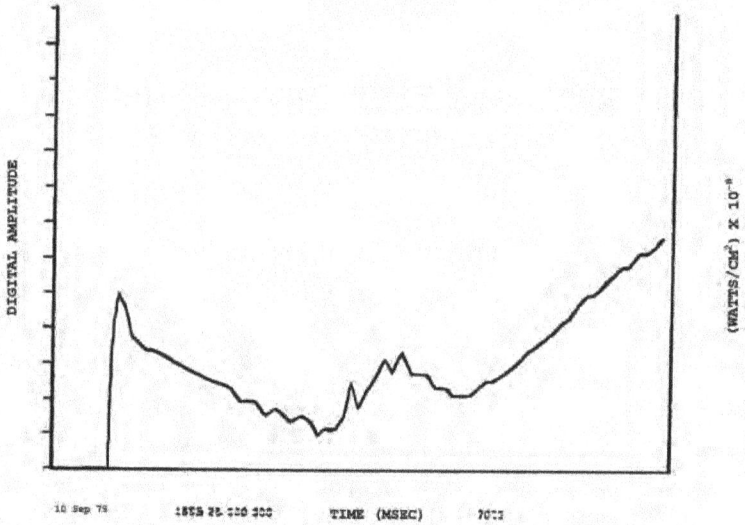

Figure 13-4A. "Zoo Event."

362

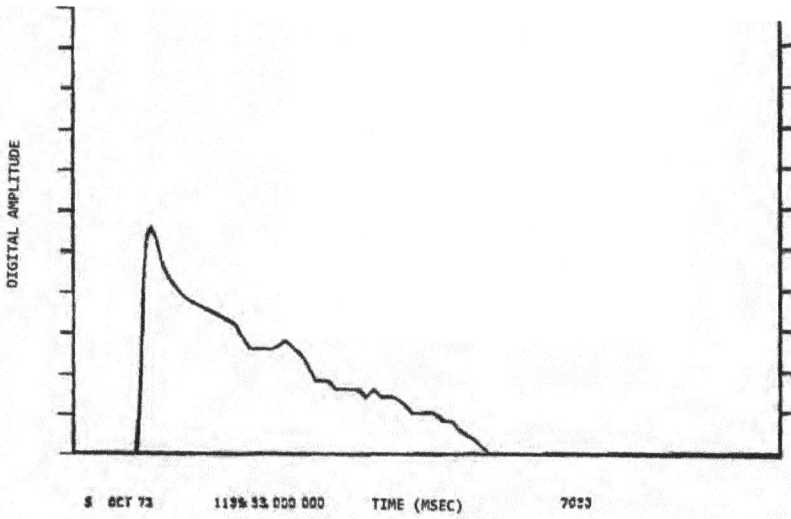

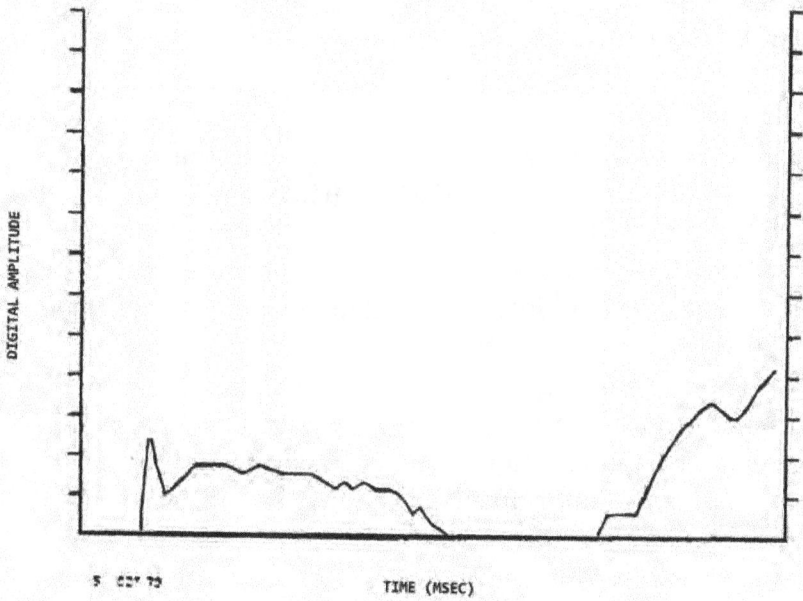

Figure 13-4B. "Zoo Event."

concerns about the fate of the SALT II treaty and the political ammunition a clandestine test would give to Republican opponents. My own experience showed that the Reagan White House was equally concerned over the prospect of a confirmed clandestine Israeli nuclear test at a time when the United States was ostensibly trying to hold the line on proliferation activities in Pakistan and Congress was considering legislation prohibiting military assistance to Pakistan in the event of a Pakistani nuclear test. Hersh also quotes a number of prominent members of the Nuclear Intelligence Panel who had examined the Vela data and concluded it was a nuclear test but were ordered not to discuss it publicly. In particular, the chairman of the panel, Donald Kerr, who had been acting director of defense programs at the Department of Energy, told Hersh, "We had no doubt it was a bomb."[28]

On April 20, 1997, an article in the Israeli newspaper, *Ha'aretz*, quoted South African Deputy Foreign Minister Aziz Pahad as confirming that the Vela event was a nuclear test. The article said that Israel had helped South Africa develop its bomb designs in return for 500 tons of uranium and other assistance. Although Pahad later claimed his statement had been taken out of context, the *Ha'aretz* article was referenced in a July 11, 1997, Los Alamos Laboratory newsletter under the headline, "Blast from the past: Lab scientists receive vindication." This referred to earlier work by the laboratory, concluding that a nuclear test had taken place on September 22, 1979. Dave Simons of the Nonproliferation and Arms Control Research and Development division said, "The whole federal laboratory community came to the conclusion that the data indicated a bomb," and that "we were quite thoroughly convinced of our interpretation."[29] Although the power of the article has been diminished some-what by Pahad's partial retraction, the latter did not result in any retraction by laboratory scientists that a nuclear test took place.

That the Vela event was the result of a cooperative effort by Israel and the apartheid regime of South Africa has been claimed or suggested many times,[30] such an effort would have been the logical result of an arms trade relationship between the two countries that included the transfer of advanced military technology and nuclear materials. It has been reported that, at one point in 1975, Israel offered to sell Jericho missiles to South Africa that could carry nuclear warheads, and it may even have offered to sell the warheads themselves.[31]

A U.S. GOVERNMENT COVER-UP AT THE TOP?

As of this time, the conclusion that the Vela event was a nuclear test is shared by the directors of the U.S. nuclear weapons laboratories, senior officials at the DIA, and many members of the scientific community.[32] Others in the intelligence community—such as the Director of Central Intelligence's Nuclear Intelligence Panel; many scientists and analysts at the Los Alamos, Livermore, and Sandia National Laboratories; and at SRI International, DIA, Mission Research Corporation, and the Aerospace Corporation—subscribe to the conclusion that the event was "most probably" a nuclear test.[33] Yet, despite this considerable body of expert opinion, the U.S. Government under both Democratic and Republican administrations still has not admitted that a nuclear test took place.

In his recently published book with diary entries, former President Jimmy Carter briefly, but revealingly, writes about the September 22, 1979, "flash." In the entry dated on the day of the flash, he writes, "There was indication of a nuclear explosion in the region of south Africa -- either South Africa, Israel using a ship at sea, or nothing."[34] In another diary entry dated October 26, Carter writes, "At the foreign affairs breakfast we went over the South African nuclear explosion. We still don't know who did it."[35] It is no coincidence that this entry occurred the day after ABC reporter John Scali revealed publicly the existence of the Vela event. Five months later, on February 27, 1980, Carter writes, "We have a growing belief among our scientists that the Israelis did indeed conduct a nuclear test explosion in the ocean near the southern end of Africa."[36] That Israel is immediately mentioned in the first entry by Carter about a possible nuclear test near South Africa is not a surprise. The intelligence agencies were watching the military relationship between Israel and South Africa, and Carter was specifically aware of the Israeli nuclear weapons program and where they might have obtained weapons materials.

In a cryptic reference to the Nuclear Materials and Equipment Corporation (NUMEC) affair,[37] his diary entry of August 2, 1979, reads as follows: "The question of lost uranium in the 1960s that may or may not have gone to Israel is a matter we have been discussing. It's going to be a public issue shortly when ERDA [the Energy Research and Development Agency] makes its report."[38] It is clear from these entries that Israel was a prime suspect in the Vela event from the beginning, and the appearance of these entries in his book strongly suggests that Carter believes the flash was indeed an Israeli nuclear test. But he did not say anything approaching that when he was President. The public path of ambiguity taken by Carter as Presi-

366

dent on the Vela event has been trod by every President since then, enabled by the refusal to declassify relevant data and documents.

Keeping important evidentiary data secret makes it difficult for outside independent investigators to evaluate critically and definitively the conclusions of the Ruina panel and the 300-page NRL analysis, among other things. One of the likely reasons that the U.S. Government is withholding the declassification of relevant documents is to assist Israel to maintain its policy of opacity in nuclear affairs, a policy which had its origin during the Lyndon Johnson presidency and was reinforced in a bargain made with the United States during the Richard Nixon presidency.[39] Its abandonment, accompanied by the admission that Israel violated the Limited Test Ban Treaty, would create some serious political fallout for both countries. But it is hard to argue that helping Israel in this way contributes to U.S. national security at a time when the U.S. demands openness in the nuclear activities of Iran, North Korea, Syria, and all other countries who may be engaged in clandestine weapon-related nuclear activities.

FINAL COMMENT

This raises a general policy question. The Iraq War has shown the harm that can result from the politicization of intelligence in order to support a desired policy outcome whose support by the public would otherwise be problematic. In the case of the Vela event, U.S. administrations on both sides of the political fence have sought to ignore or demote the value of legitimately collected and analyzed intelligence information in order to reduce or eliminate pressure to

367

take an action with unpredictable or negative political repercussions. Obfuscating or denigrating hard intelligence data in order to avoid a political problem can be as dangerous to national security and democracy as inventing bogus intelligence in order to smooth the way into a war. Both tactics are designed to mislead the public and are therefore antithetical to democratic governance. It is time for the U.S. Government to open up its files on the Vela event and end a charade that has persisted for over 30 years.

ENDNOTES - CHAPTER 13

1. See Sidney Singer, "The Vela Satellite Program for the Detection of High Altitude Nuclear Detonations," *Proceedings of the Institute of Electrical and Electronics Engineers* [IEEE], Vol. 53, No. 12, December 1965, pp. 1935-1948, available from *ieeexplore.ieee. org/stamp/stamp.jsp?tp=&arnumber=1446400*.

2. See Jeffery Richelson, *Spying on the Bomb*, New York: W. W. Norton and Co., 2006, p. 285. See also Carey Sublette, "Report on the 1979 Vela Incident," available from *nuclearweaponarchive.org/ Safrica/Vela.html*.

3. See Guy Barasch, "Light Flash Produced by an Atmospheric Nuclear Explosion," Report LASL 79-84, Los Alamos, NM: Los Alamos National Laboratory, November 1979, available from *www.fas.org/sgp/othergov/doe/lanl/docs4/00362363.pdf*.

4. Dale S. Sappenfeld, David H. Sowle, and Trella H. McCartor, "Possible Origins of Event 747 Optical Data," Prepared for Air Force Tactical Operations Center, Patrick AFB, Mission Research Corporation, Santa Barbara, CA, August 1, 1980, p. 8, available from *foia.abovetopsecret.com/VELA_SATELLITE/ THE_VELA_INCIDENT/REPORTS/POSSIBLE_ORIGINS_OF_ EVENT_747_AUG_1980.pdf*.

5. See David Albright and Corey Gay, "A Flash from the Past," *Bulletin of Atomic Scientists*, Vol. 53, No. 6, November/De-

cember 1997, p. 17, available from *books.google.com/books?id=vgwA AAAAMBAJ&pg=PA15&lpg=PA15&dq=Albright,+vela,+flash&sour ce=bl&ots=XP0fvSvR46&sig=xHyRNQdS2MiMoPP_ZLLm7Z6GG5 I&hl=en&ei=3OqPTfWkJY24sQOCkczzCA&sa=X&oi=book_result&c t=result&resnum=2&ved=0CBwQ6AEwAQ#v=onepage&q=Albright %2C%20vela%2C%20flash&f=false.*

6. Henry G. Horak, "Vela Event Alert 747," Los Alamos, NM: Los Alamos National Laboratory, May 1980, available from *foia. abovetopsecret.com/VELA_SATELLITE/THE_VELA_INCIDENT/ REPORTS/LA_8364_MS_MAY_1980.pdf.*

7. See David Albright, "The Flash in the Atlantic," *Bulletin of the Atomic Scientists,* Vol. 50, No. 4, July/August 1994, p. 42.

8. Eliot Marshall, "Navy Lab Concludes the Vela Saw a Bomb," *Science,* Vol. 209, No. 4460, August 29, 1980, pp. 996-997, available from *www.sciencemag.org/content/209/4460/996.extract.*

9. Albright and Gay, p. 17.

10. National Security Council (NSC) Memo dated October 22, 1979, PDF file, p. 2, available from *foia.abovetopsecret.com/VELA_ SATELLITE/THE_VELA_INCIDENT/NATIONAL_SECURITY_ COUNCIL/NATIONAL_SECURITY_COUNCIL_OCT_22_ 1979.pdf.*

11. National Security Council Memo dated Jan. 7, 1980, PDF file, p. 2, available from *foia.abovetopsecret.com/VELA_SATELLITE/ THE_VELA_INCIDENT/NATIONAL_SECURITY_COUNCIL/ NATIONAL_SECURITY_COUNCIL_JAN_7_1980.pdf.*

12. NSC Memo dated October 22, 1979.

13. Interagency Intelligence Memorandum, "The 22 September 1979 Event," January 21, 1980, p. 8, approved for release June 2004, available from *www.gwu.edu/~nsarchiv/NSAEBB/ NSAEBB190/03.pdf.* See also Thomas O'Toole, "New Light Cast on Sky Flash Mystery," *The Washington Post,* January 30, 1980. See also Thomas O'Toole and Milton Benjamin, "Officials Hotly Debate Whether African Event was Atom Blast," *The Washington Post,* January 17, 1980.

14. Interagency Intelligence Memorandum, p. 11.

15. Seymour Hersh, *The Samson Option,* New York: Random House, 1991, p. 274.

16. *Ibid.,* p. 277.

17. See CBS Evening News, West Coast Edition, Walter Cronkite's last day (March 6, 1981). Part 4 of 5, available from *www.youtube.com/watch?v=piLAsh6GRE0.*

18. George Oetzel and Steven Johnson, "Vela Meteoroid Evaluation," T/8503/T/PMP, *SRI International,* January 29, 1980, p. 42, available from *foia.abovetopsecret.com/VELA_SATELLITE/ THE_VELA_INCIDENT/REPORTS/SPECIAL_TECHNICAL_ REPORT_2%20JAN_29_1980.pdf.*

19. John Mansfield and Houston Hawkins, "The South Atlantic Mystery Flash: Nuclear or Not," Defense Technical Intelligence Report, Washington, DC: DIA, June 26, 1980, summarized in "The Vela Incident: Nuclear Test or Meteoroid," Jeffery Richelson, ed., *National Security Archive Electronic Briefing Book,* No. 190, May 5, 2006, available from *www.gwu.edu/~nsarchiv/NSAEBB/ NSAEBB190/index.htm.* See also Richelson, p. 305.

20. Ad Hoc Panel Report on the September 22 Event, May 23, 1980, available from *foia.abovetopsecret.com/VELA_- SATELLITE/THE_VELA_INCIDENT/REPORTS/AD_HOC_ REPORT_SEPT_23_1980.pdf.*

21. Luis Alvarez, *Alvarez: Adventures of a Physicist,* New York: Basic Books, 1987, Chap. 42.

22. Sublette, pp. 6-7.

23. Marshall, pp. 206-207.

24. Quoted by CBS News.

25. Sasha Polakow-Suransky, *The Unspoken Alliance: Israel's Secret Relationship with Apartheid South Africa,* New York: Pantheon Books, 2010, p. 140.

26. Richelson, p. 310.

27. John Pike, "The Death-Beam Gap," e-Prints, October 1992, available from *www.fas.org/spp/eprint/keegan.htm*.

28. Hersh, p. 281.

29. *Daily Newsbulletin*, Los Alamos National Laboratory, Friday, July 11, 1997, available from *nuclearweaponarchive.org/Safrica/071197.html*.

30. Richelson, p. 313.

31. Polakow-Suransky, p. 83.

32. Thomas Reed and Danny Stillman, *The Nuclear Express: A Political History of the Bomb*, Minneapolis, MN: Zenith Press, 2009, p. 177.

33. *Ibid.*

34. Jimmy Carter, *White House Diary*, New York: Norton and Co., 2010, p. 357.

35. *Ibid.*, p. 365.

36. *Ibid.*, p. 405.

37. See Victor Gilinsky and Roger Mattson, "Revisiting the NUMEC Affair," *Bulletin of Atomic Scientists*, Vol. 66, No. 2, March/April 2010, pp. 61-75.

38. Carter, p. 405.

39. See Avner Cohen, *Israel and the Bomb*, New York: Columbia University Press, 1998.

CHAPTER 14

THE NONUSE AND ABUSE OF NUCLEAR PROLIFERATION INTELLIGENCE IN THE CASES OF NORTH KOREA AND IRAN

Robert Zarate

INTRODUCTION

When it comes to stopping the spread of nuclear weapons to additional nations, much attention has focused on the so-called supply side problem of nuclear proliferation-related intelligence. Here, the main challenge is to provide policymakers with accurate and timely information about the accumulating moves of foreign governments to acquire nuclear weapons-making capability, so that they can respond appropriately with diplomacy, economic sanctions, interdiction, covert actions, military force, or other tools of statecraft.[1] Less attention, however, has focused on what might be called the demand side problem of proliferation intelligence. Here, the main challenge is that policymakers sometimes may prefer not to receive information about a foreign government's nuclear proliferation-related provocations, lest they be required to respond in ways that would complicate or fundamentally contradict their preferred policies. "Intelligence is important in dealing with proliferation, but only if you want it," former Nuclear Regulatory Commissioner Victor Gilinsky keenly observed, adding:

> it is also true that sometimes—contrary to the usual assumption—major players don't want to get reliable information at all because it would force them to act, or otherwise face uncomfortable political consequences.[2]

To illustrate why the demand side problem of proliferation intelligence is not a hypothetical one, this chapter identifies and examines key instances of the nonuse or abuse of proliferation intelligence by U.S. policymakers with regard to North Korea's and Iran's respective nuclear programs. Whereas North Korea succeeded in building and detonating a nuclear explosive device in October 2006 after repeatedly violating its international nonproliferation obligations, Iran — which, in many ways, is following North Korea's model for nuclear misbehavior — continues to violate its international obligations as it steadily improves its capability to build a nuclear explosive device on ever shorter notice.

Although the U.S. Government has declassified more proliferation-related intelligence on the North Korean case than on the Iranian case, it is nonetheless possible to arrive at some tentative and general conclusions regarding the demand side problem of proliferation intelligence in both of cases. To be sure, U.S. policymakers — in both Democratic and Republican presidential administrations — have consistently described North Korea's and Iran's respective nuclear programs over the last few decades as grave threats to American and allied security. That said, they sometimes declined to react decisively to worrisome nuclear provocations by North Korea and Iran, especially early on, because doing so would have required decisions that were too difficult, risky, or politically inconvenient. In certain instances, policymakers appeared not only to overrate their preferred policy palliatives towards North Korean and Iranian nuclear misbehavior, but also to be disinclined to receive or further pursue intelligence, suggesting that these proliferation cases had worsened or failed to be adequately addressed, on occasion even suppressing the sharing of relevant intelligence or denying its existence.

FAILURES OF INTELLIGENCE DEMAND IN THE NORTH KOREAN CASE

Despite roughly 2 decades of U.S.-led international efforts to stop the Democratic People's Republic of Korea (DPRK) from getting nuclear weapons-making capability, Pyongyang succeeded in detonating its first nuclear explosive device in October 2006. This outcome was made possible, in no small part, by the failures of policymakers to use available intelligence or demand new intelligence on the North Korean nuclear program that would have compelled a refinement of or fundamental revision to their policies.

Much of the record of intelligence related to the DPRK's march to nuclear weapons-making capability has not been declassified by the U.S. Government. However, a review of the available declassified record suggests that the most egregious nonuse or abuse of nuclear proliferation-related intelligence occurred during the mid-1990s, when policymakers apparently prioritized preserving the so-called Agreed Framework—America's controversial nuclear "grand bargain" with North Korea—over fully reckoning with what appears to be the U.S. intelligence community's classified judgment in the mid-1990s that Pyongyang already had produced "one, possibly two, nuclear weapons."[3]

A closer examination of the successive failures of intelligence demand in the DPRK's nuclear case is instructive, and it can help contemporary policymakers to avoid a North Korean-like outcome in the ongoing Iranian case and to adopt early on more effective nonproliferation and counterproliferation strategies in future cases.

Failing to Get the DPRK to Accept Nuclear Inspections.

Upon learning of North Korea's undeclared nuclear activities in the mid-1980s, the Ronald Reagan administration sought to persuade Pyongyang to join the Treaty on the Nonproliferation of Nuclear Weapons (NPT) and sign the NPT-required "full-scope" nuclear safeguards agreement with the International Atomic Energy Agency (IAEA). Although they achieved the first objective, they did not achieve the second. Given how Pyongyang's conclusion of an NPT-required IAEA safeguards agreement could have helped to fill some of the U.S. intelligence community's admitted gaps in understanding of the DPRK's nuclear program during this period, policymakers arguably should have treated North Korea's repeated refusal to conclude this safeguards agreement as a much graver violation then they did.

According to the declassified record, the U.S. intelligence community first learned in the early- to mid-1980s that North Korea was engaging in undeclared nuclear activities relevant to nuclear weapons making.[4] In particular, Pyongyang quietly had begun the construction of two nuclear reactors capable of producing weapons-grade plutonium at Yongbyon—namely, a 5-megawatt electric (MWe) research reactor in 1980, and a 50-MWe Magnox power reactor in the mid-1980s that was not connected to the country's electrical grid. However, the declassified record suggests that the intelligence community understood very little about the full extent of North Korea's nuclear efforts during this period, using carefully worded caveats to avoid strong conclusions about whether Pyongyang had a weapons program. For example, the Directorate of Intelligence in the Central Intelligence Agency (CIA) wrote in a May 1983 report: "We have very little information on North Korea's ability to conduct the non-nuclear research, particularly that involving high explosives, required for a nuclear weapons research program," adding:

> In considering whether to embark on a venture as costly, hazardous, and politically sensitive as a nuclear weapons program, P'yongyang would face complex calculation of benefits versus costs as well as uncertainty regarding the effect of such a program on its ultimate goal of reunifying the peninsula on its own terms.[5]

Even though the United States and the Soviet Union were locked at the time in a heated Cold War strategic rivalry, the Reagan administration was able to respond to the DPRK's nuclear provocations by working with Moscow, which had no small amount of influence on Pyongyang and shared a common interest in getting North Korea to join the NPT. Although the DPRK agreed to accede to the NPT in December 1985 — apparently as a *quid pro quo* for a Soviet promise to build a nuclear power plant — it failed to meet its NPT obligation to conclude, within the first 18 months of signing the treaty, a full-scope safeguards agreement with the IAEA. A complicating factor was that the IAEA reportedly provided the North Koreans with an outdated version of the safeguards agreement and thus gave them another 18 months to conclude the pact, but Pyongyang failed to meet this revised deadline.[6]

An IAEA safeguards agreement would have required the DPRK to make a full declaration of all its nuclear material and related equipment and activities, and also legally authorized the IAEA's nuclear inspectors to verify the correctness and completeness of its declarations. Implementation of IAEA safeguards in North Korea could therefore have given U.S policymakers, lawmakers, and the intelligence community critical information for better assessing the DPRK's program's weapons potential, especially in light of the intelligence community's admitted knowledge gaps.

Indeed, the declassified record shows that the intelligence community expressed repeated concern about Pyongyang's failure to conclude an IAEA safeguards agreement. For example, the CIA's Director of Intelligence issued a May 1988 report that stated there is "no evidence that North Korea is pursuing a nuclear weapons option, but we cannot rule out that possibility," adding:

> the possibility that P'yongyang is developing a reprocessing capability [that would enable it to extract weapons-usable plutonium from spent nuclear fuel] and its footdragging on implementing NPT provisions, suggest close scrutiny of the North's nuclear effort is order.

It is crucial to note that the 1988 report also stated that the South Korean government "believes P'yongyang is developing a nuclear weapon capability — a concern that Seoul has raised publicly."[7] Despite internal debates, however, the Reagan administration appears to have treated North Korea's NPT violation as relatively routine, perhaps due in part to competing priorities in foreign policy, such as America's larger Cold War rivalry with Moscow.

378

Indeed, rather than raise the ante with economic sanctions or other forms of high-profile international pressure, U.S. policymakers relied on quieter multilateral diplomacy in the hopes of eventually changing Pyongyang's behavior. In the meantime, the DPRK had already begun to produce and accumulate plutonium-laden spent nuclear fuel via its 5-MWe nuclear reactor, which had started operations in the mid-1980s. Moreover, as it continued construction of its 50-MWe nuclear reactor at Yongbyon, it also began to build a new 200-MWe plutonium-producing reactor at Taechon in January 1989. What's especially troubling—especially in retrospect—is that North Korea likely undertook additional efforts during this period not only to experiment with the non-nuclear components necessary for the construction of a nuclear explosive device, but also to develop and perhaps even use reprocessing technologies to begin separating weapons-usable plutonium from its growing stockpile of spent nuclear fuel. At any rate, the DPRK's weapons-relevant nuclear activities would metastasize in the next decade.

The Agreed Framework: Giving Up Our Ends to Preserve Our Means.

Under President George H. W. Bush, the United States eventually achieved the Reagan administration's goal of getting North Korea to conclude its NPT-required full-scope IAEA safeguards agreement in early-1992. Yet disturbing discrepancies in the DPRK's subsequent declaration of nuclear materials to the IAEA—and the weapons-related worries surrounding its suspected covert reprocessing activities— led the Bush administration and later the Bill Clinton administration to escalate diplomatically the controversy. What is problematic, though, is that when the North Koreans publicly said that international pressure—in particular, the imposition of economic sanctions—against it would amount to a "declaration of war," the Clinton administration apparently blinked. Instead, U.S. policymakers doubled-down on the diplomatic track, eventually concluding the so-called Agreed Framework, a controversial grand bargain with Pyongyang that aimed, above all, at de-escalating the North Korean crisis. Although senior officials in the Clinton administration cited the Agreed Framework as a means to achieving the goal of halting the DPRK's march to nuclear weapons-making capability, it appears that they came to see preserving the grand bargain as an end in of itself, even in the face of intelligence reports suggesting egregious North Korean nuclear violations of the agreement.

In 1989, the incoming Bush administration undertook a sustained effort to get Pyongyang to agree to put its nuclear facilities under IAEA inspections. Towards that end, U.S. policymakers wrangled, on the one hand, with North Korea's rising demand for the establishment of a nuclear-weapons-free Korean Peninsula (which would require U.S. withdrawal of forward-deployed tactical nuclear weapons), and on the other, with reassuring South Korea, Japan, and other allies of their security. At the same time, American diplomats also worked behind the scenes to persuade the IAEA's 35-nation Board of Governors to posture itself for increased diplomatic pressure on the DPRK. Although the effort had its share of controversies, it nonetheless produced certain results. On September 27, 1991, with Cold War tensions with Moscow at a low (indeed, the Soviet Union would dissolve in late-December 1991), President George H. W. Bush announced that the United States would withdraw all land- and sea-based tactical nuclear weapons from South Korea. In December 1991, Seoul and Pyongyang concluded a joint declaration to establish a nuclear-weapons-free Korean Peninsula, which entered into force in February 1992, and stipulated that "South and North Korea shall not possess nuclear reprocessing and uranium enrichment facilities." After subsequent delays, the DPRK finally signed its full-scope IAEA safeguards agreement on January 30, 1992, and ratified it on April 9, 1992.

Although the Bush administration arguably paid a steep price to persuade Pyongyang merely to conclude its NPT-required IAEA safeguards agreement, the U.S. intelligence community's increasingly grave assessments of the DPRK's nuclear program during this period reflected the need for urgent actions. For example, the CIA reported in a March 1989 special analysis:

> North Korea may be willing to risk the international censure that a nuclear weapons program would bring in order to maintain a decided military advantage over the South, the keystone of the North's national security policy. Pyongyang may believe that nuclear weapons are crucial to preserving that edge.[8]

But in the years following, evidence emerged that North Korea had escalated activities relevant to an actual nuclear weapons program. For instance, the CIA reported in February 1992:

> P'yongyang recently conducted its first high-explosive test since 1988. . . . The activity [related to reprocessing] at Yongbyon suggests the North may be trying to complete its nuclear weapons program before inspec-

tions begin; it may also have no intention of allowing inspections.[9]

The report added that these developments "suggest P'yongyang is moving forward with its nuclear weapons programs."[10] Thus, while the U.S. intelligence community had strong suspicions that the North Koreans had already engaged in some level of reprocessing activities, policymakers in the Bush administration apparently declined to publicly raise this point, for a public confrontation might have also forced them to question the validity of the North and South's December 1991 joint declaration. Right or wrong, it appears that the Bush administration — which was becoming consumed with the aftermath of Operation DESERT STORM and other foreign policy priorities — concluded that pushing for IAEA nuclear inspections offered perhaps the best way to exert international pressure on Pyongyang in a manner that could build multilateral consensus. Yet, to the extent that policymakers believed that North Korea's efforts constituted an actual nuclear weapons program, they arguably should have treated Pyongyang's nuclear provocations as a much graver threat than they did.

The Bush administration — and then the Clinton administration, at least in its first 2 years — thus relied mainly on the IAEA inspections process to raise the stakes over the DPRK's controversial nuclear program. Indeed, tensions mounted after the IAEA found disturbing discrepancies in North Korea's May 1992 initial declaration of nuclear materials. Questions emerged about whether Pyongyang had correctly and completely declared its nuclear material inventories and related activities — in particular, whether it had used so-called hot cells supplied by the Soviets in the

1960s to separate plutonium from its growing stock-pile of spent nuclear fuel. After the DPRK repeatedly denied IAEA inspectors access to two suspected nu-clear-related sites at Yongbyon during a series of visits in mid- to late-1992 then-IAEA Director General Hans Blix formally requested Pyongyang to allow more intrusive "special inspections" on February 11, 1993. In addition, the IAEA Board of Governors passed a resolution on February 24, 1993, urging North Korea to accept special inspections within 1 month. Pyong-yang defiantly responded by announcing its inten-tion to withdraw from the NPT on March 12, 1993. The IAEA Board of Governors subsequently found the DPRK to be in noncompliance with its obligations under the NPT-required IAEA safeguards agreement, and referred its case to the United Nations (UN) Secu-rity Council in April 1993. Led by the United States, the UN Security Council passed a May 1993 resolution that called upon North Korea to reconsider its inten-tion to withdraw from the NPT and comply with its IAEA safeguards agreement. However, the UN Se-curity Council resolution did not impose any actual sanctions to pressure a change in Pyongyang's nuclear misbehavior.

Over the next year, the United States and IAEA continued to engage North Korea diplomatically, hop-ing to convince it to change course. However, after the DPRK began removing spent nuclear fuel rods from its 5-MWe reactor at Yongbyon in early- to mid-1994, the U.S. State Department attempted to draw a "red line" by threatening that any removal of the spent fuel from the fuel rods themselves would lead the United States to actively seek sanctions in the UN Security Council. Nonetheless, Pyongyang responded by withdraw-ing from the IAEA in June 1994 and issuing boisterous counterthreats. For example, North Korea's First Vice Foreign Minister Kang Sok Ju definitely declared: "Sanctions would equal a declaration of war. All the people in our country and our military are gearing up now to respond to the sanctions."[11]

Rather than press the UN Security Council to impose sanctions, the Clinton administration elected to negotiate a so-called grand bargain with Pyongyang in an apparent effort, above all, to diffuse the crisis. In October 1994, U.S. and North Korean diplomats concluded the so-called Agreed Framework, which obliged the DPRK to suspend construction of its 50-MWe and 200-MWe nuclear reactors in return for heavy fuel oil and allegedly "proliferation-resistant" light water reactors (LWRs) with a total 2,000-MWe capacity. One of the Agreed Framework's key problems, however, was that it explicitly suspended the IAEA's routine and ad hoc nuclear inspections pursuant to North Korea's full-scope IAEA safeguards agreement and did not authorize the resumption of inspections until after "conclusion of the supply contract for the provision of the LWR project." The suspension of IAEA nuclear inspections had the immediate effect of reducing the Agency's access to the DPRK's overall nuclear program, and—critically—the flow of information about the country's activities. Another key problem was that the grand bargain required Pyongyang to come into full compliance with its NPT and IAEA obligations—including enabling inspectors to verify the correctness and completeness of its nuclear declarations—but only after a "significant portion of the LWR project is completed, but before delivery of key nuclear components." Such provisions effectively created the time and space necessary for North Korean nuclear violations to accumulate.

Although the stated objective of the Agreed Framework was to prevent the DPRK from getting nuclear weapons, the Clinton administration apparently came to view preserving the grand bargain as an end in and of itself, even in the face of growing evidence that Pyongyang had egregiously violated the agreement. To take one example from the mid- to late-1990s, senior policymakers sought to downplay intelligence that North Korea had potentially violated the Agreed Framework by pursuing a covert program to enrich uranium. In testimony before various congressional committees in 1997 and 1998, Secretary of State Madeleine Albright repeatedly claimed that the Agreed Framework had definitively halted the DPRK's nuclear weapons program. However, with lawmakers repeatedly confronting the Clinton administration about emerging intelligence of Pyongyang's procurement and development activities related to uranium enrichment, she reluctantly conceded to the House Committee on International Relations in February 1999: ". . . we have suspicions that North Korea has engaged in construction activities that could constitute a violation of its commitment to freeze its nuclear-related facilities under the Agreed Framework."[12]

More troubling, policymakers in the Clinton administration may have failed to reckon fully with the implications of what appear to be troubling judgments in the mid-1990s by elements of the U.S. intelligence community that North Korea had already built a nuclear weapon—a conclusion that would have undermined a key premise of the Agreed Framework, if it had a public airing at the time. In December 2001, the National Intelligence Council revealed in *Foreign Missile Developments and the Ballistic Missile Threat Through 2015*, an unclassified summary of a National Intelligence Estimate (NIE), that the intelligence community had already concluded in the mid-1990s, apparently after the Agreed Framework had been signed, that Pyongyang had built as many as two nuclear weapons:

> The Intelligence Community *judged in the mid-1990s* that North Korea had produced one, possibly two, nuclear weapons, although the North has frozen plutonium production activities at Yongbyon in accordance with the Agreed Framework of 1994 (emphasis added).[13]

In November 2002, the CIA subsequently provided to Congress an unclassified estimate on North Korea's nuclear program that repeated and elaborated the claim made by the December 2001 NIE summary:

> The U.S. has been concerned about North Korea's desire for nuclear weapons and *has assessed since the early 1990s* that the North has one or possibly two weapons using plutonium it produced prior to 1992 (emphasis added).[14]

If it is indeed true that the intelligence community arrived at this conclusion in the mid-1990s, then policymakers in the Clinton administration should have not only tasked the intelligence community to determine whether the DPRK at the time was continuing to engage in a covert nuclear program—including the development of uranium enrichment capabilities—in violation of the Agreed Framework, but also questioned the wisdom of continuing with the Agreed Framework. But they apparently chose not to do either one. As Henry Sokolski, who served as the Pentagon's Deputy for Nonproliferation under President George H. W. Bush, explained in a November 2002 article:

If North Korea already had built one or more weapons and was hiding them in violation of the 1994 deal, wouldn't it be reasonable to assume that North Korea was still conducting a covert nuclear weapons program? *The answer from the intelligence community: Probably, but since no one had yet asked the community formally to review the matter in a national intelligence estimate, it had no definitive view* (emphasis added).[15]

Sokolski added:

Why was there no such request? Almost certainly because Clinton officials knew what the answer would be — yes — and that that would spell the end of their 1994 deal.[16]

Another complicating factor may have been that the Clinton administration did not want to endanger the May 1995 review conference for the NPT, in which treaty signatories were scheduled to vote on the pact's indefinite extension. What is troublesome, however, is that during this same period, the mid- to late-1990s, A. Q. Khan, the Pakistani engineer who headed a rogue international nuclear proliferation network, repeatedly visited North Korea and allegedly provided it with components and designs related to uranium centrifuges, as well as other nuclear assistance.[17]

Legacy: Pyongyang's Final Sprint to the Bomb.

In the early years of the 21st century, the Agreed Framework began to fall apart. After President George W. Bush identified North Korea as a member of the so-called Axis of Evil in a post-September 11, 2011, State of the Union speech, U.S. diplomats confronted their

North Korean counterparts with evidence of an un-declared uranium enrichment program in September 2002. Bristling at the Bush administration's more con-frontational approach, Pyongyang began to reactivate dormant nuclear facilities and then withdrew from the NPT in January 2003.

Over the next 3 years, however, policymakers in the Bush administration changed course and pur-sued a multilateral diplomatic process with the DPRK through the so-called Six-Party Talks process that also brought Russia, China, Japan, and South Korea to the table. Although the Six-Party Talks may have enjoyed temporary—albeit limited—gains, if any questions lingered over the overall success of North Korea's long march to nuclear weapons-making capability, they were dispelled when Pyongyang detonated under-ground its first nuclear explosive device on October 9, 2006. It then exploded underground another nuclear device on May 25, 2009. Moreover, as DPRK officials revealed to former Los Alamos National Laboratory Director Siegfried Hecker and other nuclear specialists visiting the country in November 2010, North Korea had built a 2,000-centrifuge uranium enrichment facil-ity with a surprising level of technical sophistication.[18] Equally troubling, Pyongyang was also now interna-tionally spreading nuclear weapons-related technolo-gies. To take a key example, a surprise Israeli air strike on a secret Syrian nuclear facility at the al-Kibar site near Deir Alzour in September 2007 led to subsequent public revelations by the U.S. intelligence community that North Koreans had actively assisted the Assad regime in building a nuclear reactor designed to pro-duce weapons-grade plutonium.[19] Having success-fully tested a long-range missile with direct relevance to intercontinental ballistic missile (ICBM) capability on December 12, 2012, Pyongyang detonated its third nuclear explosive device on February 12, 2013, raising tensions on the Korean Peninsula and throughout the region.

In sum, after North Korea's nuclear weapons program was discovered by the U.S. Government in the early- to mid-1980s, it metastasized in the 1990s and became a full-blown proliferation nightmare in the 2000s. What is worrisome is that the available record of declassified intelligence about the North Korean nuclear case appears to suggest that policymakers in both Democratic and Republican presidential administrations — when confronted with intelligence indicators that potentially endangered their preferred policy palliatives towards Pyongyang — did not always want to know.

FAILURES OF INTELLIGENCE DEMAND IN THE IRANIAN CASE

Although the United States has helped lead international efforts to stop the Islamic Republic of Iran from developing the capability to build a nuclear weapon on increasingly short notice since 2003, Tehran has consistently refused to yield. Now Israeli officials — whose country Iranian President Mahmoud Ahmadinejad has threatened to "wipe off the map" — have warned that Iran's nuclear program is about to enter a so-called zone of immunity, a state of technological progress in which not even a military strike may prevent, with much confidence, the current Iranian regime from eventually building a nuclear weapon. Nonetheless, President Obama has repeatedly stated his belief that "there is still time and space to pursue a diplomatic solution."

Here, it is important to recall once again that the available record of declassified intelligence on Iran's nuclear program is currently much more limited than the available record on North Korea's program. As a result, conclusions about the nonuse or abuse of proliferation intelligence in the Iranian case will be inherently more tentative than in the North Korea case. That said, what is troubling is how so-called realist foreign policy analysts are now urging the United States to "handle" Iran's accelerating march to nuclear weapons-making capability "like North Korea"—which is to say, to try to negotiate a grand bargain, if nothing else, to decrease tensions, but also to be prepared to accept a nuclear-armed Iran.[20] Such advice is wrongheaded, and a closer examination of key failures of intelligence demand with regard to the Iranian nuclear program can help policymakers refine and revise their nonproliferation and counterproliferation strategies, and hopefully avoid a North Korea-like outcome in Iran's case.

Halting the Shah's March to Nuclear Weapons-Making Capability.

In the early-1970s, Iran's Shah Mohammad Rez Pahlavi announced plans to initiate a civil nuclear program in his country. Towards that end, he not only struck power reactor deals with French and West German nuclear suppliers, but also offered to buy nuclear reactors from the United States. Despite warnings from elements of the U.S. intelligence community that the Shah's pro-Western government might use a civil nuclear program to pursue specific technologies and eventually acquire a nuclear weapons-making capability, the Gerald Ford administration initially worked to conclude a bilateral agreement for civil nuclear cooperation that would "accommodate Iranian demands" for some level of access to weapons-relevant nuclear fuel-making technologies like reprocessing to separate plutonium from spent nuclear fuel. However, when President Ford announced sweeping changes to America's nuclear export and nonproliferation policies in October 1976, his announcement effectively foreclosed U.S. efforts to provide Iran with access to reprocessing. Although President Jimmy Carter apparently momentarily reversed his predecessor's decision on providing Tehran with access to reprocessing, the possibility of any U.S. civil nuclear cooperation with Iran ended with the Shah's overthrow in the country's 1979 Islamic Revolution.

In 1974, as the Shah started to ramp up efforts to building a civil nuclear program, America's nuclear export and nonproliferation policies were being rocked by Smiling Buddha, India's May 18th detonation of a nuclear bomb. What disturbed policymakers and lawmakers in Washington, DC, was that New Delhi had obtained the plutonium for the bomb using a reactor that Canada had built for India to use "for peaceful purposes only" and heavy water to moderate the reactor that the United States had supplied, again expressly "for peaceful purposes."[21] Nonetheless, officials in New Delhi attempted to use semantics to explain away their nuclear test, describing the bomb as a so-called peaceful nuclear explosive device that had not violated **their** understanding of the terms of the nuclear cooperation agreements with the United States and Canada.[22]

According to declassified records, the U.S. intelligence community worried about the Shah's long-term nuclear intentions in the aftermath of India's nuclear test. Some intelligence analyses appeared to take little comfort that Iran had signed the NPT in July 1968 and concluded an NPT-required full-scope IAEA safeguards agreement in May 1974. For example, the CIA's Director of Central Intelligence issued *Prospects for Further Proliferation of Nuclear Weapons*, a Special National Intelligence Estimate (SNIE) dated August 1974, that cautioned:

> Iran's much publicized nuclear power intentions are entirely in the planning state. . . . There is no doubt, however, of the Shah's ambition to make Iran a power to be reckoned with. If he is alive in the mid-1980s, if Iran has a full-fledged nuclear power industry and all the facilities necessary for nuclear weapons, and if other countries have proceeded with weapons development, we have no doubt that Iran will follow suit. Iran's course will be strongly influenced by Indian nuclear programs.[23]

Nonetheless, Secretary of State Henry Kissinger, concurrently serving at the time as the National Security Advisor, apparently prioritized the perceived benefits of a U.S.-Iranian nuclear cooperation agreement over the intelligence community's statements of the risks and pushed ahead with efforts to negotiate a deal with Tehran. But as the declassified record shows, the Ford administration internally debated how to respond to the Shah's communicated desire to reprocess spent nuclear fuel, or at least have some access to reprocessing technology. On April 22, 1975, Kissinger issued a National Security Decision Memorandum (NSDM) stating that Washington's negotiations for a nuclear agreement with Tehran should seek: ". . . to require U.S. approval for reprocessing U.S. supplied fuel, while indicating that the establishment of a multinational reprocessing plant would be an important factor favoring such approval." But the NSDM added:

> As a fall back, we could inform the Government of Iran that we shall be prepared to provide our approval for reprocessing of U.S. material in a multinational plant in Iran if the country supplying the reprocessing technology or equipment is a full and active participate in the plant, and holding open the possibility of U.S. participation. The standard provisions requiring mutual agreement as to safeguardability shall apply. An expression of U.S. willingness to explore cooperation in establishing such a facility at an appropriate time should Iran so desire, may be made.[24]

Another option was to "buy back" spent nuclear fuel from the Iranians at market prices. Over the next 1 1/2 years, Washington and Tehran exchanged various draft agreements and wrangled over the reprocessing issue, but they found little common ground as negotiations intermittently stalled.

Nonetheless, by October 1976, President Ford effectively had foreclosed the possibility of any U.S. assistance in helping Iran to access to reprocessing technologies when he announced a major shift in America's nuclear nonproliferation and energy policies. In particular, Ford stated that the United States would defer the pursuit of activities to reprocess spent fuel, fabricate plutonium-based nuclear fuels, and export plutonium-based fuels and related technologies:

> I have decided that the United States should no longer regard reprocessing of used nuclear fuel to produce plutonium as a necessary and inevitable step in the nuclear fuel cycle, and that we should pursue reprocessing and recycling in the future only if they are found to be consistent with our international objectives.[25]

One key motivating factor behind the President's decision was that the White House had assembled in mid-1976 an interagency panel — led by then Deputy Administrator of the Energy Research and Development Agency (ERDA) Robert Fri, and composed of representatives from the Arms Control and Disarmament Agency (ACDA), the Office of Budget and Management, the State Department, and other agencies — to examine U.S. nuclear energy and export policy. The panel's still-classified study, known as the Fri Study, apparently offered both majority and minority recommendations on policy changes that fundamentally informed President Ford's decision to prioritize nonproliferation while deferring the domestic and international promotion of reprocessing and plutonium-based nuclear fuels.

President Carter made the Ford administration deferral policy "indefinite" in April 1977 but nonetheless apparently moved to reverse the Ford administration's decision to prevent Iran from getting reprocessing during a one-on-one meeting with the Shah in December 1977. Drawing on firsthand interviews with aides to President Carter, nuclear nonproliferation expert Henry Sokolski wrote in March 2005:

> In an effort to show support for the Shah, President Carter visited Iran in late December 1977. At the time, it was U.S. policy to export U.S. reactors but not to share reprocessing or enrichment technology with any state, Iran included. Yet, when he met with the Shah, Carter, to the amazement of his aides, cast U.S. nuclear policy aside and orally assured the Shah that he could have anything nuclear he wanted from the United States, including reprocessing, if he liked.[26]

That said, the possibility of any substantive U.S. civil nuclear cooperation with Tehran ended with the fall of the Shah in the 1979 Iranian Revolution. Indeed, the new theocratic regime in Tehran would temporarily put Iran's push for a civil nuclear program on the backburner, but it would not abandon its nuclear efforts completely.

Failing to Reckon with Iran's Chinese Nuclear Connection.

Consumed by the Iran-Iraq War (September 1980 to August 1988) — the bloody conflict in which Saddam Hussein's Ba'athist regime used chemical weapons against Iran and both countries attacked each other's nuclear facilities — the Iranian regime did not prioritize efforts related to acquiring a nuclear weapons-making capability for much of the 1980s.[27] However, after the war's conclusion, Iran initiated a tenacious and often covert campaign to gain access to nuclear materials, technologies, and know-how, seeking help from entities in China, Russia, and elsewhere, to acquire elements necessary for developing the capability to make a nuclear weapon on ever-shorter notice.

In response, policymakers in the Clinton administration — who claimed to be gravely concerned about Iranian nuclear proliferation activities — attempted to pressure both Moscow and Beijing to curb their permissive nuclear policies towards Iran by dangling before each the possibility of concluding a much-coveted bilateral civil nuclear cooperation agreement with the United States. Although it appears that the U.S. intelligence community was deeply concerned about Chinese and Russian assistance to Iran's nuclear programs, the Clinton administration decided to treat the two countries differently, pushing to fully implement a U.S. civil nuclear cooperation agreement with China, while holding off on even negotiating a similar agreement with Russia.

While the U.S. Government has not yet declassified much intelligence from the 1990s related to Sino-Iranian and Russo-Iranian civil nuclear relations, various unclassified reports to Congress help to give a sense of the intelligence community's worries about such relationships during the period. For example, a September 1996 report by the Congressional Research Service (CRS) noted concerns that "[s]ince the 1980s, China has agreed to provide nuclear technology to Iran . . .," adding: "There is concern about Iran's nuclear collaboration with Pakistan, long a recipient of Chinese assistance."[28] Of note, the CRS report elaborated on aspects of Sino-Iranian nuclear collaboration that the intelligence community found problematic such as:

> *Secret Cooperation.* U.S. and European intelligence reportedly found that, since 1988, 15 Iranian nuclear engineers from Iran's nuclear research center at Esfahan have been secretly trained in China; that a secret Iranian-Chinese nuclear cooperation agreement dates from after 1985; and that China transferred designs and technology for reactor construction and other projects at Esfahan. . . .[29]

and:

> *Other Controversial Deals.* The China National Nuclear Energy Industry Corporation reportedly plans to sell Iran a facility to convert uranium ore into uranium hexafluoride gas, which could be enriched to weapons-grade material. U.S. policy is complicated by the fact that Westinghouse Electric Corporation wants to sell equipment to the Chinese company. According to

intelligence reports, the deal is proceeding with Chinese nuclear experts going to Iran to build the new uranium conversion plant near Esfahan.[30]

Indeed, only a few months earlier, Congressman John Spratt, Jr., (D-South Carolina) had publicly warned about China's alleged cooperation with Iran on a uranium conversion facility during a June 1996 floor speech: "China is assisting Iran in building a uranium hexafluoride [HEX] facility which converts uranium into a gaseous form so it can be diffused to produce highly enriched uranium."[31] But perhaps most alarmingly, the CIA's Nonproliferation Center subsequently issued a July 1997 report bluntly stating that "China [in the latter half of 1996] was the single most important supplier of equipment and technology for weapons of mass destruction [WMD]" worldwide (emphasis added).[32]

The intelligence community in the 1990s also had strong concerns about Russo-Iranian nuclear cooperation. According to news reports from the middle of the decade, the Clinton administration—concerned that Russian assistance to Iran might come to entail uranium enrichment or other nuclear fuel-making technologies—took the unprecedented step of directly sharing U.S. intelligence findings on Iran's suspected nuclear weapons program with the Kremlin, in the hopes of persuading Russia to end all nuclear assistance to Iran.[33] Although it appears that Moscow subsequently refrained from direct assistance to Iranian efforts to gain nuclear fuel-making technologies, the Director of Central of Intelligence (DCI) nonetheless warned Congress in an unclassified September 1997 report:

Russian entities continued to market and support a variety of nuclear-related projects in Iran in 1997, ranging from the sale of laboratory equipment for nuclear research institutes to the construction of a 1,000-megawatt nuclear power reactor in Bushehr, Iran, that will be subject to . . . IAEA safeguards. These projects, along with other nuclear-related purchases from abroad, helped to build Iran's nuclear technology infrastructure, which in turn would be useful in supporting nuclear weapons research and development.[34]

However, the DCI's report tentatively added:

Russia has committed to observe certain limits on its nuclear cooperation with Iran. For example, President [Boris] Yel'tsin has stated publicly that Russia will not provide militarily useful nuclear technology to Iran.

The Clinton administration decided to take markedly different approaches when it came to linking the possibility of U.S. civil nuclear cooperation to changes in Russia's and China's respective behaviors towards Iran's nuclear program. On the one hand, President Clinton—following President Bush's policy precedent—declined even to negotiate with the Kremlin for an agreement to permit bilateral civil nuclear cooperation until Russia had ended all nuclear, advanced conventional military, and missile assistance to Iran.[35] One particular sticking point was Moscow's decision to try to complete the construction of Iran's light water reactor at Bushehr.

On the other hand, the President decided to positively respond to Beijing's request that Washington fully implement the controversial U.S.-Chinese civil nuclear cooperation agreement. After the Reagan administration had negotiated and signed the bilateral agreement in July 1985, Congress passed a joint resolution (Public Law 99-183) that technically allowed the agreement to enter into force, but conditioned its full implementation—e.g., the issuance of export licenses—on the President legally certifying, among other things, that China's peaceful use of U.S. nuclear exports can and will be effectively verified and that Beijing's provision of further details about its nuclear nonproliferation policies and practices conformed with Section 129 of the Atomic Energy Act, which prohibits nuclear exports to countries that proliferate.

To lay the political groundwork for implementing the U.S.-Chinese civil nuclear cooperation agreement forward, senior policymakers in the Clinton administration began to tout reversals in China's historically troubling nuclear policies—in particular, its nuclear practices towards Iran. For example, State Department official Robert Einhorn told lawmakers in September 1997 that China had cancelled its controversial project to build a uranium conversion plant in Iran, although he conceded that the Chinese still had provided the Iranians with blueprints to build the problematic facility.[36] (Worse, the IAEA would subsequently reveal in a June 2003 report that China had also secretly exported in 1991 roughly one metric ton of uranium hexafluoride to Iran.[37] In the mid-2000s, Iran would reportedly use some of this gaseous uranium feedstock in its uranium enrichment centrifuges.) Moreover, the Clinton administration leaked the contents of a "secret" letter that Foreign Minister Qian Qichen had given to Secretary of State Madeleine Albright on the eve of a U.S.-China summit in Washington in October 1997, in which Beijing had promised not to start new nuclear projects in Iran, but only after first completing a small nuclear research reactor and a facility to fabricate zirconium cladding for encasing nuclear reactor fuel rods.[38]

In January 1998, President Clinton issued the required certifications to clear the final legal hurdles to formally begin the congressional review period for the controversial U.S.-Chinese civil nuclear cooperation agreement.[39] Although individual lawmakers moved to push a joint resolution of disapproval to block the U.S.-Chinese agreement's implementation, Congress as a whole did not act on the proposed joint resolution before the legislative branch's review period ended. As a result, the door to the nuclear deal's full implementation opened.

What is troubling about this episode is that, even though key elements of the intelligence community had singled out China as a worse WMD proliferator than Russia, it appears that the Clinton administration prioritized geopolitics and the U.S. nuclear industry's desire to sell nuclear goods and services to the Chinese over a principled policy on nuclear nonproliferation. By failing to hold China's proliferation activities towards Iran to a similar standard as Russia's proliferation activities, policymakers certainly gave up a point of powerful leverage on Beijing's nuclear behavior — one that conceivably could have been used to get China not only to divulge the full measure of its assistance to Iran, but also to take an even tougher stand on Iranian efforts to get nuclear weapons-making capability, especially as these efforts metastasized in the next decade.

Legacy: Continuing Struggle to Halt Iran's March to the Bomb.

The controversy over the Iranian nuclear program turned into a bona fide crisis in August 2002, when the

IAEA learned that Iran had engaged in a host of un-declared nuclear activities relevant to a weapons pro-gram for nearly 2 decades. Iran, as a signatory to the NPT, had obligated itself to make correct and complete declarations of its nuclear material and related activi-ties to IAEA inspectors. As a result, then-IAEA Direc-tor General Mohamed ElBaradei reported in June 2003 that Iran had "failed to meet its obligations" under its NPT-required nuclear transparency and inspections agreement with the IAEA, and he urged Iran to fully cooperate with nuclear inspectors so they could pro-vide credible assurances regarding the [current and future] absence of undeclared nuclear activities.

Yet despite nearly a decade's worth of U.S.-led in-ternational efforts to use diplomacy and pressure to change Iranian behavior, the regime in Tehran to this day has refused to take IAEA-required actions that would help allay international worries about its nucle-ar program; they instead pursued technical capabili-ties that have shrunk the amount of time that it needs to make its first nuclear weapon.[40] What is troubling is the extent to which China and Russia have acted to slow or halt Western efforts to get the UN Security Council to impose sanctions on Iran's ongoing non-compliance with its international nuclear nonprolifer-ation obligations, especially in recent years. But just as President Clinton conceded a point of leverage on Bei-jing's nonproliferation policy by fully implementing the U.S.-Chinese civil nuclear cooperation agreement in March 1998, so President Obama conceded leverage on Moscow's nonproliferation policy by successfully concluding the controversial U.S.-Russian civil nucle-ar cooperation agreement in January 2011.

Given that the U.S. Government so far has declas-sified very little intelligence related to Iranian nuclear

proliferation efforts during the 2000s, a thorough examination of any nonuse or abuse of proliferation intelligence on Iran during this period remains beyond
the scope of this chapter. However, given U.S. decisionmaking with regard to assisting the Shah's nuclear
program or holding accountable major supplier states
relevant to the Islamic Republic's nuclear efforts, it appears that policymakers have often struggled to strike
a principled balance between the objective of nuclear
nonproliferation vis-à-vis Iran, and the desire to satisfy other competing geopolitical or national aims.

CONCLUSION

Over the last few decades, U.S. policymakers tried
to use a mixture of policies short of military action—
including diplomatic negotiations, economic sanctions, interdictions, and covert actions—to deal with
North Korean and Iranian efforts to develop nuclear
weapons-making capabilities. However, U.S. policies
ultimately did not stop the DPRK from building its
first nuclear explosive device, and detonating it in
October 2006. Observers today rightly worry whether
Iran can be persuaded or prevented from following
North Korea's nuclear precedent.

As this chapter's examination of the nonuse and
abuse of proliferation intelligence in the cases of Iran
and North Korea suggests, despite a long tradition of
official statements about how nuclear proliferation
poses the gravest danger to the United States and its
allies, U.S. policymakers in both Democratic and Republican presidential administrations sometimes have
tended to subordinate nuclear nonproliferation policy
to other international or domestic concerns—even in
the face of proliferation intelligence that counseled

otherwise. In turn, this tendency has served at times to frustrate, if not also undermine, the very aims of nuclear nonproliferation policy. What is worrisome is that it was often when proliferation problems metastasized and became far less manageable that risks of subordinating nuclear nonproliferation policy came to be more fully appreciated.

The failures of intelligence demand in the North Korean and Iranian nuclear proliferation cases raise a significant and thorny issue—namely, if policymakers will not be more hardnosed and act on timely intelligence early on, when a proliferation case is still manageable and easier to respond to, then might they be even less likely to take meaningful yet more difficult actions later, when the case becomes much less manageable and much more dangerous? The answer appears to a tentative and regrettable "yes." However, there is ground for modest hope. Indeed, if policymakers in the Executive Branch, as well as lawmakers in Congress who oversee them and other interested parties, soberly examine and attempt to apply the lessons learned of these and other past instances when the demand-side problem of proliferation intelligence negatively affected U.S. policymaking, then they potentially can put themselves in a better a position to deal more effectively with current and future proliferation cases.

ENDNOTES - CHAPTER 14

1. For example, see Commission on the Intelligence Capabilities of the United States Regarding Weapons of Mass Destruction, *Final Report*, March 31, 2005, available from *www.fas.org/irp/off-docs/wmd_report.pdf*.

2. Victor Gilinsky, "Sometimes We Don't Want to Know: Kissinger and Nixon Finesse Israel's Bomb," draft paper, August 4, 2011, available from *www.npolicy.org/article_file/Sometimes_We_Do_not_Want_to_Know__Kissinger_and_Nixon_Finesse_Israeli_Bomb_-_Gilinsky.pdf*.

3. For more on this, see endnotes 12-15.

4. For example, see Central Intelligence Agency (hereinafter "CIA"), *North Korea: Nuclear Reactor,* possibly from the *National Intelligence Daily,* July 9, 1982, (formerly SECRET) declassified, available from *www.foia.cia.gov/docs/DOC_0000453456/DOC_0000453456.pdf*; and CIA, *North Korea's Nuclear Efforts,* April 28, 1987, (formerly SECRET) declassified, available from *www.foia.cia.gov/docs/DOC_0000835118/DOC_0000835118.pdf*.

5. CIA's Directorate of Intelligence, *A 10-Year Projection of Possible Events of Nuclear Proliferation Concern,* May 1983, (formerly SECRET) declassified, available from *www.foia.cia.gov/docs/DOC_0000835123/DOC_0000835123.pdf*.

6. Author's conversations with former officials in the U.S. Government and the IAEA.

7. CIA's Directorate of Intelligence, *North Korea's Expanding Nuclear Efforts,* May 3, 1988, (formerly SECRET) declassified, available from *www.foia.cia.gov/docs/DOC_0000835120/DOC_0000835120.pdf*.

8. CIA, *Special Analysis: "North Korea: Nuclear Program of Proliferation Concern,"* March 22, 1989, (formerly SECRET) declassified, available from *www.foia.cia.gov/docs/DOC_0000835122/DOC_0000835122.pdf*.

9. CIA, *North Korea: Increased Nuclear-Related Activity,* February 1, 1992, (formerly TOP SECRET) declassified, available from *www.foia.cia.gov/docs/DOC_0001085723/DOC_0001085723.pdf*.

10. *Ibid.*

11. For examples of North Korean First Vice Foreign Minister Kang Sok Ju's statements likening the imposition of sanctions to a

war declaration or war itself, see Joel S. Wit, Daniel B. Poneman, and Robert L. Gallucci, *Going Critical: The First North Korean Nuclear Crisis*, Washington, DC: Brookings Institution Press, August 2005, pp. 55, 222.

12. Quoted from transcript of the House Committee on International Relations, "Hearing on the FY 2000 Budget," February 25, 1999, via *Congressional Quarterly Transcripts*.

13. U.S. National Intelligence Council, *Foreign Missile Developments and the Ballistic Missile Threat Through 2015*, Unclassified Summary of an NIE, December 2001, UNCLASSIFIED, a reproduction of which is available from *www.fas.org/irp/nic/bmthreat-2015.htm*. In a 2003 essay, Jonathan D. Pollack wrangles with the implications of the December 2001 NIE summary:

> If the report was claiming that U.S. intelligence analysts had concluded that North Korea had produced these weapons in the mid-1990s, it reflected either reinterpretation of old data or the inclusion of new information in older estimates. If the authors were claiming that the CIA had made this determination in the mid-1990s, then the claim is patently false, or all intelligence assessments published in the 1990s were false, in as much as the December 2001 claim contradicts all intelligence assessments published during the latter half of the 1990s.

See Pollack, "The United States, North Korea, and the End of the Agreed Framework," *Naval War College Review*, Vol. 56, No. 3, Summer 2003. Quoted by Joshua Pollack, "N. Korea: Deadly in a Snowball Fight," *Arms Control Wonk* weblog, February 7, 2010, available from *pollack.armscontrolwonk.com/archive/2615/n-korea-deadly-in-a-snowball-fight*.

14. CIA, *Untitled Estimate of North Korea's Nuclear Weapons Potential Provided to Congress*, November 19, 2002, UNCLASSIFIED, a reproduction of which is available from *www.fas.org/nuke/guide/dprk/nuke/cia111902.html*.

15. Henry Sokolski, "Fool Us Once . . . The North Koreans Get Ready to Shake Us Down Again," *The Weekly Standard*, Vol. 8, No. 9, November 11, 2002, available from *www.weeklystandard.com/Content/Public/Articles/000/000/001/851znpbr.asp?nopager=1*.

16. *Ibid.*

17. For example, see David Albright and Paul Brannan, "Taking Stock: North Korea's Uranium Enrichment Program," Washington, DC: Institute for Science and International Security, October 8, 2010, available from *isis-online.org/uploads/isis-reports/documents/ISIS_DPRK_UEP.pdf.*

18. See Siegfried S. Hecker, "A Return Trip to North Korea's Yongbyon Nuclear Complex," Palo Alto, CA: Stanford University, Center for International Security and Cooperation, November 20, 2010, available from *iis-db.stanford.edu/pubs/23035/HeckerYongbyon.pdf.* See also Henry Sokolski, "Getting Serious about North Korea's Nukes," *National Review Online*, November 23, 2010, available from *www.nationalreview.com/blogs/print/253792.*

19. For example, see "Background Briefing with Senior U.S. Officials on Syria's Covert Nuclear Reactor and North Korea's Involvement," Washington, DC: Office of the Director of National Intelligence, April 24, 2008, available from *www.dni.gov/interviews/20080424_interview.pdf.*

20. For example, see Alexandre Debs and Nuno P. Monteiro, "The Flawed Logic of Striking Iran," *Foreign Affairs*, January 17, 2012, available from *www.foreignaffairs.com/articles/137036/alexandre-debs-and-nuno-p-monteiro/the-flawed-logic-of-striking-iran.* See also Will Marshall, "Yes, We Can Contain Iran," *Foreign Policy*, March 16, 2012, available from *www.foreignpolicy.com/articles/2012/03/16/yes_we_can_contain_iran?page=full;* and Kenneth Waltz, "Why Iran Should Get the Bomb," *Foreign Affairs*, July/August 2012, available from *www.foreignaffairs.com/articles/137731/kenneth-n-waltz/why-iran-should-get-the-bomb.*

21. See *Agreement on the Canada-India Colombo Plan Atomic Reactor Project*, New Delhi, India, April 28, 1956, a reproduction of which is available from *www.nci.org/06nci/04/Canada-India%20 CIRUS%20agreement.htm;* and *U.S.-India CIRUS Agreement*, March 16, 1956, a reproduction of which is available from *www.nci. org/06nci/04/US-India%20CIRUS%20agreement.htm.*

22. For a penetrating case study of how diplomatic confusion over the terms of U.S.-India nuclear cooperation during the 1950s and 1960s not only unwittingly furthered the Indian nuclear weapons program, but also facilitated New Delhi's construction and shocking test of a nuclear explosive device, see Roberta Wohlstetter, *The Buddha Smiles: Absent-Minded Peaceful Aid and the Indian Bomb*, PH-78-04-370-23, final report prepared for the U.S. Energy Research and Development Administration in partial fulfillment of E (49-1)-3747, Los Angeles, CA: PAN Heuristics, November 15, 1976, revised November 1977, available from *www. robertawohlstetter.com/writings/BuddhaSmiles.*

23. Director of Central Intelligence, *Special National Intelligence Estimate: Prospects for Further Proliferation of Nuclear Weapons*, SNIE 4-1-74, August 23, 1974, (formerly TOP SECRET) declassified, available from *www.gwu.edu/~nsarchiv/NSAEBB/NSAEBB240/ snie.pdf.*

24. Henry A. Kissinger, *U.S.-Iran Nuclear Cooperation*, National Security Decision Memorandum 292, National Security Council, April 22, 1975, (formerly SECRET) declassified, available from *www.gwu.edu/~nsarchiv/nukevault/ebb268/doc05d.pdf.*

25. President Gerald Ford, "Statement of the President on Nuclear Policy," Office of the White House Press Secretary, October 28, 1976, available from *www.presidency.ucsb.edu/ws/index.php?pid =6561&st=&st1=#axzz1usYpla7k.*

26. Henry Sokolski, "The *Washington Post* Bombs Nuclear History: Did Dick Cheney, Donald Rumsfeld, and Paul Wolfowitz try to stoke Iran's nuclear ambition in the '70s?" *The Weekly Standard*, March 28, 2005, available from *www.weeklystandard.com/Content/ Public/Articles/000/000/005/417gusvl.asp?nopager=1.*

27. For a discussion of Iran's 1980 attack on Iraq's Osirak nuclear reactor and Iraq's series of attacks on Iran's Bushehr nuclear reactor and other nuclear facilities between 1984 and 1988, see Sarah E. Kreps and Matthew Fuhrmann, "Attacking the Atom: Does Bombing Nuclear Facilities Affect Proliferation?" *Journal of Strategic Studies*, Vol. 34, No. 2, April 2011, especially pp. 176-177.

28. Shirley A. Kan, *Chinese Proliferation of Weapons of Mass Destruction: Background and Analysis*, Report 96-767, Washington, DC: Congressional Research Service, Updated September 13, 1996, p. 29.

29. *Ibid.*, pp. 29-30.

30. *Ibid.*, p. 32.

31. Congressman John Spratt, Jr., "Defense Against Weapons of Mass Destruction Act of 1996," *Congressional Record*, June 27, 1996, p. E1192, available from *www.gpo.gov/fdsys/pkg/CREC-1996-06-27/pdf/CREC-1996-06-27-pt1-PgE1192-3.pdf#page=1.*

32. CIA, *The Acquisition of Technology Relating to Weapons of Mass Destruction and Advanced Conventional Munitions*, Washington, DC: Nonproliferation Center, July 1997, p. 5.

33. For example, see Stephen Greenhouse, "U.S. Gives Russia Secret Data on Iran to Discourage Atom Deal," *The New York Times*, April 3, 1995, p. A9; Stephen Erlanger, "Russia Says Sale of Atom Reactors to Iran is Still On," *The New York Times*, April 4, 1995, p. A1; Stephen Greenhouse, "U.S. Keeps On Urging Russia On Iran Deal," *The New York Times*, April 9, 1995, p. 1; Eric Schmitt, "Republicans Warn Russia That Deal With Iran Threatens Aid," *The New York Times*, May 8, 1995, p. A7; and Michael R. Gordon, "Russia Plans to Sell Reactors to Iran Despite U.S. Protests," *The New York Times*, March 7, 1998, p. 3.

34. Director of Central Intelligence, *Report of Proliferation-Related Acquisition in 1997*, unclassified report to Congress pursuant to Section 721 of the *Intelligence Authorization Act of Fiscal Year 1997*, September 17, 1998, available from *www.cia.gov/library/reports/archived-reports-1/acq1997.html*

35. Stephen Greenhouse, "Russia and China Pressed Not to Sell A-Plants to Iran," *The New York Times*, January 25, 1995, p. A6.

36. John Pomfret, "U.S. May Certify China on Curbing Nuclear Exports," *The Washington Post*, September 18, 1997, p. A28.

37. *Implementation of the NPT Safeguards Agreement in the Islamic Republic of Iran*, Report by the Director General, GOV/2003/40, Vienna, Austria, IAEA, June 6, 2003, available from *www.iaea.org/Publications/Documents/Board/2003/gov2003-40.pdf*.

38. R. Jeffrey Smith, "China's Pledge to End Iran Nuclear Aid Yields U.S. Help; Clinton Says He'll Allow U.S. Exports of Technology; Scrutiny and Debate Are Expected," *The Washington Post*, October 30, 1997, p. A15.

39. Besides issuing the certification required by the 1985 joint resolution on the U.S-China civil nuclear cooperation agreement, President Clinton issued a second certification that waived sanctions that blocked the implementation of the U.S.-Chinese nuclear deal—sanctions which Congress imposed after the Tiananmen Crackdown through Section 902 of the *Foreign Relations Authorization Act for Fiscal Years 1990 and 1991* (P.L. 101-246).

40. For discussions about how Iran has used Western-led diplomacy to improve its capability to make a nuclear weapon on ever-shorter notice, see "FPI Fact Sheet: The False Promise of Negotiations over Iran's Nuclear Program," *Foreign Policy Initiative*, Washington, DC, April 12, 2012, available from *www.foreignpolicyi.org/content/fpi-fact-sheet-false-promise-negotiations-over-iran%E2%80%99s-nuclear-program*; and Gregory S. Jones, *Iran's Rapid Enrichment Progress Moves It Ever Closer to a Nuclear Weapons Capability: Centrifuge Enrichment and the IAEA May 25, 2012 Safeguards Update*, Washington, DC: Nonproliferation Policy Education Center, June 6, 2012, available from *www.npolicy.org/article_file/Iran_Rapid_Enrichment_Progress.pdf*.

CHAPTER 15

FACING THE REALITY OF IRAN
AS A DE FACTO NUCLEAR STATE

Gregory S. Jones

This chapter was produced for the Nonproliferation
Policy Education Center (NPEC). Though the author
is also a part-time adjunct staff member at the RAND
Corporation, this chapter is not related to any RAND
project, and RAND bears no responsibility for any of
the analysis and views expressed in it. The original
version of this chapter was published by NPEC on
March 22, 2012, under the title, "Facing the Reality of
Iran as a De Facto Nuclear State: Centrifuge Enrichment and the IAEA February 24, 2012 Safeguards Update," available from *npolicy.org/article_file/Facing_the_
Reality_of_Iran_as_a_De_Facto_Nuclear_State.pdf*.

In various papers since 2008, this author has outlined how Iran's growing centrifuge enrichment program could provide it with the ability to produce
highly enriched uranium (HEU) for nuclear weapons.[1]
On February 24, 2012, the International Atomic Energy Agency (IAEA) published its latest safeguards update. This update shows not only that Iran's centrifuge
enrichment effort has continued to be unimpeded by
Western counteraction, but that it has undergone a significant expansion. In particular, Iran has made good
on its announcement of June 2011 that it would triple
its production of 19.7 percent enriched uranium by beginning enrichment operations at the well-protected
Fordow facility. At its main enrichment facility at Natanz, Iran increased its production of 3.5 percent enriched uranium by an additional 15 percent, meaning

it has doubled production since 2009—this is in stark contrast to the popular perception that cyber attacks have crippled Iran's enrichment effort. I estimate that Iran could produce enough HEU for a nuclear weapon in 1 1/2 months to 3 2/3 months, and might be able to produce enough HEU for three nuclear weapons in just 6 months if it were to decide to do so quickly (see Appendix 15-I).

Iran's rapid progress has changed the perception of the problem of its nuclear program, even for those who disagree with my current assessments. It is now obvious that even if my assessments are not true at this moment, they soon will be. For example, Olli Heinonen, a Senior Fellow at the Harvard Belfer Center and former deputy director general of the IAEA, has estimated that if Iran were to make an all-out effort now, it could produce enough HEU for a nuclear weapon in just 6 months. However, due to Iran's rapid progress in producing 20 percent enriched uranium, by the end of 2012, Heinonen estimates that Iran could produce enough HEU in just 1 month.[2]

It was in fall 2011 that David Albright of the Institute for Science and International Security (ISIS) was promoting the cheery notion that sanctions had capped Iran's nuclear program and that with its increasingly unreliable centrifuges, Iran's enriched uranium production had reached its maximum and was beginning to decline. Clearly this is not the case, and no longer is there any pretense that direct sanctions on Iran's nuclear program will stop Iran from being able to produce the HEU for nuclear weapons.

This does not mean that most analysts (including those in the U.S. Government) are willing to accept my view stated in September 2011 that Iran, in fact, was so close to having a nuclear weapon that it is already a de facto nuclear weapons state.[3] Rather, the focus has shifted to the nonnuclear components that would be needed to detonate Iran's HEU and implausible claims that it will take Iran 1 to 3 years to develop a "deliverable" nuclear weapon. In addition, most observers still cling to the hope that somehow Iran can be stopped from acquiring nuclear weapons. The methods that they foresee for stopping Iran are some form of military strike, the effects of sanctions, diplomacy, or some combination of these elements. As we will see, none of these methods holds much promise.

NONNUCLEAR WEAPONS COMPONENTS

As I have pointed out in prior writings,[4] the viewpoint that it will take Iran years to develop the nonnuclear components required for a nuclear weapon is hard to square with the actual historical experience of the nuclear weapons states. It is well-known that for past nuclear weapons programs, the key impediment was the acquisition of fissile material (HEU or plutonium) for the weapon. The production of the nonnuclear components needed to detonate the fissile material was relatively easy, and the development of these components was usually done in parallel to the more costly and time-consuming effort to produce fissile material. After all, the nonnuclear components of a nuclear weapon rely on conventional high-explosive technology, and any country advanced enough to acquire nuclear weapons has a military large enough to have substantial high-explosive expertise.

In 1944, the United States was able to develop an implosion-type nuclear weapon (the type that Iran would produce) in just 11 months, and this should be considered an upper bound on the time that Iran

413

would require. Though today Iran would not have the talent and resources available to the Manhattan Project, it would be starting from a far better position than did the United States. In 1944, no one knew whether or how the implosion method could work. Today, it is not only well-known that such weapons work, there are also descriptions of such weapons and pictures showing their general construction. Additionally, knowledge of explosives, as well as computing power, is far superior today than it was 68 years ago when the United States undertook this effort.

Moreover, Iran would not be starting from scratch. As the IAEA described last November, prior to 2004, Iran was assisted in developing "a multipoint initiation system that can be used to initiate effectively and simultaneously a high explosive charge over its surface" by "a foreign expert" who "worked for much of his career with this technology in the nuclear weapon programme of the country of his origin."[5] According to press reports, this "foreign expert" is a Russian named Vyacheslav Danilenko. The IAEA has been told by nuclear-weapon states that the specific multipoint initiation concept is used in some known nuclear explosive devices.

This "multipoint initiation system" will allow Iran to manufacture sophisticated nuclear weapons. Iran is now in a position to build nuclear weapons that are significantly lighter and have a smaller diameter than the cruder nuclear weapons that are typical of countries' early efforts. In 2003, Iran had already conducted at least one full-scale test of its multipoint initiation system with the hemispheric shape required for a nuclear weapon and sized to be used as a missile warhead. Furthermore, since that time, Iran has continued to test this system, but it is now using scaled down versions and employing a cylindrical geometry. Such geometry is not directly applicable to a nuclear weapon, but according to the IAEA, such tests would still allow Iran to improve and optimize the multipoint initiation design. As a result, I estimate that Iran could develop the nonnuclear components for a nuclear weapon in just 2 to 6 months.

A common mistake is to assume that Iran's production of HEU and its production of the nonnuclear components for its nuclear weapons would need to occur in series. However, it is clear from the published accounts of the U.S, British, and Chinese nuclear weapons programs that this development tends to occur in parallel instead. William Penney, who led the British effort to develop nuclear weapons, outlined the process. According to the official British history:

> He said that the manufacture of an atomic bomb of present design fell naturally into two parts: firstly the production of the active material and secondly the ordnance part, that is, the manufacture and assembly of the components causing the explosion of the active material. The second part of the work could be begun and *completed* without the need to use fissile material at any stage.[6]

Therefore, not only can the production of the fissile material and the nonnuclear components of a nuclear weapon occur in parallel, the production of the nonnuclear components can occur first. This fact was demonstrated by the U.S. experience in World War II. The nonnuclear components of the Hiroshima, Japan, nuclear weapon were on the cruiser *Indianapolis* and sailing across the Pacific Ocean, while some of the HEU components for the weapon were still being manufactured. The fact that the IAEA has provided information showing that Iran is currently developing the nonnuclear components for nuclear weapons, even though Iran does not yet have any HEU, further reinforces this point.

Though some have indicated that Iran might be able to develop a nuclear weapon in a year or less, they estimate that it could take Iran 2 to 3 years to develop a "deliverable" nuclear weapon, i.e., one that could be fitted as a warhead to a ballistic missile. There are several problems with this estimate. First, Iran does not need to use ballistic missiles to deliver its nuclear weapons. Vehicle delivery of bombs (up to now all conventional) has become quite common in the region, and many such attacks have been carried out on U.S. forces. Vehicle delivery of a nuclear weapon against U.S. forces could have a devastating effect and would have the advantage of making it more difficult to attribute the source of the attack.

Second, Iran already possesses and has tested a multipoint initiation system that has been sized for a ballistic missile warhead. Therefore Iran's first nuclear weapon will probably already be small and light enough to fit on a ballistic missile. One should note, however, that given the antimissile systems of Israel and the United States, it is not clear that ballistic missiles will be Iran's preferred nuclear weapon delivery mode, even if it has the capability.

Recent U.S. Government statements on how quickly Iran could build a nuclear weapon, should it decide to do so, have also indicated that the time required has been declining. But given the now widely-held assessment that Iran can produce enough HEU for a nuclear weapon in a matter of months, the U.S. Government assessments are still surprisingly long and are inconsistent as well. Media reporting, in particular that of

416

CBS News, has further complicated the situation. On December 19, 2011, CBS News broadcast an interview with then-U.S. Secretary of Defense Leon Panetta conducted by Scott Pelley:[7]

> **Pelley:** So are you saying that Iran can have a nuclear weapon in 2012?
>
> **Panetta:** It would probably be about a year before they can do it. Perhaps a little less. But one proviso, Scott, is if they have a hidden facility somewhere in Iran that may be enriching fuel.
>
> **Pelley:** So that they can develop a weapon even more quickly . . .
>
> **Panetta:** On a faster track . . .
>
> **Pelley:** Than we believe . . .
>
> **Panetta:** That's correct.

This interview caused quite a stir, since it was the first time that someone from the U.S. Government had given a public statement to the effect that Iran could produce a nuclear weapon in less than 1 year. Prior to that time, there was the belief that Iran was at least several years away. But a month later, Panetta seemed to be saying something different in an interview that CBS News broadcast on "60 Minutes" on January 29, 2012. Again, the interview was conducted by Scott Pelley.[8]

> **Narration by Pelley:** "We were surprised to hear how far he thinks Iran has come."

> **Panetta:** The consensus is that, if they decided to do it, it would probably take them about a year to be able to produce a bomb and then possibly another 1 to 2 years in order to put it on a deliverable vehicle of some sort in order to deliver that weapon.

At first glance, in this second interview, it seems that Panetta is just backtracking, as well as making the rather dubious assumption that Iran would first produce a nuclear weapon and only then start to work on a means of delivery. But the reality is more complicated. When one watches the video of these interviews,[9] it becomes clear that this is the **same interview** that has simply been edited differently. I find it disconcerting how easily the same interview can be edited to provide a quite different sense of how quickly Iran could produce a nuclear weapon and disappointing that a news organization as distinguished as CBS should have done so. It would be of great value for CBS to publish an unedited version of this interview so that Panetta's real view of this matter could be determined.

James Clapper, the U.S. Director of National Intelligence, has presented the assessment of the U.S. intelligence community to a congressional hearing on January 31, 2012. In part, this statement said:

> We assess Iran is keeping open the option to develop nuclear weapons, in part by developing various nuclear capabilities that better position it to produce such weapons, should it choose to do so. We do not know, however, if Iran will eventually decide to build nuclear weapons. . . . Iran's technical advancement, particularly in uranium enrichment, strengthens our assessment that Iran has the scientific, technical, and industrial capacity to eventually produce nuclear weapons, making the central issue its political will to do so. These advancements contribute to our judg-

ment that Iran is technically capable of producing enough highly enriched uranium for a weapon, if it so chooses.[10]

A few comments are in order. The first sentence contradicts itself since, if Iran is developing various nuclear capabilities that better position it to produce nuclear weapons, it is doing significantly more than just "keeping open" the option to develop nuclear weapons. Rather, Iran is either further developing the option or exercising it.

Absent any concrete time estimates, many of these statements are devoid of meaning. After all, any country (Belize, for example) has the "scientific, technical, and industrial capacity" **eventually** to produce nuclear weapons. For Belize, the time required would be many decades, but for Iran, it is presumably a good deal shorter. With the repeated use of the word "eventually," this intelligence assessment gives the impression that Iran is many years away, but when pressed on this issue at congressional hearings in February, Clapper said that Tehran could produce a nuclear weapon in 1 or 2 years.[11] Not only is this estimate much more immediate than the term "eventually" implies, it is not consistent with Panetta's estimate of 1 year (or perhaps less than 1 year).

Panetta's statement has also placed a great deal of reliance on a semantic distinction that, upon further examination, turns out to have no significance. This relates to the question of whether Iran is developing a nuclear weapons capability or a nuclear weapon. On January 8, 2012, he said:

Are they [Iran] trying to develop a nuclear weapon? No. But we know that they're trying to develop a nuclear capability. And that's what concerns us. And our red line to Iran is to not develop a nuclear weapon. That's a red line for us.[12]

But what is the difference? To build a nuclear weapon, Iran (or any country) needs sufficient fissile material (in Iran's case, HEU) and the nonnuclear components to detonate the fissile material. Iran is developing both of these elements. How is this not developing a nuclear weapon?

In addition, as was discussed previously, **any** country is "nuclear capable" in the sense that, given enough time, it can build a nuclear weapon. Yet most discussions (and not just those of U.S. Government officials) use "nuclear capable" without reference to any time element and thus render the term meaningless.

A related factor is the oft repeated statement that Iran has yet to decide to build a nuclear weapon. The implication seems to be that Iran cannot be building a nuclear weapon if it has not decided to do so. But many current nuclear states had nuclear weapons programs before there was a specific decision to build a nuclear weapon, and these programs helped enable the decision to build nuclear weapons by allowing countries to get close to acquiring nuclear weapons before any explicit decision was required. As I have written before:

> Though Iran's leadership may have not yet specifically decided to develop nuclear weapons, the U.K., France, India and Nazi Germany at one time all had nuclear weapons programs before their governments had decided specifically to produce nuclear weapons. The U.K., France and India all went on to make such a decision and have produced nuclear weapons. This underscores the point that as Iran moves closer to having a nuclear weapons capability, it becomes increas-

ing likely that Iran will make the decision to produce nuclear weapons.[13]

MILITARY STRIKE ON IRAN

The possibility of an Israeli military strike to "take out" Iran's enrichment facilities has been much in the news lately. Though not explicit, there seems to be a general view that this would be a one-time strike, similar to the ones that Israel carried out on nuclear reactors in Iraq in 1981 and in Syria in 2007. Concerns have been raised about the progress of the Iranian program and whether, with the partial move of its centrifuge enrichment activities to the underground site near Qom, Iran may be entering a "zone of immunity," whereby the Iranian centrifuge enrichment program can no longer be successfully attacked in a single strike.

In fact, attacking centrifuge enrichment facilities is quite different from attacking single nuclear reactors, and Iran's enrichment program is already well into a zone of immunity with regard to a single air strike. Iran has between 32 and 52 cascades operating in parallel at its main enrichment facility at Natanz.[14] An air strike on Natanz that scored multiple bomb hits would shut down the entire facility, but the majority of the cascades would be undamaged and unable to operate only due to damage to piping and the loss of utilities. It would only take a few months of repairs before these undamaged cascades were back in operation. Even for the cascades that suffered bomb hits, the majority of the centrifuges would still be undamaged. Iran could pull out the undamaged centrifuges and use them to build new cascades. It would only take 4 to 6 months before Iran would have returned to close-to-full production.

Iran's current stockpiles of about 3,000 kilograms of 3.5 percent enriched uranium and 67 kilograms of 19.7 percent enriched uranium are also a problem. These stockpiles represent years of centrifuge plant operation, but they would be very difficult to destroy by air attack. The combined volume of these two stockpiles is less than one cubic yard, making it easy to hide or protect.

It is small wonder that U.S. officials, when discussing possible attacks on Iran's centrifuge enrichment program, have begun to talk of bombing campaigns rather than single strikes.[15] By bombing Iran's facilities every few months, it would be possible to keep Iran's enrichment facilities shut down. Such a campaign would also have the advantage in that the question of whether U.S. large bunker-buster bombs can actually penetrate and hit Iran's underground enrichment facility near Qom would largely become moot: No matter how deep and well protected a bunker is, it is always possible to collapse the entrance tunnels and cutoff the utilities from the outside.

There are two problems with such an air bombing campaign. First, Iran could respond by dispersing its centrifuges. Indeed, centrifuge enrichment with its many parallel cascades would be ideal for such dispersal. The United States would be able to find and bomb some of these dispersed enrichment sites, but many would continue in operation undetected. Second, such a prolonged bombing campaign would run a serious risk of turning into a large-scale war with Iran. Though no doubt the United States would eventually win such a war, I think that, given the financially exhausted and war-weary condition of the United States, such a war would be ill-advised.

SANCTIONS

A key element of U.S. policy is to impose increasingly severe sanctions on Iran. The latest round of sanctions is designed to affect Iran's overall economy significantly by making it more and more difficult for Iran to export its oil. However, these sanctions are not authorized by the United Nations (UN) but rather imposed unilaterally by the U.S. and the European Union (EU). The reason for this is that, despite the IAEA's revelations last November of Iran's efforts to develop the nonnuclear components for a nuclear weapon, both Russia and China have refused to support any additional sanctions against Iran. Indeed, both countries have continued trading with Iran, and China continues to purchase oil from Iran.

Nor are China and Russia the only countries that have not adopted these sanctions. India, with its important economy, has actually increased its purchases of Iranian oil, as has South Korea. India has gone so far as to change its tax code to facilitate a method of payment that involves using rupees rather than dollars. Pakistan and Turkey have also continued trading with Iran. Pakistan has even proposed deals based on a straight barter arrangement. Japan has cut back on its oil purchases but is expected to ask the United States for an exemption from a requirement to eliminate all Iranian oil purchases.

With all of these important economies not complying with the sanctions on Iran, it is unclear that the sanctions will be enough to compel Iran to change its current policies. Even if they can, the real problem is that Iran can resolve all of its outstanding issues with the IAEA, and still, due to the laxity of IAEA safeguards, maintain its drive towards the production of nuclear weapons.

DIPLOMACY

Many have continued to hope that negotiations with Iran could provide a means to prevent it from obtaining nuclear weapons. But no one has outlined how any realistic agreement with Iran can achieve this goal. President Obama has said that the United States will not allow Iran to obtain nuclear weapons. But as I have written elsewhere, Iran has no need actually to produce a nuclear weapon unless it wants to test or use such a weapon.[16] Therefore, it is likely to be many years before Iran does so. The real issue in the near term is not preventing Iran from obtaining nuclear weapons, but rather stopping Iran from moving ever closer to being able to build nuclear weapons. Similarly, President Obama has said that Iran must "make a decision to forsake nuclear weapons," but since many U.S. Government officials have said that Iran has not yet made a decision to produce nuclear weapons, the Iranians can argue that they have already complied with this requirement.

Iran has outstanding issues with the IAEA regarding its nuclear weapons development program. But most of these issues relate to events from 2003 and before. Though domestically it would be politically difficult for Iran, if it were to admit to these prior transgressions, it would be able to end its disputes with the IAEA while not having to give up any of its current centrifuge enrichment program. Indeed, given the laxity of IAEA safeguards, Iran could go on to produce HEU with the blessings of the IAEA.

Most of those who believe that Iran's nuclear weapons program can be stopped diplomatically have suggested that, in order to reach an agreement,

Iran should be allowed to keep its centrifuge uranium enrichment program. Those who hold this view realize that this poses a risk of allowing Iran to obtain the HEU needed for nuclear weapons, but they believe that with the proper controls, this risk can be obviated. In particular, it is often suggested that Iran be limited to producing uranium with an enrichment level of less than 5 percent and reduce its stockpile of 19.7 percent enriched uranium to zero by exporting all of this material, and that Iran's future enrichment program be limited to that which can be justified by its peaceful nuclear needs. However, Iran's current enrichment facilities are very small compared to those needed for most peaceful nuclear activities (such as providing fuel for a single nuclear power reactor), and such an agreement would provide Iran with the justification for greatly expanding its current enrichment facilities. These greatly expanded facilities would provide Iran easy access to the HEU needed for nuclear weapons.

For example, even if Iran produced only 4.1 percent enriched uranium and expanded its enrichment capacity by a factor of 20, it would only produce about 15 metric tons of enriched uranium per year. This amount would still be less than that needed to fuel a single large power reactor, yet, using batch recycling, these enrichment facilities could produce enough HEU for a nuclear weapon in just 2 weeks or enough HEU for five weapons in just 5 weeks (see Appendix 15-II). One might argue that, using its own resources, it would take Iran a very long time to expand its enrichment facilities by a factor of 20, but such a diplomatic agreement would serve the function of legitimizing Iran's enrichment activities. This would lead to the removal of sanctions that are designed to prevent Iran from importing the materials needed to build additional centrifuges and, in addition, Iran might receive assistance to expand its enrichment facilities (from China or Pakistan, for instance) as part of normal nuclear commerce.

As it is, Iran appears to be laying the groundwork to make such an agreement impossible and to present the P5+1[17] with a lose-lose situation. In the middle of February 2012, Iran announced a set of three advances in its nuclear program: It had manufactured and installed a fuel element using 20 percent enriched uranium into the Tehran Research Reactor (TRR), increased the number of centrifuges operating at Natanz from 6,000 to 9,000, and successfully developed more advanced and efficient centrifuges than the type Iran currently uses. This announcement was generally met with derisive comments to the effect that these advances were not special, and the Iranian Government was playing to its domestic audience.

At the same time, Iran indicated it was interested in restarting negotiations with the P5+1 regarding its nuclear program. Some observers were puzzled by this seemingly schizoid behavior. Others were so eager for negotiations that they did not care about any contradictory indications, including Iran's assault on the British Embassy in November 2011. However, very few seem to have recognized the significance of the two events when considered together.

By announcing its nuclear advances, Iran is throwing down markers for negotiations. By loading a 20 percent fuel element into its research reactor, Iran can now argue it has a legitimate need to produce such enriched uranium and that it will not stop its production. By claiming to have 9,000 centrifuges in operation, Iran is establishing a base below which it will refuse to go. By claiming to have finished developing advanced centrifuges, Iran is putting itself in a position to be able to significantly upgrade and expand its current uranium enrichment capacity.

This, then, is the P5+1's no-win situation. It can refuse to allow Iran to keep its centrifuge enrichment facilities, in which case Iran can break off the talks, claiming that the P5+1 are being unreasonable, and then use this claim to help break sanctions by playing to on-the-fence countries such as India. Or, if the P5+1 should be so foolish as to agree to allow Iran to keep its current enrichment facilities, then Iran will have legitimized these facilities and its ability to quickly produce the HEU for nuclear weapons whenever it decides to do so.

The only negotiated solution that would prevent Iran from being able to quickly produce HEU would be for Iran to permanently shut down its enrichment facilities and export its stockpiles of enriched uranium. By saying that the P5+1 must accept continued Iranian uranium enrichment, advocates of a negotiated solution are essentially admitting that no satisfactory negotiated solution is possible.

NONPROLIFERATION AFTER IRAN

If Iran is already a de facto nuclear weapons state, where should the United States go from here with regards to its nonproliferation policy? The key will be to learn from our failure with Iran and prevent additional countries from acquiring nuclear weapons. This will require a two-pronged approach.

First, as President Obama has indicated, Iran's de facto nuclear status will motivate a number of other countries to try to emulate Iran's success. The United States needs to take decisive action to head off these efforts on a county by country basis as soon as the first steps towards acquiring the fissile material for nuclear weapons are detected. Taking early action runs counter to normal government instinct, which is to try to "kick the can down the road" and avoid taking any unpleasant actions unless it has to. The lack of early action has been a hallmark of U.S. nonproliferation policy since the Reagan administration and has allowed Pakistan, India, North Korea, and now Iran to acquire the fissile material required for nuclear weapons.

Yet as we saw with Libya, early action can be quite effective. Many believe that Muammar Gaddafi made a mistake by giving up a nuclear weapons program, but he had no choice. His effort was discovered early, before Libya had even begun to enrich uranium, and Gaddafi had no other option.

Second, there needs to be a change to the IAEA's safeguards regime to prevent countries from acquiring the fissile material needed for nuclear weapons with the IAEA's approval. Some in the U.S. Congress have called for military action against Iran if it starts to enrich uranium to levels greater than 20 percent, but under current IAEA rules, such Iranian actions would be perfectly acceptable as long as Iran declared the activity to the IAEA. Similarly, the IAEA permits non-nuclear weapons states to produce pure compounds of plutonium by reprocessing spent fuel. Informally, the IAEA does require that the country carrying out these activities provide some rationale as to how these activities are related to some peaceful nuclear activity, but the rationale does not have to be very plausible. For example, a country can say that it is stockpiling the plutonium for use in a breeder reactor, even if it is now more than 40 years since such reactors were first supposed to come into operation, and that such reactors are still decades away.

Much of providing the proper rationale involves learning to play the game properly. As discussed previously, Iran got itself into trouble by conducting clandestine nuclear activities prior to 2004. More recently, Iran did a better job and explained that its production of 20 percent enriched uranium was required to produce research reactor fuel. This activity, which is generally agreed to be carrying Iran close to the possession of the fissile material for a nuclear weapon, has not caused the IAEA to say that Iran is violating safeguards even though Iran is currently producing more 20 percent enriched uranium in 1 month than the research reactor uses in 1 year.

The U.S. Government has recognized this problem, and in its Nuclear Cooperation Agreement with the United Arab Emirates (UAE), it requires that the UAE not possess facilities that can be used for uranium enrichment or the reprocessing of spent fuel, which could produce plutonium, HEU, or U-233 (another material that can be used to produce nuclear weapons). However, the U.S. administration has discovered the drawback of attempting to handle this problem though bilateral nuclear cooperation agreements. In the face of competition from Russia and France, the United States has proposed nuclear cooperation agreements with Vietnam and Jordan that lack these provisions on enriching and reprocessing. Only if the issue is approached by the IAEA will there be uniform standards without commercial pressures undercutting nonproliferation.

Furthermore, even the standards for the UAE are not enough. Non-nuclear weapons states need to be prohibited from possessing any materials or facilities that can quickly provide fissile material for nuclear weapons. This includes prohibiting not only enrichment and reprocessing facilities, but also HEU, plutonium, or U-233 that has either been separated from spent fuel and/or HEU, plutonium, or U-233 contained in unirradiated reactor fuel (such as HEU fuel for research reactors or mixed oxide fuel for power reactors).

The IAEA does not have the legal authority to prohibit countries from possessing such materials or facilities, but it does have the responsibility to safeguard these materials and facilities. As I have discussed elsewhere,[18] IAEA safeguards are supposed to be more than just an accounting system; they should provide "timely warning" of diversions of nuclear materials. However, the IAEA cannot safeguard these facilities and materials in a timely warning sense. The IAEA needs to admit this fact and make clear that any such facilities and materials in non-nuclear weapons states are not being effectively safeguarded. This issue is significantly larger than just Iran and, at a minimum, includes Japan, Germany, the Netherlands, and Brazil. It will be up to these countries to explain why they need to continue to possess these materials and facilities since they cannot effectively be safeguarded. Given the state of nuclear power in a post-Fukushima world, this could be difficult.

The United States needs to urge the IAEA to be clear about what materials and facilities it can effectively safeguard and which it cannot. At the same time, the United States needs to take early action to ensure that any countries that attempt to follow Iran's successful path are prevented from gaining access to the fissile material required for nuclear weapons. Otherwise, the number of nuclear-armed countries will continue to grow until the catastrophe of nuclear use occurs. Just one nuclear weapon detonated in a city could kill hundreds of thousands of people—roughly 100 times as many as were killed on September 11, 2001.

ENDNOTES - CHAPTER 15

1. The author's most recent report is Gregory S. Jones, "Iran's Enriched Uranium Stocks Can Produce Enough HEU for 3 to 5 Nuclear Weapons," September 10, 2013, available from *nucle-arpolicy101.org/wp-content/uploads/2013/09/Iran-Enrichment-Up-date-09-2013.pdf*.

2. Olli Heinonen, "The 20 Percent Solution," *Foreign Policy*, January 11, 2012.

3. Gregory S. Jones, "No More Hypotheticals: Iran Already Is a Nuclear State, *New Republic*, September 9, 2011, available from *www.tnr.com/article/environment-and-energy/94715/jones-nuclear-iran-ahmadinejad*.

4. Gregory S. Jones, "Iran's Efforts to Develop Nuclear Weapons Explicated, Centrifuge Uranium Enrichment Continues Unimpeded, The IAEA's November 8, 2011, Safeguards Update," December, 6, 2011, pp. 10-13, available from *npolicy.org/article.php?aid=1124&rid=4*; and Gregory S. Jones, "An In-Depth Examination of Iran's Centrifuge Enrichment Program and Its Efforts to Acquire Nuclear Weapons?" August 9, 2011, pp. 23-25, available from *npolicy.org/article.php?aid=1092&rid=4*.

5. IAEA, *Implementation of the NPT Safeguards Agreement and relevant provisions of Security Council resolutions in the Islamic Republic of Iran*, GOV/2011/65, November 8, 2011, Annex, pp. 8-9.

6. Emphasis added. At the time, the memo was so highly classified that Penney had to type it himself. See Margaret Gowing, assisted by Lorna Arnold, *Independence and Deterrence: Britain and Atomic Energy, 1945-1952*, Vol. I, Policy Making, New York: St. Martin's Press, 1974, p. 180.

7. Leon Panetta, interview by Scott Pelley, *60 Minutes*, CBS News, December 19, 2011, transcript available from *www.cbsnews.com/8301-18563_162-57345322/panetta-iran-will-not-be-allowed-nukes*.

8. Leon Panetta, interview by Scott Pelley, *60 Minutes*, CBS News, January 29, 2012, transcript available from *www.cbsnews.com/8301-18560_162-57367997/the-defense-secretary-an-interview-with-leon-panetta/?tag=contentMain;cbsCarousel*.

9. Available at the same locations as the transcripts.

10. James R. Clapper, "Unclassified Statement for the Record on the Worldwide Threat Assessment of the U.S. Intelligence Community for the Senate Select Committee on Intelligence," January 31, 2012, pp. 5-6.

11. Associated Press, "Panetta says Iran enriching uranium but no decision yet on proceeding with a nuclear weapon," *The The Washington Post*, February 16, 2012.

12. "Face the Nation transcript: January 8, 2012," *Face the Nation*, CBS News, January 8, 2012, available from *www.cbsnews.com/8301-3460_162-57354647/face-the-nation-transcript-january-8-2012*.

13. Jones, "An In-Depth Examination of Iran's Centrifuge Enrichment Program and Its Efforts to Acquire Nuclear Weapons," p. 35.

14. Iran has declared to the IAEA that it has 52 cascades in operation, but its enriched uranium production is only equivalent to about 32 cascades operating at full capacity.

15. Joby Warrick, "Iran's Underground Nuclear Sites Not Immune to U.S. Bunker-Busters, Experts Say," *The Washington Post*, February 29, 2012.

16. Jones, "No More Hypotheticals: Iran Already Is a Nuclear State":

> That's not to say that I expect Iran to divert nuclear material from IAEA safeguards anytime soon. After all, why should it? It can continue to move ever closer to the HEU required for a nuclear weapon with the blessing of the IAEA. Iran would only need to divert nuclear material from safeguards when it would want to test or use a nuclear weapon. Recall that the U.S. was unable to certify that Pakistan did not have nuclear weapons in 1990, but it was only in 1998 that it actually tested a bomb. Similarly, though it could be many years before Iran becomes an overt nuclear power, it needs to be treated as a de facto nuclear power simply by virtue of being so close to having a weapon.

17. The United States, UK, France, Russia, China, and Germany.

18. Jones, "An In-Depth Examination of Iran's Centrifuge Enrichment Program and Its Efforts to Acquire Nuclear Weapons," pp. 16-23.

CHAPTER 15

APPENDIX I

DETAILED ANALYSIS OF THE INTERNATIONAL ATOMIC ENERGY AGENCY FEBRUARY 24, 2012, SAFEGUARDS REPORT AND METHODS WHEREBY IRAN COULD PRODUCE HIGHLY ENRICHED URANIUM FOR NUCLEAR WEAPONS IRANIAN CENTRIFUGE ENRICHMENT OF URANIUM

Iran has three known centrifuge enrichment facilities. It's main facility is the Fuel Enrichment Plant (FEP) at Natanz. The basic unit of Iran's centrifuge enrichment effort is a cascade that originally consisted of 164 centrifuges, but Iran has now modified the majority of the cascades by increasing the number of centrifuges to 174 (all centrifuges installed up to now have been of the IR-1 type). Each cascade is designed to enrich natural uranium to 3.5 percent enriched uranium. As of February 19, 2012, Iran had installed a total of 54 cascades, 30 of which each contain 174 centrifuges, with the remaining 24 cascades each containing 164 centrifuges. This results in a total of 9,156 centrifuges. Of these 54 cascades, 52 (containing 8,808 centrifuges) were declared by Iran as being fed with uranium hexafluoride and, therefore, were producing 3.5 percent enriched uranium, though the International Atomic Energy Agency (IAEA) has indicated that not all of these 8,808 centrifuges may be operational.[1] Indeed, given the amount of enriched uranium that was actually being produced at the FEP, it seems likely that Iran's declaration was simply a negotiating ploy so as to be able to claim it has this number of centrifuges in

435

operation and that the real number of centrifuges in operation was significantly less.

Iran began producing 3.5 percent enriched uranium at the FEP in February 2007, and as of February 4, 2012, Iran had produced a total of 3,685 kilograms (kg) in the form of 5,451-kg of uranium hexafluoride. Since 666-kg of this enriched uranium has already been processed into 19.7 percent enriched uranium (see the Pilot Fuel Enrichment Plant [PFEP] and Fordow Fuel Enrichment Plant [FFEP]), and a further 21-kg was converted into uranium dioxide for use as fuel in the Tehran Research Reactor (TRR), Iran's current stockpile of 3.5 percent enriched uranium is 2,998-kg. Iran's current production rate of 3.5 percent enriched uranium is about 115-kg per month.[2] This production rate represents about a 15 percent increase from 2011, when the production rate was about a steady 100-kg per month, and represents a doubling of the rate since 2009 (see Table 15-AI1). From the production rate of 3.5 percent enriched uranium, it is easy to calculate that the FEP has a separative capacity of about 5,000 separative work units (SWU) per year.[3]

Iran also has the PFEP at Natanz, which is used to test a number of more advanced centrifuge designs. These are usually configured as single centrifuges or small 10- or 20-centrifuge test cascades. However, Iran has installed a cascade of 164 IR-2m centrifuges, and, although this cascade appears ready to begin to produce enriched uranium, it has yet to do so. Iran has also installed 58 IR-4 centrifuges in a separate cascade but has not yet begun feeding them with uranium hexafluoride.

IAEA Reporting Interval	Average 3.5 Percent Enriched Uranium Production Rate (Kilograms Uranium per Month)
11/17/08-1/31/09	52
2/1/09-5/31/09	53
6/1/09-7/31/09	57
8/1/09-10/31/09	57
11/22/09-1/29/10	78
1/30/10-5/1/10	81
5/2/10-8/6/10	80
8/7/10-10/17/10	95
10/18/10-2/5/11	88
2/6/11-5/14/11	105
5/15/11-8/13/11	99
8/14/11-11/1/11	97
11/2/11-2/4/12	115

**Table 15-AI1. Average Iranian Production Rate
of 3.5 Percent Enriched Uranium
Late-2008 to Early-2012.**

In addition, there are two full cascades, each with 164 IR-1 type centrifuges, at the PFEP. These two cascades are interconnected and are being used to process 3.5 percent enriched uranium into 19.7 percent enriched uranium. In February 2010, Iran began producing 19.7 percent enriched uranium at the PFEP using one cascade. It added the second cascade in July 2010. As of February 11, 2012, Iran had produced 64.5-kg of 19.7 percent enriched uranium (in the form of 95.4-kg of uranium hexafluoride) at this facility. Iran's production rate of 19.7 percent enriched uranium at the PFEP has been fairly steady over the past year and is currently about 3.05-kg per month. The centrifuges at this facility are each producing about 0.9 SWU per year.

Finally, Iran has constructed an enrichment facility near Qom. Known as the FFEP, Iran clandestinely started to construct this plant in violation of its IAEA safeguards. Iran only revealed the existence of this plant in September 2009, when it believed that the West had discovered the plant.

Iran has installed two sets of two interconnected cascades at the FFEP (each cascade contains 174 centrifuges, IR-1 type) in order to produce 19.7 percent enriched uranium from 3.5 percent enriched uranium as is being done at the PFEP. The first of these two sets began production on December 14, 2011, and the second set began operation on January 25, 2012. As of February 17, 2012, Iran had produced 9.3-kg of 19.7 percent enriched uranium (in the form of 13.8-kg of uranium hexafluoride) at this facility. This facility is producing 19.7 percent enriched uranium at the rate of 6.45-kg per month. As with the centrifuges at the PFEP, the individual centrifuges at the FFEP are producing about 0.9 SWU per year.

With the start of these two sets of interconnected cascades at the FFEP, Iran made good on its announcement of June 2011 that it would triple its production rate of 19.7 percent enriched uranium. Currently, Iran is producing a total of about 9.5-kg of 19.7 percent enriched uranium per month. As of mid-February 2013, Iran had produced a total of about 74-kg of 19.7 percent enriched uranium. Since Iran has converted about 7-kg of this uranium into a uranium oxide compound for use as fuel in the TRR, Iran's current stockpile of 19.7 percent enriched uranium is about 67-kg.

Iran has installed the piping and centrifuge casings for an additional 2,088 centrifuges (12 cascades) at the FFEP. Iran has informed the IAEA that these additional cascades, when completed, will be used to produce either 3.5 percent or 19.7 percent enriched uranium without specifying how many cascades will be producing what type of enriched uranium. This opens the possibility that Iran could further increase its rate of 19.7 percent enriched uranium. Given Iran's current production rate of 3.5 percent enriched uranium at the FEP, Iran could run two additional sets of two interconnected cascades to produce 19.7 percent enriched uranium without the need to drawdown its stockpile of 3.5 percent enriched uranium. If Iran were to construct and start to operate these two additional sets of cascades, its overall production rate of 19.7 percent enriched uranium would be about 16-kg per month.

IRANIAN OPTIONS FOR PRODUCING HIGHLY ENRICHED URANIUM

Given that it currently has an enrichment capacity of 5,000 SWU per year at the FEP and stockpiles of about 3,000-kg of 3.5 percent enriched uranium and 67-kg of 19.7 percent enriched uranium, Iran has a number of options for producing the 20-kg of highly enriched uranium (HEU) required for a nuclear weapon.

The most straightforward method Iran could use to produce HEU would be batch recycling at the FEP. In this process, no major modifications are made to the FEP but rather enriched uranium is successively run though the FEP in batches until the desired enrichment is achieved. In the past, I have calculated that Iran could use a two-step process to produce HEU. In the first step, 3.5 percent enriched uranium would be enriched to 19.7 percent enriched uranium. Iran has already demonstrated this step by producing 19.7 percent enriched uranium at the PFEP and FFEP. In the second step, 19.7 percent enriched uranium would be enriched to 90 percent enriched uranium. My calculations for this second step rely on work by Alexander Glaser, which demonstrated that, by reducing the flow through the cascade, it was possible to achieve the production of 90 percent enriched uranium from 19.7 percent enriched uranium in one step without a significant loss of separative capacity.[4] This process is illustrated for Iran's current situation in Table 15-AI2.

Cycle	Product Enrichment and Quantity	Feed Enrichment and Quantity	Time for Cycle (Days)
First	19.7 percent 91.2-kg	3.5 percent 1,080-kg	32
Second	90.0 percent 20-kg	19.7 percent 153.2-kg*	11
Total			47**

* Includes 67-kg of 19.7 percent enriched uranium that Iran has already stockpiled.
**Includes 4 days to account for equilibrium and cascade fill time.

(The Second Step is Based on Glaser's Analysis.)

Table 15-AI2. Time, Product, and Feed Requirements for the Production of 20-Kg of HEU by Batch Recycling at the FEP (5,000 SWU Per Year Total).

Two steps are required. In the first step, Iran needs to produce 158.2-kg of 19.7 percent enriched uranium (including 5-kg for the plant inventory in the second step). However, since it has already produced 67-kg of 19.7 percent enriched uranium, Iran needs only to produce an additional 91.2-kg. This step requires 1,080-kg of 3.5 percent enriched uranium as feed, but Iran's current stockpile well exceeds this figure. In the second step, the 19.7 percent enriched uranium is further enriched to the 90 percent level suitable for a nuclear weapon. Using Iran's currently operating centrifuges at the FEP, the batch recycling would take about 1 and 1/2 months.

As stated previously, this calculation depends on Glaser's published calculations of the effectiveness of reduced cascade flow so that uranium can be enriched from 19.7 percent to 90 percent in one step. I am not the only analyst who has relied on Glaser's work, as both Levi[5] and the International Institute for Strategic Studies (IISS)[6] have based their calculations on Glaser's calculations. However, as I wrote in my last paper, questions have been raised about the validity of Glaser's work, and I have had to examine methods whereby Iran could produce the 20-kg of HEU required for a nuclear weapon without relying on his calculations.[7]

Iran could still produce HEU by batch recycling at the FEP, but the process would require three steps. Each pass would produce the feed required for the next cycle, which would include the plant inventory (in this case, 2-kg for each cycle). Iran would need to produce sufficient 19.7 percent enriched uranium from 3.5 percent enriched feed, then further enrich this 19.7 percent enriched uranium to 55.4 percent enriched uranium, and finally enrich the 55.4 percent enriched uranium to 86.3 percent enriched uranium. I have increased the amount of HEU required from 20-kg to 21-kg to keep the quantity of U-235 in the product about the same.

The results for the first step can be found using separative work calculations, but for the other two steps, an SWU calculation would not produce accurate results. Since the plant at Natanz is designed to produce 3.5 percent product from natural uranium, its cascade is more tapered than is optimal for the upper stages of an enrichment plant designed to produce HEU. As a result, some of the SWU output of the plant cannot be utilized during the latter cycles of the batch production process. The plant is restricted by the flow at the product end of the cascade. Therefore, the time required per cycle is determined by the amount of product required and the amount of product the plant can produce per day, not by an SWU calculation.

The results (see Table 15-AI3) show that this method of batch recycling would take just under 5 months, in contrast to the 1 1/2 months required in Table 15-AI2. In addition, Iran would need to start with 3,840-kg of 3.5 percent enriched uranium, much more than the 1,080-kg required by the calculations in Table 15-AI2 and significantly more than the 3,000-kg that Iran currently possesses. At current production rates, it would take about 7 months before Iran would possess enough 3.5 percent enriched uranium to start the batch recycling process.

Cycle	Product Enrichment and Quantity	Feed Enrichment and Quantity	Time for Cycle (Days)
First	19.7 percent 325-kg	3.5 percent 3,840-kg	114
Second	55.4 percent 68.4-kg	19.7 percent 390-kg*	18
Third	86.3 percent 21-kg	55.4 percent 66.4-kg	6
Total			144**

* Includes 67-kg of 19.7 percent enriched uranium that Iran has already stockpiled.
**Includes 6 days to account for equilibrium and cascade fill time.

(Does Not Rely on Glaser's Analysis).

Table 15-AI3. Time, Product, and Feed Requirements or the Production of HEU by Batch Recycling at the FEP (5,000 SWU Per Year Total).

Iran, however, has additional options for producing the HEU required for a nuclear weapon. As stated earlier, in addition to the FEP, Iran is producing 19.7 percent enriched uranium at the PFEP and recently tripled its production of 19.7 percent enriched uranium by starting two sets of two interconnected cascades at the FFEP. Iran can use its 19.7 percent production capacity to carry out the final step of the three step batch recycling process. The results are shown in Table 15-AI4.

As in the previous case, the times for the second and third steps are determined by the cascade product production rate and not by SWU calculations. The total time required is about 3 2/3 months, which is over a month shorter than the prior case where all three batch recycling steps were carried out at the FEP. In addition, this method has the advantage of reducing the required amount of 3.5 percent enriched uranium feed from 3,840-kg to 1,640-kg, which is smaller than Iran's current 3,000-kg stockpile and therefore could be carried out today, if Iran so desired.

Cycle and Enrichment Plant	Product Enrichment and Quantity	Feed Enrichment and Quantity	Time for Cycle (Days)
First FEP	19.7 percent 139-kg	3.5 percent 1,640-kg	37
Second FEP	55.4 percent 39.2-kg	19.7 percent 223-kg	10
Third PFEP & FFEP**	89.4 percent 20-kg	55.4 percent 39.0-kg	64
Total			111**

* Includes 67-kg of 19.7 percent enriched uranium that Iran has already stockpiled, and 19 kilograms of 19.7 percent enriched uranium from the tails of the PFEP and FFEP.
**Plant
***Includes 6 days to account for equilibrium and cascade fill time.

(Does Not Rely on Glaser's Analysis).

Table 15-AI4. Time, Product, and Feed Requirements for the Production of HEU by Batch Recycling at the FEP (5,000 SWU Per Year Total).

Final step at PFEP and FFEP.

If Glaser's calculations are incorrect, the only way that Iran could currently produce the HEU for a nuclear weapon in just 2 months would be to use batch recycling at the FEP, combined with a clandestine "topping" enrichment plant. Since Iran continues to refuse to implement the Additional Protocol to its safeguards agreement, the IAEA would find it very difficult to locate a clandestine enrichment plant—a fact that the IAEA confirmed.[8] While this has been a theoretical possibility since 2007, its salience increased with the discovery in September 2009 that Iran was actually building such a clandestine enrichment plant (the FFEP near Qom).

In this case, the clandestine enrichment plant could be designed as an ideal cascade to enrich 19.7 percent enriched uranium to the 90 percent enriched uranium needed for a nuclear weapon. By starting from 19.7 percent enriched uranium, this clandestine enrichment plant need only contain about 1,400 IR-1 type centrifuges to be able to produce the 20-kg of HEU required for a nuclear weapon in just 2 months. Furthermore, since Iran already has a stockpile of 19.7 percent enriched uranium, the production of the 19.7 percent enriched uranium at the FEP and the 90 percent enriched uranium at the clandestine enrichment plant could be carried out **simultaneously**.

The results of this process are shown in Table 15-AI5. As can be seen, the production of the 19.7 percent enriched uranium needed (including 0.5-kg for the plant inventory at the clandestine plant) to produce 20-kg of HEU at the clandestine enrichment plant now requires only 325-kg of 3.5 percent enriched feed. Since the cycle time at the FEP is shorter than that at the clandestine enrichment plant and the cycles are carried out simultaneously, the time required at the FEP has no impact on the overall time required to produce the HEU.

Cycle and Enrichment Plant	Product Enrichment and Quantity	Feed Enrichment and Quantity	Time for Cycle (Days)
First FEP	19.7 percent 27.5-kg	3.5 percent 325-kg	12**
Second Clandestine	90.0 percent 20-kg	19.7 percent 106.8-kg*	63**
Total			63***

* Includes 67-kg of 19.7 percent enriched uranium that Iran has already stockpiled. Processing the tails of the clandestine plant at the PFEP and FFEP produces an additional 12.8-kg of 19.7 percent enriched uranium.
** Includes 2 days to account for equilibrium and cascade fill time.
***Cycle times are not additive since cycles are simultaneous.

(Does Not Rely on Glaser's Analysis).

Table 15-AI5. Time, Product, and Feed Requirements for the Production of HEU by Batch Recyclingat the FEP (5,000 SWU Per Year Total).

Final Step at 1,400-Centrifuge Clandestine Plant (0.9 SWU per Centrifuge-Year). Cycles Carried Out Simultaneously.

Further, since Iran would have a substantial quantity of 3.5 percent enriched uranium left over (about 2,700-kg), it could continue the process and produce additional HEU. An additional 20-kg of HEU would require 1,109-kg of 3.5 percent enriched uranium feed, so with its current stockpile, Iran could produce a total of about 68-kg of HEU, which is enough for about three nuclear weapons. Since the clandestine enrichment plant has been sized to produce about 10-kg of HEU per month, Iran could produce enough HEU for a nuclear weapon at successive 2-month intervals.

Batch recycling of enriched uranium isn't the only pathway for Iran to produce the fissile material required for nuclear weapons, though it is the process that allows Iran to produce HEU most quickly. Iran could produce HEU at a clandestine enrichment plant designed to produce 90 percent enriched uranium from natural uranium feed.

A clandestine enrichment plant containing 3,800 centrifuges (0.9 SWU per centrifuge-year) could produce around 20-kg of HEU (the amount required for one nuclear weapon) each year using natural uranium as feed. Since this option does not require any overt breakout from safeguards, the relatively slow rate of HEU production would not necessarily be of any concern to Iran. Such production could be going on right now, and the West might well not know. A clandestine enrichment plant would need a source of uranium, but Iran is producing uranium at a mine near Bandar Abbas.[9] Since Iran has refused to implement the Additional Protocol to its IAEA safeguards, this uranium mining is unsafeguarded, and the whereabouts of the uranium that Iran has produced there is unknown. A significant drawback to this stand-alone clandestine enrichment plant is that it requires many more centrifuges than would the 1,400-centrifuge clandestine plant discussed earlier. It is not clear whether Iran could provide this number of centrifuges to a clandestine plant, and the larger any clandestine enrichment plant is the more likely it is that it will be discovered.

Iran, then, has a number of methods whereby it could produce the HEU required for a nuclear weapon. If Glaser's previously published calculations are correct, then batch recycling at the FEP alone could produce enough HEU for a weapon in just 1 1/2 months. If Glaser's calculations are incorrect, then the most threatening cases are those involving clandestine enrichment plants. If Iran were to produce 19.7 percent enriched uranium at the FEP and simultaneously enrich 19.7 percent enriched uranium to HEU at a clandestine enrichment plant, then it could produce a weapon's worth of HEU in 2 months and enough HEU for three weapons in 6 months. Alternatively, Iran might build a stand-alone clandestine plant to enrich natural uranium to HEU. Such a plant would only produce enough HEU for one weapon a year, but, since the plant could go undetected for many years, Iran could produce a sizable stockpile before detection.

If Glaser's calculations are incorrect, and one does not want to posit the existence of a clandestine enrichment plant, then the fastest way Iran could produce HEU would be to carry out batch recycling at the FEP and the final enrichment step at the PFEP and FFEP. In this fashion, Iran could produce sufficient HEU for a weapon in about 3 2/3 months, which is longer than the 1 1/2 months that would be required if Glaser's calculations are correct. Clearly, it would be helpful to resolve the uncertainties regarding Glaser's calculations. However, even if these uncertainties are not resolved, it is obvious that clandestine Iranian enrichment facilities pose a serious threat.

ENDNOTES - CHAPTER 15 - APPENDIX I

1. "Not all of the centrifuges in the cascades that were being fed with UF6 may have been working." IAEA, *Implementation of the NPT Safeguards Agreement and relevant provisions of Security Council resolutions in the Islamic Republic of Iran,* GOV/2012/9, February 24, 2012, p. 3.

2. To avoid problems with the fact that the length of a month is variable, we have adopted a uniform month length of 30.44 days.

3. Assuming 0.4 percent tails. A separative work unit (SWU) is a measure of the amount of enrichment a facility can perform. The SWU needed to produce a given amount of enriched uranium product can be calculated if the U-235 concentration in the product, feed, and tails are known.

4. Alexander Glaser, "Characteristics of the Gas Centrifuge for Uranium Enrichment and Their Relevance for Nuclear Weapon Proliferation," *Science and Global Security,* Vol. 16, No. 1-2, 2008, pp. 1-25. In particular, see Table 3 on p. 16.

5. Michael A. Levi, "Drawing the Line on Iranian Enrichment," *Survival,* Vol. 53, No. 4, August-September 2011, pp. 180-181.

6. *Iran's Nuclear, Chemical and Biological Capabilities, A Net Assessment, an IISS Strategic Dossier,* London, UK: The International Institute for Strategic Studies, February 2011, p. 73.

7. Gregory S. Jones, "Iran's Efforts to Develop Nuclear Weapons Explicated, Centrifuge Uranium Enrichment Contin-

ues Unimpeded, The IAEA's November 8, 2011 Safeguards Update," December, 6, 2011, p. 5, available from *npolicy.org/article.php?aid=1124&rid=4*.

8. IAEA, GOV/2012/9, pp. 10-11.

While the Agency continues to verify the non-diversion of declared nuclear material at the nuclear facilities and LOFs declared by Iran under its Safeguards Agreement, as Iran is not providing the necessary cooperation, including by not implementing its Additional Protocol, the Agency is unable to provide credible assurance about the absence of undeclared nuclear material and activities in Iran, and therefore to conclude that all nuclear material in Iran is in peaceful activities.

9. IAEA, *Implementation of the NPT Safeguards Agreement and relevant provisions of Security Council resolutions in the Islamic Republic of Iran*, GOV/2011/7, February 25, 2011, p. 9.

CHAPTER 15

APPENDIX II

LIMITING IRAN TO PRODUCING AND STOCKPILING LESS THAN 5 PERCENT ENRICHED URANIUM DOES NOT PREVENT EASY ACCESS TO HIGHLY ENRICHED URANIUM

As was discussed in the text, many who propose a diplomatic solution with Iran have suggested that it should be allowed to continue to enrich uranium as long as this activity is subject to proper controls. In particular, they propose that Iran should not enrich uranium to more than 5 percent and that its current stockpile of nearly 20 percent enriched uranium should be removed from Iran. Further, they propose that the size of Iran's enrichment effort be determined by the needs of its peaceful nuclear program.

But Iran's current enrichment effort is quite small compared to those needed for most peaceful nuclear activities, such as providing fuel for a single nuclear power reactor. A diplomatic solution could provide Iran with the justification for greatly expanding its current enrichment facilities as well as removing sanctions. Under these circumstances, Iran might receive assistance to expand its enrichment facilities (from say China or Pakistan) as part of normal nuclear commerce. These greatly expanded facilities would provide Iran easy access to the highly enriched uranium (HEU) needed for nuclear weapons.

For example, even if Iran produced only 4.1 percent enriched uranium[1] and expanded its enrichment capacity by a factor of 20 (100,000 SWU/year), it would only produce about 15 metric tons of enriched uranium per year. This amount would still be less than that needed to fuel a single large power reactor, yet, using batch recycling, these enrichment facilities could produce enough HEU for a nuclear weapon in just 2 weeks. This process is shown in Table 15-AII1.

Cycle	Product Enrichment and Quantity	Feed Enrichment and Quantity	Time for Cycle (Days)
First	20.2 percent 304 kg	4.1 percent 1,990 kg	7.5
Second	60.2 percent 69.5 kg	20.2 percent 274 kg	1.7
Third	90.0 percent 20 kg	60.2 percent 39.5 kg	0.5
Total			16*
* Includes 6 days to account for equilibrium and cascade fill time.			

(Does Not Rely on Glaser's Analysis).

Table 15-AII1. Time, Product, and Feed Requirements for the Production of 20-Kg of HEU by Batch Recycling at a Centrifuge Enrichment Plant Designed to Produce 4.1 Percent Enriched Uranium (100,000 SWU Per Year Total).

In the first step, 4.1 percent enriched uranium is processed into 20.2 percent enriched uranium. In the second step, this uranium is processed into 60.2 percent enriched uranium, and the third step completes the process by producing the 20-kg of 90 percent enriched uranium needed for a nuclear weapon. Each step produces not only the material needed to be processed in the next step, but also the material needed for the plant inventory, which in this case is 30-kg per step.

Instead of just producing enough HEU for one nuclear weapon, Iran could produce enough HEU for five nuclear weapons (100-kg) in a single batch recycling campaign. This process would take about 5 weeks and is shown in Table 15-AII2. This process would require starting with 6,090-kg of 4.1 percent enriched uranium, but since the plant will be producing about 15,000-kg per year, it would not be hard for Iran to stockpile this quantity of enriched uranium.

Cycle	Product Enrichment and Quantity	Feed Enrichment and Quantity	Time for Cycle (Days)
First	20.2 percent 929-kg	4.1 percent 6,090-kg	23
Second	60.2 percent 228-kg	20.2 percent 899-kg	5.6
Third	90.0 percent 100-kg	60.2 percent 198-kg	2.5
Total			37*
* Includes 6 days to account for equilibrium and cascade fill time.			

(Does Not Rely on Glaser's Analysis).

Table 15-AII2. Time, Product, and Feed Requirements for the Production of 100-Kg of HEU by Batch Recycling at a Centrifuge Enrichment Plant Designed to Produce 4.1 Percent Enriched Uranium (100,000 SWU Per Year Total).

ENDNOTES - CHAPTER 15, APPENDIX II

1. With tails of 0.2 percent.

PART V:

SERIOUS RULES FOR
NUCLEAR NONPROLIFERATION

CHAPTER 16

SERIOUS RULES FOR NUCLEAR POWER
WITHOUT PROLIFERATION

Henry Sokolski
Victor Gilinsky

In this chapter, we try to step back from the day-to-day struggles in Washington, DC, over nuclear non-proliferation policy to ask what measures we would need to have in place to be reasonably confident that expanding nuclear power globally will not increase the number of nuclear weapons-armed states. We recognize that, since the start of the Atoms for Peace Program in the mid-1950s, the United States has supported worldwide use of nuclear power. It also has opposed the spread of nuclear weapons and supported measures to control the nuclear weapons proliferation risks inherent in spreading nuclear technology for civilian purposes. The principal administrative elements of this nonproliferation effort are the Nuclear Nonproliferation Treaty (NPT) and the associated inspection activities of the International Atomic Energy Agency (IAEA), as well as various national and international export controls.

In practice, the success of our policies promoting the global use of nuclear energy have raced ahead of the means available to control the associated nuclear weapons proliferation risks, leaving a broad security gap. What passes for U.S. nuclear nonproliferation policy—the perennial pushing and pulling over the details of nuclear export controls and agreements—does not begin to address that broad gap.

Unless the members of the NPT agree to deal with its fundament deficiencies by interpreting the treaty in a way that sharply limits access to fuels that are also nuclear explosive materials and agree to universal enforcement of that interpretation, increased worldwide nuclear energy use will carry with it the inevitable risk of the further nuclear weapons proliferation.

NUCLEAR WEAPONS PROLIFERATION AND NUCLEAR POWER: WHAT'S THE WORRY?

In any effort to assess our current nonproliferation policies, we must remind ourselves why we still resist the spread of nuclear weapons. In fact, it has become fashionable in some industry and academic circles to discount the dangers on the grounds that, chiefly, proliferation has proceeded more slowly than once feared.

The usual reference is to U.S. President John F. Kennedy's 1962 statement that 15 to 25 countries could obtain nuclear weapons. But this was a warning, not a prediction, and a useful one that led to non-proliferation efforts that slowed the process. In view of our experiences with countries falsely claiming to be conducting "peaceful" nuclear programs and later using their facilities for illicit activities or conducting clandestine bomb activities — in India, Iran, Iraq, Israel, North Korea, Pakistan, South Africa, and Syria — it is time to heed these warnings again.

There is also a school of thought that, even if some more countries obtained nuclear weapons, it would not make much difference because they would just serve as deterrents.[1] There is a troubling disconnect between this cheerful theorizing — which is not without an element of self-interest — and any awareness of the devastating possibility of nuclear war. Just because the weapons are supposed to be for deterrence does not mean they will not be used. Such use is, after all, implied in the threat that underlies deterrence. And if they are used, they are likely to change profoundly the way the world is organized, with unpredictable but likely unhappy consequences.[2] A few years ago, former U.S. Secretary of State Henry Kissinger wrote:

> If one imagines a world of tens of nations with nuclear weapons and major powers trying to balance their own deterrent equations, plus the deterrent equations of the subsystems, deterrence calculation would become impossibly complicated. To assume that, in such a world, nuclear catastrophe could be avoided would be unrealistic.[3]

Happily, we have not reached this state. No such weapons have exploded in anger since World War II, and it has been a long time since people have seen the results of atmospheric tests. But this has also meant there is not the gut-level consciousness about proliferation dangers that there is about the dangers of nuclear accidents. Whereas everyone agrees that expanded use of nuclear power has to be predicated on tough safety rules, there is no corresponding agreement when it comes to rules to protect against nuclear weapons spread, especially when it comes to restrictions on nuclear power programs.

One often hears from nuclear industry sources that "civilian" nuclear programs are not a proliferation worry because they are an unlikely source of nuclear explosive materials for would-be bomb makers. They argue that, just as current nuclear weapons states relied on dedicated military programs, so would any future would-be weapons country.[4]

Our view is different. Leaving aside the correctness of the assumptions about past weapons programs, the past is not a good guide to the future because conditions have changed fundamentally. Today all non-weapon states are members of the NPT. If one of these countries should decide to obtain weapons, it would have to withdraw or cheat, both courses risking a military response until the would-be bomb maker had weapons comfortably in hand. This would put a very high premium on traversing the period of vulnerability as quickly as possible. Kissinger made this point in the previously cited 2006 Trilateral Commission report: "A policy of using preventive force against aspiring nuclear powers, however, creates incentives for them to acquire nuclear weapons as rapidly as possible. . . ."[5]

That means drawing on bomb material and know-how where it is most quickly obtainable, which would mean tapping a nuclear power program if there is one, unless, of course, there are strict measures in place to prevent that happening. If there is any doubt about this conclusion, consider the following counterhistorical: Suppose each of the major World War II belligerents already had civilian nuclear power programs before the war started. Would they not have tapped them rather than start anew to develop independent nuclear weapons programs? The answer suggests why strict nonproliferation measures are important.

THE NUCLEAR NONPROLIFERATION TREATY'S DEFICIENCIES

In this regard, no one believes that we have adequate preventive antiproliferation measures in place today. Otherwise, we would not be discussing endlessly various international fuel supply schemes to mitigate the risks that national uranium enrichment and fuel reprocessing might be used to produce nuclear explosives. Everyone understands that the NPT as it has been interpreted up to now has basic deficiencies:

- The treaty allows withdrawal on 3 months' notice.
- It does not delineate the limits on permissible "peaceful" technology, with respect to fuels that are immediately usable to make nuclear explosives.
- It sharply restricts IAEA inspections.
- The treaty lacks an established enforcement system, so that each violation requires an improvised response.
- The treaty's universality is undermined by India, Israel, North Korea, and Pakistan, which remain as examples of what a country can get away with doing.

The advance of technology since the treaty went into force has exacerbated these problems by lowering the technological barriers between civilian nuclear activities and nuclear weapons. The prime example is the spread of centrifuge enrichment technology, which can be used to produce low enriched uranium to fuel power reactors but also can bring states within weeks of acquiring weapons-grade uranium to make a bomb. More generally, worldwide advances in materials, manufacturing, and computing skills put weapons design and manufacture within reach of a larger group of countries.

NUCLEAR POWER EXPANSION REMAINS GOAL OF MAJOR NUCLEAR SUPPLIERS AND OF THE IAEA

Despite these acknowledged basic inadequacies of current anti-proliferation protections, the U.S. Government has supported worldwide use of nuclear power since U.S. President Dwight Eisenhower's Atoms for Peace Program and continues to do so today. The rationale, however, has evolved.

In proposing the program, Eisenhower said that starting with small projects had:

> the great virtue that it can be undertaken without irritations and mutual suspicions incident to any attempt to set up a completely acceptable system of world-wide inspection and control.[6]

In time, however, the projects got bigger and much more significant from the point of view of international security. Meanwhile, the effectiveness of the IAEA inspection system did not keep up. This arguably mattered less when the two Cold War camps expected to keep their client states in line mainly through their own intelligence and intervention. But now we really do need the "completely acceptable system of world-wide inspection and control" that President Eisenhower spoke about, especially if there is to be a major expansion in use of nuclear power plants.

At the moment, a major, global nuclear expansion is not in play, mainly because of unfavorable economics and, since the March 11, 2011, Fukushima, Japan, accident, also because of increased safety concerns.[7] Nevertheless, such expansion remains the goal, or at least the expectation, of key nuclear exporters—the United States included—and of the IAEA in Vienna, Austria.

U.S. President Barack Obama has consistently supported an expanded role for nuclear power both abroad and at home. In a March 2012 speech at Hankuk University, South Korea, almost exactly a year after Fukushima, the President said the world needed nuclear power.[8] He predicted that "nuclear energy will only become more important," and that remains the operational assumption in the U.S. Government.[9]

The IAEA also announced optimism about nuclear power expansion post-Fukushima. At the 2012 IAEA General Conference, Director General Yukiya Amano said he expected "a steady rise in the number of nuclear power plants in the world in the next 20 years." His low case for 2030 projected a nuclear power capacity increase of about a quarter, and his high case projected a doubling of current capacity.[10] The projections are significant as expressions of the Agency's sentiments and those of the national nuclear bureaucracies it represents. However unrealistic, these projections find their way into official and semi-official nuclear establishment reports and bolster support for nuclear power.

SECURITY IMPLICATIONS OF THE CLIMATE ARGUMENT FOR NUCLEAR POWER

In his January 28, 2013, inaugural address, Obama reiterated his support for building "a new generation of safe, clean nuclear power plants in this country." In this he is following in his predecessors' footsteps.[11] But his rationale—that nuclear power is necessary to deal with climate change—is significantly different, and it has far-reaching security implications.

Since we are talking about a global rather than a local effect, the climate benefit of nuclear power installations only accrues if there are very many of them. In resting the case for nuclear power on the need for them to deal with global warming, the proponents are therefore saying that we must build a very large number of nuclear plants. The experts say it would take well over 1,000 plants just to make a **dent** on the climate problem.[12] But an increase of that size would likely involve nuclear power programs in dozens more countries, including many in the rougher parts of the world — most of the Middle Eastern countries have already expressed interest in building nuclear plants — a worrisome prospect from a security point of view.

The putative climate imperative for nuclear power has made it easy for U.S. nuclear officials to argue that, yes, they would like to see effective antiproliferation protection, but at the end of the day, we have to settle for what we can get because we must have lots of nuclear power to deal with climate change, no matter what. However, that is exactly the case where that antiproliferation protection is needed most.

More important, there are environmentally acceptable energy alternatives to nuclear power, including ones superior for coping with climate change. An obvious example is natural gas, which allows faster and cheaper reductions in carbon.[13] We certainly do not accept the notion that the world is locked into eventually relying on large numbers of nuclear power plants to cope with global warming.

PUTTING SECURITY FIRST

For this reason, we believe it would make more sense to reverse the current policy priorities under which "nuclear" trumps protection, and instead insist on adequate protection against proliferation as a condition for nuclear trade. We would try to persuade others to accept that standard. If prospective customers are unwilling to agree to such protection, our answer is to stop encouraging nuclear expansion until such protection is available.

We certainly do not buy the rationalization that the United States has to sell into the international nuclear market, even if the agreements covering the trade are not as tight as we would like, because if we do not, others will, and "you have to play the game if you want to participate in setting the rules." We think the United States would be more convincing by setting a principled example. Nor do we go along with that ultimate cop-out—that technology controls are not important because proliferation is really a political, and not a technical problem. It is obviously both, and an essential aspect of nonproliferation is to keep it difficult, both technically and politically, for countries to join the weapons game, so difficult that we can reasonably exclude the possibility that civilian nuclear programs will contribute to weapons development.

We are very much aware of the positions of the majority of countries on NPT issues and their negative reactions to further restrictions and even existing restrictions on access to nuclear technology.[14] It is clear they are in no mood to accept a major tightening of the rules, and we are under no illusion that the United States can by itself impose such a major tightening. But we can start to talk about the issues in a more straightforward manner than we have been doing, and we can seek to persuade the international community of the value to all countries of effective antiproliferation protection.

In fact, we have been moving in the opposite direction by promoting the "three-pillars" interpretation of the NPT, which hobbles our nonproliferation efforts.[15] This reinterpretation of the treaty puts sharing of nuclear technology on a par with nonproliferation. It is easy to see why U.S. diplomats find this path of least resistance appealing — technology sharing is easy and pleasant for all parties, while imposing nonproliferation restrictions is just the opposite. The trouble is, taking this approach leads to a markedly weakened NPT.

The argument is made that acceptance of the three-pillars formulation is necessary to give the United States the bona fides to conduct nonproliferation policy, and, anyhow, it is so firmly entrenched that there is no point in talking about it. The result is a kind of zero-sum game in which the nonproliferation obligations of the majority of members are held hostage to technology sharing by the main nuclear states. The loose interpretation of the latter by the hopeful recipients has been especially problematic and is what creates the proliferation problem in the first place.

That is not all. By putting the technology sharing obligation on a par with nonproliferation, the three-pillars formulation singles out nuclear energy as the internationally politically anointed energy source, irrespective of its real economic value. This reflects energy-of-the-future thinking of the 1960s, but makes no sense today.

Unfortunately, Obama's comments on the foundations of the NPT have not been helpful. On the treaty's 40th anniversary in 2010, the President described it as standing on "three pillars — disarmament, nonproliferation, and peaceful uses."[16] We need to get back to viewing the NPT, as primarily about the nonproliferation of nuclear weapons.

FIVE PRINCIPLES FOR STRENGTHENING
THE NPT REGIME

We propose here to examine what effective antiproliferation protection would entail. We express this
in terms of five principles addressed to the main deficiencies of the NPT. So far as we can tell, nowhere
in the nonproliferation literature is there a clear statement of a policy goal. There is an abundance, in fact a
superabundance, of discussions of the value or attain-
ability of this or that agreement provision or requirement, but nowhere do we find a statement of what it is
we want in the way of protection.[17] In effect, the principles we detail here are an outline of what it takes to
have an international environment in which nuclear
power can thrive without providing an easy target for
would-be bomb makers.

1. Make Withdrawals from the NPT Effectively Impossible.

We need to make it much more difficult—in fact,
essentially impossible—to exercise the NPT's withdrawal provision. This is vital, not least because the
member states' safeguards agreements with the IAEA
remain in force so long as the states remain parties to
the treaty.[18] The U.S. position is that safeguards continue in perpetuity, but it is unclear what would actually happen if another state announced its withdrawal.
The international responses to North Korea's 1993
withdrawal threats and ultimate 2003 withdrawal announcement were deficient in that, however much everyone deplored North Korea's action, no one made
the case at the time that a country cannot, while in

467

a state of violation, legally leave the NPT. It was a question of legitimacy, which, while intangible, remains important in international affairs, even to the North Koreans.

There seems finally to be general agreement on this point — that a country in violation of the NPT cannot relieve itself of its responsibilities by announcing withdrawal, and that the international reaction to a similar case would be more forceful.[19] Nevertheless, ambiguity still remains over whether North Korea is still obligated by its NPT membership. Obama's statement deploring North Korea's February 12, 2013, nuclear explosion did not so much as mention the NPT. It would have been helpful if the President had said that North Korea's NPT withdrawal was not valid and that it stands in continued violation of the treaty for current actions as well as past ones.[20]

But even this, in our view, would not go far enough. A country should not be allowed to gather the production tools for making a bomb while a member and then free itself of its treaty responsibilities by withdrawing, even if it is in good standing at the time. What this means is that a country should not be allowed to leave the treaty legally with technology it obtained as a member, because it did so with the forbearance of other members on the assumption that it was doing so for peaceful uses. This should apply whether the technology was imported or developed indigenously.

The position that NPT safeguards apply in perpetuity, even if it gained universal acceptance, would not entirely deal with this issue. In practice, there is no way to erase the advantages that a country bent on nuclear weapons gains from its nuclear power program, an advantage that lies in part in equipment and

468

materials, but perhaps most importantly in the training of scientific and technical personnel.

It would be useful to introduce these NPT arguments into the current discussions that have arisen in Japan and South Korea over the advisability of obtaining nuclear weapons.[21] The voices for weapons certainly have become much louder after the February 12, 2013, North Korean nuclear test.

2. Limit NPT Members' Access to and Production of Nuclear Weapons-Usable Materials.

The NPT cannot be a vehicle for legally coming overly close to a weapons capability. There has to be a technological safety margin between genuinely peaceful and potentially military applications to make it impossible to surprise the world with a bomb. As a consequence, the "inalienable right" language in the treaty has to be interpreted in terms of the treaty's overriding objective, and thus there have to be restrictions on the kinds of technology that are acceptable for nonmilitary use. Nuclear power needs to develop in a way that does not provide easy access to nuclear explosive materials. Where to draw the line is now coming to a head in the context of Iran's nuclear program.

In the early days of nuclear power — the 1950s and 1960s — technology control was not on the international agenda. Enthusiasm for plutonium as the fuel of the future in fast breeder reactors overrode any official concern about its weapons potential. Under Atoms for Peace, the United States shared its reprocessing technology with all countries, as it was seen as an essential part of nuclear power programs.[22] This complacency about easy access to plutonium was jolted by India's 1974 nuclear explosion, which led the principal ex-

469

porters to organize the Nuclear Suppliers Group to exercise some control over the spread of what were euphemistically called "sensitive" nuclear technologies. The resort to euphemisms is itself telling. Neither the United States nor the other exporters ever publicly addressed the tensions in the NPT between prohibitions on bombs and liberal promises of technology — between the NPT Article IV's "inalienable right of all the Parties" to "nuclear energy for peaceful purposes" and the qualification that this activity must be "in conformity with Articles I and II," which prohibit acquisition of nuclear weapons. The conflict remains to be resolved. That, after all, is what the struggle over Iran's enrichment program is about.

The well-known root problem is that separated plutonium and highly enriched uranium (HEU) can be converted to weapons use too quickly for international inspection to provide protection against that possibility.[23] The IAEA refers to its inspections and related protection systems as "safeguards." This introduced some confusion because the IAEA uses "safeguards" as a term of art for all the inspections it conducts, whether or not the inspections in question actually achieve their purpose of providing the timely warning needed to safeguard against military diversions.

It is therefore important that the IAEA's basic NPT safeguards document states the purpose of IAEA safeguards is to deter diversion by the threat of early detection.[24] If sufficient early detection cannot be counted on — as it realistically cannot for the nuclear explosive materials plutonium and HEU — there is no deterrence related to inspection, and therefore no effective safeguarding in terms of the INFCIRC/153 standard. In our view, the commercial use of materials that cannot be safeguarded in this sense should not be permitted.

The same concerns, once removed, apply to the enrichment and reprocessing facilities capable of producing nuclear explosives. As the *Strategic Plan* for the Bush administration's Global Nuclear Energy Partnership (GNEP) program put it, "*. . . there is no technology 'silver bullet' that can be built into an enrichment plant or reprocessing plant that can prevent a country from diverting these commercial fuel cycle facilities to non-peaceful use.*"[25] This explains why we do not want to have them there in the first place.

Reprocessing Plutonium and Recycling Spent Fuel.

In principle, plutonium recycling should be easier to deal with because it is widely recognized that there is no economic case for plutonium reprocessing or recycling spent fuel.[26] The fast breeder programs that were the original incentives for separating plutonium have almost all receded into the indefinite future.[27] Nor are recent claims valid that reprocessing and recycling facilitate waste disposal. That has not, however, prevented national laboratories and nuclear fuel firms from clinging to plutonium technology as a link to the original dream of an all-nuclear future, regardless of its current impracticality. To make it pay, the industrial supporters are counting on heavy subsidies — following the principle that, to make money, you do not need an economic product, you just need someone to pay for it.

The last refuge for plutonium recycling supporters is the claim that it serves arms control purposes by consuming plutonium, thus reducing long-term risks.[28] It does so, however, by separating plutonium and exposing it, in a number of fuel cycle stages, to considerably increased near-term risks. In view of the dangers and lack of economic benefits of reprocessing plutonium and recycling spent fuel, it makes sense to ban them altogether.

Such a ban would be no more than what U.S. President Gerald Ford proposed in 1976.[29] While the nuclear community never wholly accepted the view, for many years, these fuel activities were restrained by the unfavorable economics. But in 2007, the recycling adherents convinced the George Bush administration to launch the GNEP, a futuristic reprocessing and re-cycle crash program. The advertised purpose was to "solve" the nuclear waste and proliferation problems simultaneously by having the United States and other major nuclear supplier countries provide a full range of fuel services. In reality, it was a poorly thought out scheme, based on technology that did not exist, to re-kindle the nuclear dream of a fast reactor future.[30] The Obama administration cut back the Bush program but kept much of it going, and when the President spoke in South Korea in 2012, he called for "an interna-tional commitment to unlocking the fuel cycle of the future."[31] It is not surprising that nuclear bureaucra-cies in other countries, and especially South Korea, have been emboldened in expressing their interest in plutonium technology.

A key pending policy issue concerns the so-called Gold Standard for civilian nuclear cooperation be-tween the United States and other countries. This is the standard established in the 2009 agreement with the United Arab Emirates (UAE) that permits reactor sales but rules out nuclear fuel activities. Unfortu-nately, the Obama administration has been ambigu-ous about whether it will apply this standard to all such agreements.[32] Without an American commit-ment to it, there is no chance for the standard to gain international acceptance.

We should be clear that to restrict fuel cycle activities to a small number of countries would mean not only that states that have not yet gotten into these activities would forego doing so, but also that some states already involved in such activities would have to give them up. While giving up reprocessing facilities would not involve economic penalties, as reprocessing is uneconomic and broadly understood to be so, that does not mean that it would be an easy thing.

Uranium Enrichment.

Gaining agreement for restricting enrichment is an even more difficult proposition, as enrichment is a necessary part of the nuclear fuel cycle. It is really centrifuge enrichment that is the main concern, rather than enrichment per se, because it lends itself to small-scale operation, and so is relatively accessible for many countries.

A number of countries already have enrichment facilities or development programs.[33] An aura has developed around enrichment that goes beyond any economic rationale, one that ultimately relates to the connection with nuclear weapons. That is not necessarily what the operators have in mind, but we can be sure they are aware of the weapons potential, and, at minimum, the political leverage this provides.

One way or another, to gain broad agreement on limiting access to this technology, there would have to be a reasonably common rule for all (the "inalienable right" phrase in Article IV of the NPT is immediately followed by the words "without discrimination"). To grasp the magnitude of the change that would be required from the current ad hoc approach, consider that to be consistent, the effort to restrict Iran's enrichment program would have to be matched by an effort to restrict Brazil's roughly comparable program. But that is part of what it would take to restrict enrichment activities to a small number of countries in order for nuclear power to operate with a reasonable safety margin from the point of view of proliferation. The technology has spread beyond the point where grandfathering the existing enrichment programs is a workable solution. That would be a recipe for a creeping retreat to an increasingly unstable state of affairs—a "Perils of Pauline" world in which many countries are a short step from nuclear weapons, one they could likely take before their neighbors or the international community could react.

Coming up with a satisfactory answer on how to limit access to this technology is a vexing problem. One possible approach would be to assess a safeguards fee to compensate for the very considerable IAEA effort involved in monitoring centrifuge facilities. If, as seems likely, the monitoring effort involved would not be much less for smaller plants, a large fee would discourage small operations and thus restrict the number.

3. Expand Inspections.

The IAEA's inspections, when first instituted, were infrequent and specifically limited to listed facilities, and the inspectors approached their tasks wearing blinders. The IAEA inspection system is a very different one today, especially after the upgrades that followed the first Gulf War, which revealed the ineffectiveness of the original system. Nevertheless, there are intrinsic limitations on what can be achieved through inspections alone.[34] The most recent addition, the so-called Additional Protocol (AP), further expands the agency's inspection rights.[35] It marks an important advance. But the AP remains voluntary, and a number of countries have yet to accept it. Also, to encourage acceptance of this protocol, the IAEA agreed to reduce the frequency of routine inspections for countries that accepted the AP, so that, for example, the inspection goal for reactor spent fuel inspections would be once a year instead of once a quarter.

To be effective, the IAEA's inspection system has not only to provide timely assessments at known facilities within a country, but also to be able to rule out the existence of clandestine facilities.[36] The reason for concern about clandestine facilities is that reprocessing and centrifuge enrichment lend themselves to small-scale operation. Such facilities may be insignificant in commercial terms, but they can be very significant in military terms. This means, for example, that the commonly-held view that light water reactors (LWRs) by themselves are a safe proposition (in terms of proliferation) is correct only if we can rule out clandestine reprocessing, not only contemporaneously, but also in the future.[37]

Ironically, this was pointed out in 1977 by an Oak Ridge National Laboratory group that opposed anti-proliferation restrictions. Specifically, they wanted to show that a country that wished to produce plutonium for bombs could easily get around a ban on commercial reprocessing. To drive the point home, the Oak Ridge team made public a design for a small reprocessing plant that a country with a minimal industrial base could build quickly and secretly to obtain enough plutonium for dozens of bombs.[38] The memorandum describing the design included equipment lists, process sheets, and drawings.

The Oak Ridge experts said that most equipment would be available from local industries such as wineries or dairies or could be fabricated in a small shop. The described "quick and dirty" plant is not something anyone can put together in a garage. But it is entirely credible that experienced reprocessing experts could do it.

Although this was not the objective of the Oak Ridge exercise, it also put in doubt the assumed benign character of nuclear power plants in the absence of commercial reprocessing. If a country with LWRs but no commercial reprocessing could quickly build a small "quick and dirty" plant to obtain enough plutonium for bombs, then power reactors in the wrong places could be much more dangerous propositions than previously thought.

Especially to deal with the problem of clandestine plants, two former IAEA Deputy Directors General have urged that the IAEA make greater use of "special inspection" rights it already has under the comprehensive NPT safeguards agreements. For historical and bureaucratic reasons, the threshold within the Agency for invoking such inspections rights is exceedingly high. The Agency's right to conduct a special inspection has only been invoked once in adversarial circumstances (North Korea in 1993).[39]

Finally, a problem that is not immediate but is bound to arise in the future concerns the NPT safeguards exclusion for materials used in "nonprohibited military applications," which principally means naval reactor fuel. This is a potential gaping loophole that, at a minimum, needs to be narrowed, and perhaps eliminated altogether—just as "peaceful nuclear explosives" were, in effect, read out of the NPT.

4. Ensure Enforcement.

In the initial years of the NPT, there was an implicit assumption that each of the two Cold War blocs would police its sphere. The United States did that, for example, in the mid-1970s, when it forced Taiwan and South Korea to dismantle their clandestine nuclear weapons efforts. But now the NPT needs an established enforcement mechanism to deal with treaty violation in a predictable way. At present, each violation calls for improvization by the powerful states. While the logic of "safeguards" over plutonium and HEU is that the international system will react to evidence of violations rapidly, the natural response time of the international system is more often measured in years.

It is vital that future would-be bomb makers be disabused of any notion that they could evade tough international sanctions. What is needed—as Pierre Goldschmidt, former IAEA Deputy Director General for Safeguards, recommended—is a country-neutral, reasonably predictable, more-or-less automatic sanction regime that puts all countries on notice in advance of NPT violation, including violations of any IAEA safeguards agreements.[40] Improving the ability to detect possible violations will not deter violators if they know that little will be done quickly enough to stop their bomb making. A permanent secretariat attached to the treaty would help make such a process work.

The record of U.S.-led ad hoc enforcement is decidedly mixed. The trouble is that political considerations inevitably intrude: sometimes to deflect U.S. interest from pursuing enforcement of nuclear agreements, sometimes even to look the other way.

President Obama had this to say about enforcement in his 2009 Prague, Czech Republic, speech:

> We need real and immediate consequences for countries caught breaking the rules or trying to leave the treaty without cause. . . . We must go forward with no illusions. Some countries will break the rules. That's why we need a structure in place that ensures when any nation does, they will face consequences.

More recently, the President's NPT ambassador stated:

> There is no greater threat to the integrity and vitality of the treaty than the unresolved cases of noncompliance. Because of the corrosive effect of noncompliance on international confidence in the NPT, we must redouble our efforts to encourage full compliance with treaty obligations.[41]

The present situation in which a violator does not, by any means, face immediate sanction significantly undercuts the disincentives to violations. What would actually happen in any particular instance would depend on who were the violator's friends and enemies, and what else is going on to distract the world from the violation.

Iran has been sanctioned (in a long, drawn out process) not because it has been charged with specific NPT treaty violations, but rather because it is seen as a potential violator of IAEA safeguards requirements that threatens the interests of other powerful states, mainly Israel and the United States. By contrast, North Korea's flagrant 1992 violation in refusing required IAEA inspections evoked an entirely different response, mainly because the United States feared a

North Korean withdrawal from the NPT would undermine the then-upcoming 1995 NPT review conference. Investigation of its violation was postponed, and, in return for not operating its small indigenous plutonium production reactors, North Korea was offered two large LWRs worth about $5 billion. The deal eventually broke down, but the precedent remains that, in the right circumstances, blackmail can work.

It also undermines worldwide respect for the NPT when enforcement is farmed out to a nonmember, as was the case when the United States acquiesced in Israel's 2007 bombing of Syria's clandestine reactor, instead of bringing Syria's violation before the IAEA.[42] The point applies as well to cooperation with Israel in sabotaging Iran's nuclear program.

One cannot talk about NPT enforcement today without addressing the U.S. reaction to the February 12, 2013, North Korean nuclear test. As observed earlier, Obama's statement, calling for "swift and credible" action in response to violation of United Nations (UN) Security Council resolutions, did not mention the NPT.[43] It leaves the impression the government regards the NPT as of marginal significance.

5. Bring Nonmembers into the NPT Process.

The most difficult NPT-related issue concerns what to do about the three NPT holdouts — India, Israel, and Pakistan — and the member-in-violation, North Korea. Under the NPT, there are only two classes of countries: the five nuclear weapon states, and nonweapons states, which includes the three holdouts.[44] Although discriminatory, the original intent of only recognizing five nuclear states was to make sure that number did not grow larger. To now create a new class of members, in addition to these five, that would have nuclear weapons would therefore undermine the treaty's original intent.

However implausible or even impossible it may now seem, it is probable that the only way that all states can be brought under the NPT system is if all reduce their nuclear weapons to zero. All nuclear weapons states have to participate in weapons reductions. The United States and Russia have made substantial reductions, but the continuation of that process is predicated on all nuclear states participating in further weapons reductions. Without such reductions, it does not seem likely that the non-nuclear NPT members would agree to necessary restrictive measures on the use of nuclear energy. Ultimately, the weapons reduction will have to include India, Israel, and Pakistan, and, of course, North Korea.

We would universalize the treaty—that is, regard it as applicable to all states, including the three holdouts, which would then be in noncompliance.[45] Of course, as a legal matter, you cannot force a country to join a treaty. But as a practical matter, if the 190 NPT members so decided, they could treat the three holdouts, and, of course, North Korea, as countries in noncompliance, with the appropriate disadvantages that would follow from that decision.

To take a positive view, if these countries agreed to join in the weapons reductions process under adequate monitoring, and so could be considered on their way toward compliance, any adverse treatment as a consequence of their noncompliance could be moderated. We are under no illusions about the current practicability of this proposal. But we are also convinced that it is an essential element of an effective system to bar proliferation. It simply makes no sense to accept President Obama's goal of zero nuclear weapons but, at the same time, dismiss any notion of applying that goal to India, Israel, and Pakistan. As Obama said in Prague: "This goal will not be reached quickly—perhaps not in my lifetime. It will take patience and persistence." Yet he set a goal. In the same way, we are trying to set a goal for nonproliferation.

WHAT NOW?

To sum up, it is our view that for a nonproliferation system that is reasonably able to cope with the kind of nuclear expansion implicit in U.S. nuclear energy policy, there need to be firm measures: (1) to prevent NPT withdrawals, (2) to restrict access to nuclear explosive materials, (3) to ensure adequate IAEA inspections, (4) to guarantee enforcement, and (5) to deal with NPT holdouts. We obviously are nowhere close to this standard, and, in fact, NPT members presently resist change in the direction of meeting it.

The official U.S. response to this unsatisfactory state of affairs has been to focus on incremental measures, ones on which some progress could be made through relatively low-level negotiations.[46] Even the nonproliferation academics and nongovernmental organization analysts reach no further, and in fact dismiss anything beyond this incremental approach as unrealistic or positively harmful. Not surprisingly, the nuclear power community regards any suggestion to go beyond the working-at-the-margins approach as dangerous, if not tantamount to "anti-nuclear," even when it comes to reining in enrichment and reprocessing.[47]

Along this line of thinking, to gain the necessary bona fides for tightening antiproliferation protection, the United States needs to encourage worldwide nuclear expansion and to involve itself heavily in selling nuclear technology. There is in this, of course, more than a touch of self-interest, as there has always been in such advice going back to Atoms for Peace. Historically, nuclear sales have invariably come before protection against their misuse.

To keep active on nonproliferation and yet avoid roiling the diplomatic waters, diplomats and academics spend a great deal of time discussing inoffensive — and ineffectual — schemes, such as fuel banks and multinational fuel centers.[48] Fuel banks, seemingly simple and catchy, so beloved by diplomats, are a nonsolution to a nonproblem.

There is a competitive international fuel market, and no country, other than one obviously bent on weapons, has to worry about getting fuel. In fact, no country, including one obviously seeking weapons, has had to shut reactors because it was denied nuclear fuel.

As to banking fuel, it is utterly impractical to store fuel assemblies for individual reactors as these vary significantly in their technical specifications. One could bank low enriched uranium (LEU), but even that poses problems, as the reactors' fuel enrichment levels vary. In any case, a country could much more easily bank its own LEU.

Multinational facilities would make some sense if they were coupled with a requirement for participants to forego indigenous fuel cycle facilities. But the usual arguments for the multinational projects have been that they would reduce the participants' incentives to pursue their own, which is not the same thing.[49]

Another escape from dealing with proliferation is to embrace nuclear terrorism (as opposed to weapons acquisition by states) as the immediate primary problem, which is convenient because there is not a lot of push back when it comes to opposing nuclear terrorism.[50]

The truth is that the incremental, least-common-denominator approach is never going to get us to where we need to be, and serious people responsible for security know it. To cope with proliferation hazards in the face of weak international controls over nuclear programs, the world seems to be slipping — witness the case of Iran — into relying on greatly increased national intelligence operations backed up in the last instance by bombing and even assassinations. It is difficult to imagine that this is a workable solution for the long term.

Nor is it workable to continue in the mode where nonproliferation is seen as an American obsession, and the role of the diplomats in the great majority of countries is to extract what they can from the United States for each incremental concession. They have to see it as their problem, too. To begin the process of change to a sounder long-term approach, the United States, the major nuclear countries, and the IAEA need to begin by speaking clearly about the risks of proliferation that attach to a major expansion of nuclear power use and what it takes to check them. George Orwell taught us that euphemisms and bureaucratic boilerplate can corrupt, and that they are used to make it easier to defend what otherwise cannot be defended in public.

We have to stop pretending that reprocessing and enrichment facilities can be "safeguarded" by international inspections and that they are only a problem if they are located in countries currently on the "rogue" list. We also have to end the charade that protection will come from multinational facilities, or fuel banks, or futuristic reprocessing schemes. We have to agree to abide by common standards. Finally, we have to stop downplaying the NPT by describing it as standing on "three pillars," of which nonproliferation is only one pillar. The results might surprise us. There is a lot of persuasive power in sensible thought expressed clearly.

We conclude on this optimistic note even while recognizing that countries are increasingly finding pretexts for narrowing the anti-proliferation safety margins, and major states, like the United States, are finding rationalizations for yielding to friendly states. For this chapter, our working assumption is that this process has not gone so far that it cannot be reversed. To be sure, it would require the President to make doing so and engaging other leaders a top priority, as then-President Eisenhower did in launching Atoms for Peace.

We do not, however, discount the possibility that our critics are right in saying that there is no longer any prospect for gaining the level of proliferation protection we outlined here, so there is no realistic prospect of closing the policy gap between nuclear advocacy and anti-proliferation protection. In that case, the answer is not, as they say, to fall in with the nuclear crowd in the hope of merely making minor adjustments in its rules—but to think about closing the policy gap from the other end.

ENDNOTES - CHAPTER 16

1. For an example from the nuclear industry, see Steven Kidd, "Nuclear Proliferation Risk—Is It Vastly Overrated?" *Nuclear Engineering International*, July 23, 2010, available from *www.neimagazine.com/story.asp?storyCode=2056931*. Mr. Kidd argues that:

It is likely that more countries will foolishly choose to acquire nuclear weapons. If they are really determined to do so, there is little really that the world can do to prevent them—the main effort has to be in dissuading them from this course of action. How many countries will have nuclear weapons by 2030 is hard to say, but there could well be a total of 15 by then.

Mr. Kidd then cites the work of Professor John Mueller who argues that this increase, in itself, will neither prevent nor cause wars, but will impose "substantial costs on the countries concerned." Kidd works for the World Nuclear Association, and Mueller is a professor of international relations at Ohio State University.

2. For more on these points, see the thinking of the Chairman of the International Institute for Strategic Studies, François Heisbourg, "How Bad Would the Further Spread of Nuclear Weapons Be?" Chap. 2 in this volume.

3. See Henry Kissinger, "Foreword," *Nuclear Proliferation: Risk and Responsibility*, Trilateral Commission, 2006, available from *www.scribd.com/doc/30082942/60-Nuclear-Proliferation-Risk-and-Responsibility-2006*.

4. See, e.g., Kidd; and "Assessing the Proliferation Risks of Civilian Nuclear Programs," *Nuclear Programmes in the Middle East: In the Shadow of Iran*, London, UK: International Institute for International Studies (IISS), 2008, pp. 141-50, available from *www.iiss.org/en/publications/strategic%20dossiers/issues/nuclear-programmes-in-the-middle-east--in-the-shadow-of-iran-5993*.

5. See *Nuclear Proliferation: Risk and Responsibility*.

6. See Dwight Eisenhower, "Atoms for Peace Speech before UN General Assembly," New York, December 8, 1953, available from *www.eisenhower.archives.gov/research/online_documents/atoms_for_peace/Binder13.pdf*.

7. Economics and safety are tied together. The "regulatory costs" that critics of the regulatory systems complain about are mainly the costs of providing adequate safety. There is no getting around the fact that nuclear power is intrinsically dangerous and that it takes great care in design, construction, and operation to

keep it within safe bounds, all of which are expensive. The safety of light water reactors (LWRs) is difficult to assess, as the technology is especially complex. The oxide fuel melts in hours if cooling stops. In the end, it may not prove to be the best power reactor choice. A struggle is now underway in Japan, with the newly empowered safety regulators imposing conditions that the industry claims will make it too expensive to restart Japan's reactors. See e.g., *www.japantimes.co.jp/news/2013/02/02/national/nra-safety-rules-may-keep-reactors-offline-for-years/#.UQ7OA0qjdY4*. As to alternatives to current designs, the U.S. Department of Energy (DoE) and industry are touting small modular reactors, basically ordinary reactors in smaller sizes. If they are small enough, their nuclear systems could be assembled in a factory—hence modular. Being smaller, SMRs would cost less to build than larger cousins but are unlikely to produce electricity that would cost less per kilowatt. You also have to wonder if smaller reactors are such a good idea, why they were not proposed earlier. Gaining economies of scale through factory construction would only work if very large numbers were sold, for which there is little prospect. There are efforts to breathe life into other formerly rejected designs such as the Integrated Fast Reactor, but they are unlikely to get off the ground. What is needed for safety is a reactor whose fuel will not melt in an accident. To assure significant proliferation resistance and economic competitiveness, meeting other additional, yet-to-be-achieved cost and technical attributes would also be desirable.

8. See The White House, Office of the Press Secretary, "Remarks by President Obama at Hankuk University," Seoul, Republic of Korea, March 12, 2012," available from *www.whitehouse.gov/the-press-office/2012/03/26/remarks-president-obama-hankuk-university*. Meanwhile, his former energy secretary, Steven Chu, consistently championed this argument in support of loan guarantees for nuclear power plant construction. Chu wrote on his personal Facebook page: "If we want to make a serious dent in carbon dioxide emissions -- not to mention having cleaner air and cleaner water -- then nuclear power has to be on the table." See Dr. Chu's Facebook entry, available from *www.facebook.com/notes/steven-chu/why-we-need-more-nuclear-power/336162546856*.

9. Earlier in President Obama's 2009 Prague speech, he said: "We must harness the power of nuclear energy on behalf of our efforts to combat climate change, and to advance peace oppor-

tunity for all people." Meanwhile, his remarks at Hankuk University emphasized "a renewed commitment to harnessing the power of the atom not for war, but for peaceful purposes." President Obama also praised the Koreans for showing "the progress and prosperity that can be achieved when nations embrace peaceful nuclear energy," thus glorifying the role of nuclear power in promoting economic development. See The White House, Office of the Press Secretary, "Remarks by President Obama," Hradcany Square, Prague, Czech Republic, April 5, 2009, available from *www.whitehouse.gov/the_press_office/Remarks-By-President-Barack-Obama-In-Prague-As-Delivered*.

10. See Yukiya Amano, "Introductory Statement to Board of Governors," Vienna, Austria, September 10, 2012, available from *www.iaea.org/newscenter/statements/2012/amsp2012n011.html*. At this gathering, Director General Amano noted that:

> it remains clear from the Agency's latest projections that nuclear power will remain an important option for many countries, despite the Fukushima Daiichi accident. Our new low projection is for nuclear power capacity to grow by nearly 25 percent *from* current levels to 456 gigawatts by 2030. Our high projection is 740 gigawatts, which is twice current levels.

11. See "President Obama's Second Inaugural Address (Transcript), *The Washington Post*, January 21, 2013, available at *articles.washingtonpost.com/2013-01-21/politics/36473487_1_president-obama-vice-president-biden-free-market*.

12. See, e.g., Massachusetts Institute of Technology (MIT), *The Future of Nuclear Power: An Interdisplinary MIT Study*, Cambridge, MA: MIT Press, 2009, available from *web.mit.edu/nuclearpower/pdf/nuclearpower-full.pdf*. The study analyzes a future of 1,000-Gwe of nuclear power in 2050 on grounds that "such a deployment would avoid . . . about 25% of the increment in carbon emissions otherwise expected in a business-as-usual scenario." For additional analysis, see Sharon Squassoni, "The Reality of Nuclear Expansion," testimony given before The House Select Committee for Energy Independence and Global Warming, Washington, DC, March 12, 2008, available from *carnegieendowment.org/files/squassoni_testimony_20080312.pdf*.

13. Duke Energy's announced closing of its Crystal River nuclear power plant in view of cheaper gas power is a straw in the wind not only for future U.S. plants, but also for existing ones. Modern "fracking" extraction methods for natural gas also pose environmental problems that have to be balanced against the advantages of using this fuel, but the point is that we have a choice and are not locked into depending on nuclear power. On these points, see "Exelon Cuts Nuclear Expansion Projects Amid Low Natgas Prices," *Reuters,* February 7, 2013, available from *www.reuters. com/article/idUSL1N0B76ZW20130207?irpc=932;* "Duke Energy to Close Troubled Crystal River Nuclear Power Plant," *Associated Press,* February 5, 2013, available from *www.wtsp.com/news/top-stories/article/296225/250/Crystal-River-nuke-plant-to-close;* and Rebecca Smith, "Can Gas Undo Nuclear Power?" *Wall Street Journal,* January 29, 2013, available from *online.wsj.com/article/SB10001424 127887323644904578272111885235812.html.*

14. For an excellent review of the main points of contention within the NPT regime and the issues that have made reform so difficult, see Steven E. Miller, *Nuclear Collisions: Discord, Reform & the Nuclear Nonproliferation Regime,* Cambridge, MA: American Academy of Arts and Sciences, 2012, available from *www.amacad. org/pdfs/nonproliferation.pdf.*

15. At the Preparatory Committee for the 2015 NPT Review Conference, U.S. Ambassador Susan F. Burk, the Special Representative of the President for Nuclear Nonproliferation, stated:

> It is imperative, therefore, that all NPT Parties recommit themselves to ensuring the health and vitality of this essential international agreement by advancing each of the Treaty's three pillars together. *The Treaty and the regime cannot thrive unless each pillar thrives.* All Parties must accept responsibility for taking appropriate steps to contribute to the achievement of each of the Treaty's fundamental objectives, whether collectively or individually. (Emphasis added.)

See Statement by Ambassador Susan F. Burk, Special Representative of the President for Nuclear Nonproliferation, Department of State, General Debate, First Sess. of the Preparatory Committee, as prepared on April 30, 2012, available from *vienna.usmission. gov/43014.html.* The problem here is one of emphasis. The NPT, of course, includes these other elements, which were added as conditions for the acceptance of the first, but the treaty is fundamentally about nonproliferation. The IAEA goes even further in promoting the notion that nonproliferation is not the key goal of the NPT. Vilmos Cserveny, Assistant Director General for External Relations and Policy Coordination, said at the NPT Preparatory Committee, New York, May 4, 2009: "The NPT consists of three *equally important* pillars—nuclear non-proliferation; peaceful nuclear cooperation; and nuclear disarmament—and the premise that progress in any one pillar strengthens the integrity of the whole." (Emphasis added.) He went on to say the IAEA's work is based on these three pillars. See IAEA, "Road to Disarmament, IAEA Safeguards: A Fundamental Pillar of the NPT Regime," excerpt from Vilmost Cserveny's statement at the General Debate of the NPT Preparatory Committee, New York, May 4, 2009, available from *www.iaea.org/Publications/Magazines/Bulletin/Bull511/51103570609.html*.

16. See The White House, Office of the Press Secretary, "Statement by President Obama on the 40th Anniversary of the Nuclear Nonproliferation Treaty," March 5, 2010, available from *www.whitehouse.gov/the-press-office/statement-president-obama-40th-anniversary-nuclear-nonproliferation-treaty*.

17. An arguable exception is President Ford's 1976 nuclear policy statement in which he proposed reliance on a "once-through" fuel cycle, thus eliminating reliance on plutonium until there was adequate international control to cope with proliferation risks. At that time, misuse of plutonium was seen as the main proliferation worry. HEU was not yet a principal concern. Unfortunately, the international nuclear community, with a heavy assist from our own nuclear bureaucracy, rejected this once-through standard because it seemed to bar the way to the Holy Grail of nuclear power—the plutonium-fueled fast breeder. In practice, commercial reprocessing required enormous subsidies. Although the United States has the right to control the reprocessing of nuclear fuel covered by agreements for cooperation (except with Euratom), it has since given blanket approval to Japan and, more recently, India. On the occasion of signing the 2006 U.S.-India nuclear agreement, President George W. Bush said, "You can advocate nuclear power, in order to take the pressure off of our own economy, for example, without advocating technological development of reprocessing,

. . ." See U.S. Department of State Archive, "Remarks by President Bush and Prime Minister Manmohan Singh of India," New Delhi, India, March 2, 2006, available from *2001-2009.state.gov/p/sca/rls/rm/2006/62426.htm*. The issue of reprocessing approvals is now coming to a head in the renegotiation of the nuclear cooperation agreement with South Korea. The Koreans, of course, want to be treated on a par with the Japanese, and it will be difficult to refuse them. This is the natural consequence of a compartmentalized approach to proliferation policy as a sequence of special cases, and not looking ahead at the obvious consequences. We can only climb back up this slippery slope by applying rules in a consistent and fair manner.

18. See IAEA, *The Structure And Content Of Agreements Between The Agency And States Required In Connection With The Treaty On The Non-Proliferation Of Nuclear Weapons*, IAEA INFCIRC/153, par. 26, Vienna, Austria: IAEA, available from *www.iaea.org/Publications/Documents/Infcircs/Others/infcirc153.pdf*: "The Agreement should provide for it to remain in force as long as the State is party to the Treaty on the Non-Proliferation of Nuclear Weapons."

19. See U.S. Department of State, Office of the Spokesperson, "Third P-5 conference: Implementing the NPT," a joint statement issued by China, France, Great Britain, Russia, and the United States of America, June 27-29, 2012, Washington, DC, available from *www.state.gov/r/pa/prs/ps/2012/06/194292.htm*. This statement included the following:

> As a further follow-up to the 2010 NPT Review Conference, the P5 shared their views on how to discourage abuse of the NPT withdrawal provision (Article X), and how to respond to notifications made consistent with the provisions of that article. The discussion included modalities under which NPT States Party could respond collectively and individually to a notification of withdrawal, including through arrangements regarding the disposition of equipment and materials acquired or derived under safeguards during NPT membership. The P5 agreed that states remain responsible under international law for violations of the Treaty committed prior to withdrawal.

It would have been more supportive of the NPT if the statement said such withdrawal while in violation was invalid.

20. See The White House, Office of the Press Secretary, "Statement by the President on North Korean Announcement of Nuclear Test," February 12, 2013, available from *translations.state.gov/st/english/texttrans/2013/02/20130212142430.html#ixzz2L0EWdcwt.*

21. See, e.g., Steve Herman, "Rising Voices in S. Korea, Japan Advocate Nuclear Weapons," *Voice of America,* February 15, 2013, available from *www.voanews.com/content/rising-voices-in-south-korea-japan-advocate-nuclear-weapons/1604309.html;* and Howard LaFranchi, "N. Korea Nuclear Test: Will It Spoil Obama's Disarmament Plans?" *The Christian Science Monitor,* February 15, 2013, available from *www.csmonitor.com/USA/Politics/2013/0215/N.-Korea-nuclear-test-Will-it-spoil-Obama-s-disarmament-plans.*

22. In the 1950s and 1960s, there were still a good many power reactors using natural uranium fuel. Enrichment was an essential part of the fuel cycle for programs based on U.S.-type LWRs, both pressurized water reactors which were gaining in popularity. The United States withheld enrichment data to protect the effective enrichment monopoly it had in the West because of the huge capacity it had built up to supply HEU for weapons. At the same time the U.S. Atomic Energy Commission (AEC) permitted the export of research reactors fueled with HEU and large quantities of HEU to fuel them. Until the late-1960s, the AEC did not even bother to verify the amounts shipped abroad by private firms. Up to 1996, the United States exported over 25 tons of HEU. On this point, see the Nuclear Threat Initiative, "Civilian HEU: United States" available from *www.nti.org/analysis/articles/civilian-heu-united-states.*

23. Aside from military explosive and naval propulsion applications, HEU fuel is used principally in research reactors. These can be fueled with lower enrichment fuels, albeit with some diminution in performance at higher power reactors. On the whole, the operators of these reactors, including those at universities, have dragged their feet about converting to lower enrichment fuel. Two prominent examples are the research reactors at MIT and Munich Technical University, which, despite it being U.S. and German government policy to convert such reactors, have managed to wrest delay after delay. Overall, about 50 research

reactors around the world have converted to lower enrichment fuel as a result of three decades of international effort (started at the U.S. Nuclear Regulatory Commission) to reduce HEU use, but this has resulted in only about a one-quarter reduction in annual HEU use by research reactors, which previously was about a ton per year. See Ole Reistad and Styrkaar Hustveit, "HEU Fuel Cycle Inventories and Progress On Global Minimization," *The Nonproliferation Review*, Vol. 15, No. 2, July, 2008, pp. 265-287. Reducing civilian HEU use to zero has got to be an important component of any nonproliferation policy.

24. INFCIRC/153, para. 28: ". . . the objective of safeguards is the timely detection of diversion of significant quantities of nuclear material from peaceful nuclear activities to the manufacture of nuclear weapons or of other nuclear explosive devices or for purposes unknown, and deterrence of such diversion by the risk of early detection."

25. See DoE, *Global Nuclear Energy Partnership Strategic Plan*, January 2007 (Italics in original), pp. 3-10, available from *www.fas. org/programs/ssp/_docs/GNEPStratPlanJan07.pdf*. It is often forgotten, even by the DoE, that the sophisticated reprocessing schemes proposed under GNEP—schemes that, in principle, always kept plutonium mixed with other materials—were intended as antiterrorism rather than antiproliferation protection. The program's strategic plan, however, acknowledged that there was no way to make reprocessing proliferation-proof.

26. Even the reprocessing vendors do not make a resource case for plutonium. Mixed oxide (MOX) fuel for LWRs costs several **times** what uranium fuel costs. In fact, recycling in LWRs never made any economic sense. It has been forgotten that the whole notion of thermal recycle was invented in the 1970s because the AEC's fast breeder reactor project was falling behind, and the agency needed a rationale for keeping its reprocessing efforts going "until the breeder caught up." It was never expected to be economic. By the time the breeder project collapsed, MOX acquired its own support group and took on a life of its own, leaping from one rationale to another.

27. At the start of nuclear power programs, it was generally believed that as uranium was scarce, LWRs would be a transition-

al generation that produced enough plutonium to fuel the ultimate in nuclear power plants, the fast breeders (so-called because they used fast neutrons and produced more plutonium in a surrounding uranium "blanket" than they consumed in the core). In practice, prototype fast breeders turned out to have problems and were too expensive to pursue. The Bush administration tried to revive the concept in the form of fast "burners," plutonium-fueled reactors without the surrounding blanket. This did not make any sense, either.

28. DoE made an agreement with Russia for each country to recycle 34 tons of "surplus" weapons plutonium. It has become a boondoggle. DoE contractor Shaw Areva MOX Services is constructing a mixed oxide fuel fabrication facility at DoE's Savannah River Site in South Carolina, in principle to use weapons plutonium to produce power reactor fuel that, once heavily irradiated in a commercial power reactor, would contain plutonium that was less useful for weapons. It looks as if DoE's real purpose is to use the rubric of arms control to get plutonium recycle underway. The plant is one of the largest construction projects in the southeastern United States. The out-of-control costs are said to have multiplied to $7 billion. The plant has no customers lined up for the MOX fuel it is supposed to produce, which has attracted the attention of federal budget cutters but does not seem to have diminished the project's appeal for DoE. See Sammy Fretwell, "Critics Fear $7 Billion SRS Boondoggle" *The State*, January 27, 2013, available from *www.thestate.com/2013/01/27/2606562/critics-fear-7-billion-srs-boondoggle.html#.URBjPUqjdY4*. See also "Budget Cutters Eye Nuclear Reprocessing Plant" *Roll Call*, February 5, 2013, available from *www.taxpayer.net/media-center/article/budget-cutters-eye-nuclear-reprocessing-plant*:

> It isn't expected to be completed by its 2016 target date, and the Department of Energy has found little interest from commercial power plant operators in buying the fuel, which would require costly reactor modifications . . . [Former Rep. David L.] Hobson, [R-Ohio] described the project as a jobs program for South Carolina. In addition to the 2,600 employees now working on it, the completed facility will require permanent workers to operate it for up to two decades.

29. See Gerald Ford, "Statement on Nuclear Policy," October 28, 1976, available from *www.presidency.ucsb.edu/ws/index.php?pid=6561*: "I have concluded that the reprocessing and recycling of plutonium should not proceed unless there is sound reason to conclude that the world community can effectively overcome the associated risks of proliferation." There is also a precedent for banning an entire technology in the context of the NPT—so-called peaceful nuclear explosives were read out of the NPT even though there is an entire NPT article (Article V) devoted to them. On this later point, see Robert Zarate, "The NPT, IAEA Safeguards and Peaceful Nuclear Energy: An 'Inalienable Right' but Precisely to What?" Henry Sokolski, ed., *Falling Behind: International Scrutiny of the Peaceful Atom,* Carlisle, PA: Strategic Studies Institute, U.S. Army War College, 2008, pp. 252-255, available from *www.npolicy.org/userfiles/image/Peaceful%20Nuclear%20Energy,%20an%20Inalienable%20Right%20to%20What_pdf.pdf.*

30. For a discussion on the program's technical flaws see "A Minority Opinion of Gilinsky and Macfarlane," Review of DoE's Nuclear Energy Research and Development Program, Washington, DC: National Academy Press, 2008, available from *www.nap.edu/openbook.php?record_id=11998&page=73.*

31. See President Obama's remarks at Hankuk University, March 26, 2012. DoE undoubtedly inserted the seemingly innocuous words into the speech. To the nuclear bureaucracies around the world, they spell plutonium recycle.

32. See, e.g., Elaine Grossman, "U.S. May Land Key Asian Nuclear Trade Deals in 2013," *Global Security Newswire,* January 11, 2013, available at *www.nti.org/gsn/article/us-could-secure-key-asian-nuclear-trade-deals-2013;* "Obama Team Reveals Nuclear Trade, Nonproliferation Decision on Capitol Hill," *Global Security Newswire,* January 11, 2012, available from *www.nti.org/gsn/article/obama-team-reveals-nuclear-trade-nonproliferation-decision-capitol-hill;* and "U.S. Envoy takes Issue with Nonproliferation Lingo for Nuclear Trade Pacts," *Global Security Newswire,* August 10, 2012, available from *www.nti.org/gsn/article/us-envoy-takes-issue-nonproliferation-lingo-nuclear-trade-pacts.*

33. Wikipedia lists Argentina, Brazil, China, France, Germany, India, Iran, Japan, the Netherlands, North Korea, Pakistan, the UK, and the United States in the first category, and South Africa and Australia (and probably Israel) in the second. See "Enriched uranium," Wikipedia, available from *en.wikipedia.org/wiki/Enriched_uranium.*

34. It is useful to keep in mind the verdict of the 1946 Acheson-Lilienthal Report: "A system of inspection superimposed on *an otherwise uncontrolled exploitation of atomic energy* by national governments will not be an adequate safeguard. . . . If nations or their citizens carry on intrinsically dangerous [nuclear] activities it seems to us that the chances for safeguarding the future are hopeless." In other words, the allowed activities have got to be restricted to those that can be safeguarded, in the dictionary sense, by inspection. See *Report on International Control of Atomic Energy,* Washington, DC: U.S. Government Printing Office, March 16, 1946, pp. 21-22, available from *www.learnworld.com/ZNW/LW-Text.Acheson-Lilienthal.html.*

35. See *Model Protocol Additional to The Agreement(s) Between State(s) And The International Atomic Energy Agency For The Application Of Safeguards,* INFCIRC/540, Vienna, Austria: IAEA, 1997, available from *www.iaea.org/Publications/Documents/Infcircs/1997/infcirc540c.pdf.*

36. Clandestine facilities are an obvious concern in North Korea and Iran. Iran failed to notify the IAEA of a planned enrichment facility, as it had promised to do in 2003. It later retreated from this promise. Whether this amounts to a violation of Iran's obligations or a deficiency in the IAEA system depends on legal issues that remain unclear. The IAEA has not released relevant documents and correspondence. See Board of Governors of the IAEA, *Implementation of the NPT Safeguards Agreement and Relevant Provisions of United Nations Security Council Resolutions in the Islamic Republic of Iran,*GOV/2012/50, available from *www.iaea.org/Publications/Documents/Board/2012/gov2012-50.pdf.*

37. For a detailed analysis of this proposition, see Victor Gilinsky, Harmond Hubbard, and Marvin Miller, *A Fresh Examination of the Proliferation Dangers of Light Water Reactors,* Washington, DC: Nonproliferation Policy Education Center, 2008, available from

www.npolicy.org/article_file/A_Fresh_Examination_of_the_Prolifera-tion_Resistance_of_Light_Water_Reactors.pdf. It is worth keeping in mind that many states' view of the fuel cycle is still tied to the use of LWRs with oxide fuel. It is not at all clear—for both safety and proliferation reasons and perhaps economic reasons, as well—that this is best way to extract nuclear energy. Other new types of reactors with different fuels, especially fuels that do not melt as easily, may make moot many of the issues that concern us today, possibly including reprocessing.

38. See D. L. Ferguson to F. L Culler, "Simple Quick Repro-cessing Plant," Inter-Laboratory Correspondence, Oak Ridge Na-tional Laboratory, August 30, 1977, available from *www.npolicy. org/article_file/Simple_Quick_Processing_Plant_Culler.pdf.*

39. North Korea refused to allow it. In 1992, Romania invited a special inspection. See John Carlson and Russell Leslie, "Spe-cial Inspections Revisited," paper presented at INMM 2005 Sym-posium, Phoenix, AZ, July 2005, available from *www.dfat.gov.au/ asno/publications/inmm2005_special_inspections.pdf.*

> [I]t is obvious that special inspections cannot become *regular* occurrences, and cannot substitute for complementary ac-cess, it can be questioned whether the very high threshold assumed in the 1992 board deliberations is consistent with contemporary expectations for the safeguards system and for the level of cooperation that states extend to the agency.

40. See Pierre Goldschmidt, "Looking beyond Iran and North Korea for Safeguarding the Foundations of Nuclear Nonprolif-eration," Washington, DC: Carnegie Endowment for Internation-al Peace, November 3, 2011, available from *carnegieendowment. org/2011/11/03/looking-beyond-iran-and-north-korea-for-safeguard-ing-foundations-of-nuclear-nonproliferation/8ktn.*

41. See U.S. Mission to the International Organizations in Vi-enna, Austria, "Statement by Ambassador Susan F. Burk, Special Representative of the President for Nuclear Nonproliferation, Department of State, General Debate, First Session of the Prepa-ratory Committee , as prepared, April 30, 2012," available from *vienna.usmission.gov/43014.html.*

42. See Elliott Abrams, "Bombing the Syrian Reactor: the Untold Story," *Commentary*, February 2013, available from *www. commentarymagazine.com/article/bombing-the-syrian-reactor-the-untold-story*. In Abrams's account, President Bush was initially inclined to bring the case before the IAEA but was persuaded to let the Israelis bomb the reactor.

43. See President Obama's statement on the North Korean nuclear test, February 12, 2013.

44. There is no possibility under the NPT to expand the number of weapons states "For the purposes of this Treaty, a nuclear weapon State is one which has manufactured and exploded a nuclear weapon or other nuclear explosive device prior to 1 January, 1967." NPT, Article X.

45. The Egyptian Foreign Ministry, in reaction to the February 12 North Korean explosion, emphasized the necessity of the "internationalization" of the NPT. See Egypt State Information Service, "Egypt Stresses Necessity of NPT Internationalization," February 15, 2013, available from *allafrica.com/stories/201302170072. html*. Egypt obviously has Israel in mind here, but that does not detract from the idea.

46. This outlook is captured in 2008 report of the State Department's International Security Advisory Board:

> We concluded that the current international climate is quite unpropitious for gaining support from non-nuclear weapon states to accept stricter measures against proliferation . . . we believe that incremental measures, rather than revolutionary or comprehensive changes, will be far more likely to succeed in the near term.

The report is from an earlier administration, but it captures the sense of current approaches as well. See U.S. State Department International Security Advisory Board, *Report on Proliferation Implications of the Global Expansion of Civil Nuclear Power*, Washington, DC: Department of State, April 7, 2008, available from *2001-2009. state.gov/documents/organization/105587.pdf*.

47. See, for example, Fred McGoldrick, *Nuclear Trade Controls: Minding the Gaps*, Washington, DC: Center for Strategic and International Studies, January 22, 2013, available from *csis.org/files/publication/130122_McGoldrick_NuclearTradeControls_Web.pdf*:

> Finally, the U.S. has to avoid overreach in instituting new nuclear export controls. Recent well-intentioned efforts by some in Congress and the Executive Branch to pressure other states to forswear enrichment and reprocessing capabilities could seriously damage the prospects for U.S. nuclear exports and deprive the United States of the nonproliferation influence that comes with nuclear cooperation. Some have suggested other steps that would cause similar, if not more severe, damage to U.S. influence in international nuclear affairs. Suppliers are not going to require such extreme export conditions, and most consumer states are likely to reject U.S. demands they believe deny them their rights or legitimate peaceful commercial opportunities.

McGoldrick refers the reader to an earlier version of this paper for examples of "overreach" that would cause severe damage to U.S. influence (a former Los Alamos director went further to heatedly label that paper as deliberately anti-nuclear). McGoldrick basically proposes tidying up the nonproliferation controls but doing nothing to upset any other countries, especially developing countries. He accepts that we have to offer positive incentives if we ask them to forego "sensitive" technologies. There is no larger sense here, or in the many similar reports, that the United States could, or should try to, persuade other countries of the common security advantage in agreeing to a higher level of protection against bomb making.

48. In this 2012 South Korea speech, the President said, ". . . we're creating new fuel banks, to help countries realize the energy they seek without increasing the nuclear dangers that we fear."

49. See Alan Hanson, "Nuclear Fuel Banks: Are They a Reality," presentation at the Monterey Institute, Center for Nonproliferation Studies, Monterey, CA, December 12, 2011, available from *www.youtube.com/watch?v=EYmT6ftCPhg*.

50. This includes a strained effort to include "dirty bombs" in the category of serious threats. Consider Director General Amano's October 17, 2012, speech at Chatham House, London, UK, in which he said: "One of the key risks we face is that terrorists could detonate a so-called dirty bomb, using conventional explosives and a quantity of nuclear or other radioactive material, to contaminate a major city." On proliferation, Director-General Amano dealt with Iran and North Korea, but said nothing about tightening antiproliferation protections overall.

ABOUT THE CONTRIBUTORS

RICHARD CLEARY is a student at Columbia Law School, where he is a Kent Scholar and an editor of the *Columbia Law Review*. In 2013, he was named the Charles L. Brieant Jr. fellow in the U.S. District Court for the Southern District of New York. Prior to law school, Mr. Cleary was a research assistant in the Marilyn Ware Center for Security Studies at the American Enterprise Institute (AEI). He has interned for the Senate Foreign Relations Committee and in the personal office of Senator Richard Lugar. Mr. Cleary holds a B.A., *summa cum laude*, from Washington and Lee University and an M.Phil. in international relations from Trinity College, University of Cambridge. His M.Phil. dissertation examined the influence of Mohamed El-Baradei as Director General of the International Atomic Energy Agency (IAEA) on the Iraq and Iran cases, and his undergraduate thesis considered the use of civilian nuclear energy in French diplomacy.

MATTHEW FUHRMANN is an assistant professor of political science at Texas A&M University. He is a former Stanton Nuclear Security Fellow at the Council on Foreign Relations and research fellow at the Belfer Center for Science and International Affairs at Harvard University. Dr. Fuhrmann is the author of *Atomic Assistance: How "Atoms for Peace" Programs Cause Nuclear Instability*. His research has been published in a number of journals, including the *American Journal of Political Science, British Journal of Political Science, International Organization, International Security, Journal of Conflict Resolution, Journal of Peace Research,* and *Journal of Politics*, among others. He has published opinion pieces in outlets such as the *Christian Science Monitor, Slate,* and *USA Today.*

501

VICTOR GILINSKY is an independent consultant on matters primarily related to nuclear energy. He was a two-term commissioner of the U.S. Nuclear Regulatory Commission from 1975-84, and before that Head of the Rand Corporation Physical Sciences Department. He is a member of the American Physical Society and the Institute of Electrical and Electronics Engineers. Dr. Gilinsky holds a bachelor's of engineering physics degree from Cornell University and a Ph.D. in physics from the California Institute of Technology, where he received the Distinguished Alumni Award.

PIERRE GOLDSCHMIDT is a nonresident senior associate at the Carnegie Endowment for International Peace. From 1999 to 2005, he was deputy Director General and head of the Department of Safeguards at the IAEA. Before joining the IAEA, Dr. Goldschmidt was director general of SYNATOM, which manages the fuel supply and spent fuel of Belgian nuclear plants and was a member of the directory of the French uranium-enrichment company EURODIF. He is also a member of the European Nuclear Society's High Scientific Council and has headed numerous European and international committees on nuclear energy. He is the author of over 100 publications and has received a number of cultural and scientific awards, including the 2008 Joseph A. Burton Forum Award of the American Physical Society. He was nominated Chevalier of the French Legion of Honor and was knighted by the King of Belgium.

OLLI HEINONEN is a Senior Fellow at the Belfer Center for Science and International Affairs at Harvard University. Dr. Heinonen served 27 years at the

IAEA, where he was the Deputy Director General and head of the Department of Safeguards, director of various operational divisions, and an inspector. His assignments included leading teams of investigators to examine nuclear programs of concern around the world, leading the Agency's efforts to identify and dismantle proliferation networks, and overseeing efforts to monitor and contain Iran's nuclear program. Prior to joining the IAEA, Dr. Heinonen was a senior research officer at the Technical Center of Finland's Reactor Laboratory and co-authored patents on radioactive waste solidification. He is the author of several articles, books, chapters of books, various IAEA publications, and has been published in numerous scholarly and popular periodicals around the world.

FRANÇOIS HEISBOURG is Chairman of the Council of the Geneva Centre for Security Policy and of the London-based International Institute for Strategic Studies (IISS). He is a special advisor at the Paris-based Fondation pour la Recherché Stratégique. He has also sat on a number of national and international blue-ribbon bodies, including the French Defence and National Security White Paper, the International Commission on Nuclear Non-Proliferation and Disarmament, the International Commission on the Balkans, and the European Union Commission's group of personalities on security research and development. Mr. Heisbourg has written extensively on defense and security issues and is a frequent contributor to both specialist and mainstream media on such matters.

GREGORY S. JONES is a senior researcher at the Nonproliferation Policy Education Center in Arlington, VA, and an adjunct staff member at the RAND Corporation. Mr. Jones has been a defense policy analyst for the last 40 years. A major emphasis of his work has been on the study of the potential for terrorists and hostile countries to acquire and use nuclear, chemical, biological, and radiological weapons, and the formulation of policies and actions to control and counter these weapons. Since 2008, Mr. Jones has analyzed Iran's centrifuge uranium enrichment program and charted Iran's inexorable movement towards a nuclear weapons capability.

R. SCOTT KEMP is an assistant professor of nuclear science and engineering at the Massachusetts Institute of Technology, a research scholar with the Program on Science and Global Security at Princeton University, and an affiliate of Harvard University's Program on Managing the Atom. From 2010-11, he served as Science Advisor in the Office of the Special Advisor to the Secretary for Nonproliferation and Arms Control where he was the State Department's resident centrifuge expert and was responsible for developing packages for negotiating with Iran. Dr. Kemp is also a member of the American Physical Society's Panel on Public Affairs. His work has been used in studies by the National Academies of Sciences, the American Physical Society, and the JASON advisory group. Prior to coming to Princeton, Dr. Kemp was a Fulbright Fellow in London and a member of the research staff at the Council on Foreign Relations in New York.

MATTHEW KROENIG is an associate professor of government and foreign service and international relations field chair at Georgetown University and a nonresident senior fellow at the Brent Scowcroft Center on International Security at the Atlantic Council.

From July 2010 to July 2011, he was a Special Advisor in the Office of the Secretary of Defense on a Council on Foreign Relations International Affairs Fellowship; and he worked as a strategist in the Office of the Secretary of Defense in 2005. Dr. Kroenig has held fellowships from the Council on Foreign Relations, the National Science Foundation, the Belfer Center for Science and International Affairs at Harvard University, the Center for International Security and Cooperation at Stanford University, and the Institute on Global Conflict and Cooperation at the University of California. He is a life member of the Council on Foreign Relations.

PATRICK S. ROBERTS is an associate professor in the Center for Public Administration and Policy in the School of Public and International Affairs at Virginia Tech in Alexandria, Virginia. He is the Associate Chair and Program Director CPAP, Northern Virginia. He holds a Ph.D. in government from the University of Virginia, and he spent 2 years as a post-doctoral fellow, one at the Center for International Security and Cooperation at Stanford University and another at the Program for Constitutional Government at Harvard University. He spent 2010-11 as the Ghaemian Scholar-in-Residence at the University of Heidelberg's Center for American Studies in Germany. Mr. Roberts is the author of *Disasters and the American State: How Politicians, Bureaucrats, and the Public Prepare for the Unexpected* (Cambridge, 2013).

HENRY SOKOLSKI is Executive Director of the Nonproliferation Policy Education Center in Arlington, VA, and adjunct professor at the Institute of World Politics in Washington, DC. He previously served as a military legislative aide and special assistant for nuclear energy affairs in the U.S. Senate, as Deputy for Nonproliferation Policy in the Pentagon, and as a member of the Central Intelligence Agency's Senior Advisory Group. Mr. Sokolski also was appointed by Congress to serve on the Deutch Weapons of Mass Destruction (WMD) Commission and the Commission on the Prevention of WMD Proliferation and Terrorism and has authored and edited numerous books on proliferation, including *Best of Intentions: America's Campaign against Strategic Weapons Proliferation.*

SUSAN VOSS is the President and co-founder of the Global Nuclear Network Analysis, LLC, a small, woman-owned engineering firm in Los Alamos, NM. She was previously employed at Los Alamos National Laboratory (LANL), where she worked on projects related to nuclear power, the nuclear fuel cycle, nuclear nonproliferation, counterproliferation, and arms controls for the Departments of State, Energy, Defense, and Homeland Security. Prior to her career at LANL, she worked for the Air Force Weapons Laboratory in Albuquerque on space nuclear power. Ms. Voss is currently a consultant to the Hyperion Power Generation Company's small nuclear reactor design team, focusing on advanced nuclear fuel design and safety. She holds a master's in nuclear engineering from the University of New Mexico.

LEONARD WEISS is an affiliated scholar at Stanford University's Center for International Security and Co-operation and a national advisory board member of the Center for Arms Control and Non-Proliferation in Washington, DC. He worked for over 2 decades for Senator John Glenn as the staff director of both the

Senate Subcommittee on Energy and Nuclear Proliferation and the Committee on Governmental Affairs.
He was the chief architect of the Nuclear Nonproliferation Act of 1978 and led investigations of the Indian and Pakistani nuclear programs. Dr. Weiss has
published numerous articles for the *Bulletin of Atomic
Scientists*, *Arms Control Today*, and *The Nonproliferation
Review*. He holds tenured professorships at Brown
University and the University of Maryland.

ROBERT ZARATE is policy director at the Foreign
Policy Initiative (FPI) in Washington, DC. Before joining FPI, he served as legislative assistant for foreign
affairs to U.S. Representative Jeff Fortenberry (R-NE)
and as a legislative fellow for the majority staff in the
House Foreign Affairs Subcommittee on Terrorism,
Nonproliferation, and Trade. He was previously a research scholar at the Nonproliferation Policy Education Center, an independent consultant on nuclear issues, a reporter for *Wired News*, and a policy analyst at
Steptoe & Johnson LLP. He also served on the Project
on Nuclear Initiatives Working Group on U.S.-China
Nuclear Issues convened by the Center for Security
and International Studies. Mr. Zarate co-edited *Nuclear Heuristics: Selected Writings of Albert and Roberta
Wohlstetter* and has published essays and articles in
periodicals such as *Time*, *The Weekly Standard*, *National
Review*, and *U.S. News and World Report*.

U.S. ARMY WAR COLLEGE

Major General Anthony A. Cucolo III
Commandant

*****

STRATEGIC STUDIES INSTITUTE
and
U.S. ARMY WAR COLLEGE PRESS

Director
Professor Douglas C. Lovelace, Jr.

Director of Research
Dr. Steven K. Metz

Editor
Mr. Henry Sokolski

Editor for Production
Dr. James G. Pierce

Publications Assistant
Ms. Rita A. Rummel

*****

Composition
Mrs. Jennifer E. Nevil

www.ingramcontent.com/pod-product-compliance
Lightning Source LLC
Chambersburg PA
CBHW082349270326
41935CB00013B/1562